Lecture Notes in Computer Science 16010

Founding Editors

Gerhard Goos
Juris Hartmanis

Editorial Board Members

Elisa Bertino, USA
Wen Gao, China

Bernhard Steffen, Germany
Moti Yung, USA

FoLLI Publications on Logic, Language and Information
Subline of Lecture Notes in Computer Science

Subline Editors-in-Chief

Valentin Goranko, *Stockholm University, Sweden*
Michael Moortgat, *Utrecht University, The Netherlands*

Subline Area Editors

Nick Bezhanishvili, *University of Amsterdam, The Netherlands*
Anuj Dawar, *University of Cambridge, UK*
Philippe de Groote, *Inria Nancy, France*
Gerhard Jäger, *University of Tübingen, Germany*
Fenrong Liu, *Tsinghua University, Beijing, China*
Eric Pacuit, *University of Maryland, USA*
Ruy de Queiroz, *Universidade Federal de Pernambuco, Brazil*
Ram Ramanujam, *Institute of Mathematical Sciences, Chennai, India*

Valentin Goranko · Chenwei Shi · Wei Wang
Editors

Logic, Rationality, and Interaction

10th International Conference on Logic
Rationality and Interaction, LORI 2025
Xi'an, China, October 16–19, 2025
Proceedings

 Springer

Editors
Valentin Goranko 🆔
Stockholm University
Stockholm, Sweden

Chenwei Shi 🆔
Tsinghua University
Beijing, China

Wei Wang 🆔
Xi'an Jiaotong University
Xi'an, China

ISSN 0302-9743   ISSN 1611-3349 (electronic)
Lecture Notes in Computer Science
ISBN 978-981-95-2480-8   ISBN 978-981-95-2481-5 (eBook)
https://doi.org/10.1007/978-981-95-2481-5

Preface

This volume collects the extended abstracts of the invited talks and the full papers accepted and presented at LORI 2025, the 10th International Conference on Logic, Rationality and Interaction, held in Xi'an, China, during October 16–19, 2025, and hosted by the Department of Philosophy of Xi'an Jiaotong University.

The conference received 47 submissions, 36 full papers and 11 short papers. During the single-blind review, each submission was read by at least three reviewers. The programme committee decided to accept 14 full papers and 8 as short presentations in the final program. The topics addressed in this program effectively showcase the breadth and depth characteristic of the LORI conference series, including contributions on dynamic epistemic logic, nonmonotonic reasoning, belief revision theory, decision theory, causal inference, social epistemology and so on. Additionally, the conference's program was enhanced by insightful invited presentations from distinguished speakers:

Zoé Christoff (University of Groningen, the Netherlands),
Tim French (University of Western Australia, Australia),
Aybüke Özgün (University of Amsterdam, the Netherlands),
François Schwarzentruber (ENS Lyon, France),
Marija Slavkovik (University of Bergen, Norway),
Hongjun Zhou (Shaanxi Normal University, China).

Since the first LORI event, which took place in August 2007, nine LORI conferences have been held, one every two years. As a platform for nurturing an East-Asian research community and attracting scholars from outside the region to interact and exchange ideas, the LORI series has been productive and invaluable. A full history of the series can be found at www.golori.org.

As chairs of the Organizing and Program Committees, we first wish to express our sincere appreciation to our PC members and additional reviewers for their remarkable dedication and efficient work under considerable time pressure - their contributions formed the foundation of this program. We are also grateful to the LORI Standing Committee members Fenrong Liu and Johan van Benthem, along with several former LORI PC chairs, for their advice and support. Our heartfelt thanks also go to the School of Humanities and Social Science and the Department of Philosophy at Xi'an Jiaotong University for their generous sponsorship, to the Tsinghua University - University of Amsterdam Joint Research Center for Logic for their ongoing partnership, and to the Su Tianfu Fundation of Southwest University for supporting the Best Student Paper Award. Finally, we recognize the tireless efforts and support of our Xi'an Jiaotong University colleagues and students.

July 2025

Valentin Goranko
Chenwei Shi
Wei Wang

Organization

Organizing Committee Chair

Wei Wang   Xi'an Jiaotong University, China

Organizing Committee

Xiaojun Ding   Xi'an Jiaotong University, China
Jiaxin Wang   Xi'an Jiaotong University, China
Zhensong Wang   Xi'an Jiaotong University, China

Program Committee Chairs

Valentin Goranko   Stockholm University, Sweden
Chenwei Shi   Tsinghua University, China

Program Committee

Natasha Alechina   Utrecht University, the Netherlands
Maria Aloni   University of Amsterdam, the Netherlands
Thomas Ågotnes   University of Bergen, Norway, Shanxi University,
                                    China
Alexandru Baltag   University of Amsterdam, the Netherlands
Gaia Belardinelli   Stanford University, USA
Johan van Benthem   University of Amsterdam, the Netherlands,
                                    Stanford University, USA, Tsinghua
                                    University, China
Thomas Bolander   Technical University of Denmark, Denmark
Ilaria Canavotto   University of Maryland, College Park, USA
Zoé Christoff   University of Groningen, the Netherlands
Hans van Ditmarsch   University of Toulouse, CNRS, IRIT, France
Huimin Dong   TU Wien, Austria
Rustam Galimullin   University of Bergen, Norway
Malvin Gattinger   University of Amsterdam, the Netherlands
Sujata Ghosh   Indian Statistical Institute, Chennai, India

Nina Gierasimczuk	Technical University of Denmark, Denmark
Fengkui Ju	Beijing Normal University, China
Dominik Klein	Utrecht University, the Netherlands
Sophia Knight	University of Minnesota Duluth, USA
Barteld Kooi	University of Groningen, the Netherlands
Dazhu Li	Chinese Academy of Sciences, China, University of Chinese Academy of Sciences, China
Beishui Liao	Zhejiang University, China
Emiliano Lorini	IRIT, Toulouse, France
Krzysztof Mierzewski	Carnegie Mellon University, USA
Munyque Mittelmann	University of Naples Federico II, Italy
Sara Negri	University of Genoa, Italy
Aybüke Özgün	University of Amsterdam, the Netherlands
Eric Pacuit	University of Maryland, College Park, USA
Katsuhiko Sano	Hokkaido University, Japan
Sonja Smets	University of Amsterdam, the Netherlands
Fernando Velázquez-Quesada	University of Bergen, Norway
Yanjing Wang	Peking University, China
Xuefeng Wen	Sun Yat-sen University, China
Kaibo Xie	Wuhan University, China
Jialiang Yan	Tsinghua University, China
Francesca Zaffora Blando	Carnegie Mellon University, USA

Additional Reviewers

Aleksi Anttila	Olivier Roy
Edoardo Baccini	Sayantan Roy
Tim French	Sunil Simon
Avijeet Ghosh	Pietro Vigiani
Søren Brinck Knudstorp	Zuojun Xiong

Invited Talks

How Popular is This Opinion? From Majority Illusions to Threshold Models of Collective Actions

Zoé Christoff

Bernoulli Institute for Mathematics, Computer Science and Artificial Intelligence, University of Groningen, Netherlands
`z.l.christoff@rug.nl`

In social networks, the popularity of an opinion in one's direct circles is not necessarily a good indicator of its popularity in one's entire community. For instance, when most people are confronted with a majority of opposing opinions in their circles, they might get the impression that they belong to a minority. In this sense, network structures make local information about global properties of the group potentially misleading. The way a social network is wired constrains to what extent such information distortions can occur. In this talk, I focus on when agents can correctly assess the popularity of opinions based on the information that is locally available, and when they cannot.

In the first part (based on joint work with Maaike Venema-Los and Davide Grossi), I use graph-theoretical tools to characterise networks allowing for specific types of distortion. In particular, I discuss which networks allow for a majority of agents to have locally the wrong impression about the global majority, that is, to be under 'majority illusion', before generalising to other types of illusions.

In the second part (based on ongoing joint work with Thomas Ågotnes), I will consider, from an epistemic logic perspective, the related case of threshold models of collective behaviour, where agents join a collective action if and only if they know that sufficiently many others in the community join too.

Disclosure of Interests. The author has no competing interests to declare that are relevant to the content of this abstract.

Rational Guessing and the Emergence of Reason

Tim French

The University of Western Australia, Perth, Western Australia
`tim.french@uwa.edu.au`

In this work we present an open discussion of the minimal elements required for rational thought, and particularly consider the conditions required for the emergence of reasoning from these elements. This investigation is meta-logical, in that our interest is in the possibility of logics, and we aim to be agnostic to any biological, psychological, sociological, or computational context.

The criteria we set ourselves when developing such a system is that:

1. it is *emergent*, in the sense that there is very little design or a priori structure given.
2. it is *formal*, in the sense that it has the capacity for an algebraic semantics.
3. it is *adaptive* so that it permits learning and the progressive development of reason.

We incrementally build a system with such properties, based on a simple probabilistic semantics, related to aleatoric logic. We discuss the assumptions implicit in these semantics and the challenges that remain.

1 Rational Thought

Rationality if the application reason to draw understanding from experience and a priori knowledge. There are many aspects to the process of thought, including creativity, memory, language, imagination, and these have been studied and discussed extensively in psychology [5], philosophy [8], and computer science [1]. We will take rationality to be a reasoned process to understand sensation, and anticipate consequences of those sensations.

1.1 Related Work

Reidl [7] presents a biologically inspired analysis of the evolution of knowledge, and considers the abstract processes used in the foundation of rational thought, Reidl particularly advocates the necessity of probabilistic reasoning [4, 6] to account for novel experiences and the acquisition of knowledge. In this work we will consider aleatoric logic [3] as a semantic foundation for this investigation.

In *Knowing by Imagining* [8] Williamson presents an argument for knowledge *as* imagination. The thesis presents a notion of imagination as synthesised from experience, and provides a mechanism for a primitive agent to make judgements when there is insufficient reason.

2 A Model of Emergent Rationality

We will consider a layered model of rationality, and an implementation of this model via aleatoric logic. The model consists of: *Sensation* (the recognition of a signal on a channel); *Imagination* (the hypothesis of a sensation); *Synthesis* (the composition of imagined hypotheses into a complex); *Identification* (the assertion of equivalences between complex hypotheses); *Nomination* (the understanding of things as objects, and the ability to hypothesise a thing); *Characterisation* (the association of complex properties with things); and *Induction* (the formation of imagined hypotheses based on experience).

2.1 Sensation

A sensation is a signal received on some channel. The channel provides the context (temperature, velocity), while the signal gives a value to that context (warm, fast). At its simplest we might suppose that the signal is Boolean, so that it is warm or it is not, but it could also be measured by degrees. This is the minimum for experience. We do not require any special context or even the notion of an environment. The sensation might be entirely internal, as concepts of pain, or intrusive thoughts. In a basic context we might consider a simplest single celled organism that finds sensations such as *warmth*, *nutrition*, and these sensations trigger processes to occur. It would be a stretch to identify this process as thought, but it provides a necessary first step.

2.2 Imagination

Sensations, as experiences of the world, come unbidden, reflecting different channels and events outside the agent's control. The consideration of sensation, on the other hand is internal to the agent and applies an element of focus or attention to the possibility of sensation. We call this imagination, and an agent may imagine freely creating hypothetical sensations, or events. The imagination allows the agent to model the *world* (whatever it is that generates these sensations) as some internal object that may be queried, and queried in such a way as to not only recall sensations, but to anticipate novel events.

2.3 Synthesis

Beyond simply recalling sensations and experiences, the imagination can construct complexes consisting of multiple sensations in different forms. Combinations might include the conjunction of two sensations, one sensation or another, or the absence of a sensation.

Combinations might also include the repeated imagining of the same sensation (pain, three out of four times). This provides a simple set of operations to build a language of sensation and experience, and elements of this language might feed into decision making processes: on imagining pain or hunger four out of five times, it should move to a safer location.

2.4 Identification

With a method to define experiences as complexes of sensations, there comes the possibility to assign meaning to these sensations. That is an agent to build explicit associations between imagined sensations. We might say "happiness is being fed and warm, or being fed, or being warm". This creates a correspondence between the sensation *happiness* and the sensations *fed* and *warm*. Importantly, this provides an objective element to the thought process. The sensations discussed so far are inherently subjective and tacit, in that they are easy to recognise internally but hard to communicate externally: who can say whether what they consider to be warm will meet anther's definition, or whether colours appear similarly to us all? However, we can share definitions such as the definition of happiness, or an observation that warmth and food go together, by stating the equivalence, and this can create a common understanding of experience.

2.5 Nomination

So far we have just considered sensation, which as properties experienced by the agent, are essentially properties of the agent. This is an egocentric model where effectively the agent is the only *thing* in the world. This might be adequate for simple organisms with evolved responses to stimuli, but it is clear that rational thought requires more than this. The agent should recognise things, *name* things, and understand properties of these things. More importantly, an agent should be able to imagine things. This provides a much richer and more useful imagination for the agent. An agent can imagine going *outside*, where *outside* is a thing, and it is a thing with both *danger* and a thing with *food*. Now an agent may imagine outside carefully, imagining dangerous predators and environmental hazards, and imagining food and substance. A decision then may be made that weighs all of these imaginings together, much like the process described by Williamson in [8].

2.6 Characterisation

Implicit in the process of recognising and nominating things is characterising the properties of things. An agent might imagine some mushrooms (as *things*), but imagine them as poisonous, or other times imagine them as nutritious, recalling past experiences. Properties of these things, like the colour or size, might help distinguish these cases, and there may be more complex properties, like the proximity to a tree (another *thing*). This is a large increase in the language of the agent, and is far beyond the simple language of sensations we began with with. The properties may be thought of as predicates over the set of things. This provides the agent with a language to describe properties of things, and relationships between things.

2.7 Induction

A key feature of thought, is that it should predict or in some sense align with the world that generates the experience. This alignment may be given (partially) by a designer or some ratiomorphic configuration [7] but the power of rational though is to accept new experience and adapt the imagination and characterisation to incorporate these experiences. This is a process of learning or induction. The characterisation of properties of things describes a memory or dictionary assigning properties to things, while the imagination given a subjective belief about what is likely or possible. These two elements evolve with new experience to *define* new concepts or relations and *refine* the likelihoods of different events, to imagine with greater fidelity.

3 Aleatoric Logic

Aleatoric logic [2, 3] is a many valued logic where propositions correspond to chance events, like the sampling of marbles from an urn. These events (imagined things) are drawn from a probability space, with predicates defined over that space (the characterisation of things). The operations are the basic Boolean operations, along with an expectation operator that gives the likelihood a sampled event (or imagined thing) would satisfy a proposition, and a fixed point operation that represents iteration. In [3] a scheme is presented to allow induction where the probability distribution of events is conditioned on observation

The model we have presented conditions the imagination to produce "best guesses" of unknowns. Reason comes through the probabilistic combination of these guesses in a rational process. We will discuss how aleatoric logic can represent each of the layers in this model, examine how this may allow the emergence of rationality, and consider potential future directions of this work.

Disclosure of Interests. The author has no competing interests to declare that are relevant to the content of this article.

References

1. Fagin, R., Halpern, J.Y., Moses, Y., Vardi, M.: Reasoning About Knowledge. MIT Press, Cambridge (2004)
2. French, T.: Aleatoric propositions: reasoning about coins. In: Hansen, H.H., Scedrov, A., de Queiroz, R.J.G.B. (eds.) Logic, Language, Information, and Computation. WoLLIC 2023. LNCS, vol. 13923, pp. 227–243. Springer, Cham (2023). https://doi.org/10.1007/978-3-031-39784-4_14
3. French, T.: Aleatoric predicates: reasoning about marbles. In: AAMAS 2024: Proceedings of the 23rd International Conference on Autonomous Agents and Multiagent Systems, pp. 2267–2269 (IFAAMAS) (2024)
4. Halpern, J.Y.: Reasoning About Uncertainty. MIT Press, Cambridge (2017)
5. Papineau, D.: The evolution of knowledge. Evolution and the Human Mind. Modularity, Language and Meta-Cognition, pp. 170–206 (2000)
6. Ramsey, F.P.: The Foundations of Mathematics. Oxford University Press, Oxford (1925)
7. Reidl, R.: Biology of Knowledge: The Evolutionary Basis of Reason. Wiley, Hoboken (1984)
8. Williamson, T.: Knowing by imagining. In: Knowledge Through Imagination. Oxford University Press, Oxford (2016)

Rethinking Logics of Imagination

Aybüke Özgün

ILLC, University of Amsterdam, Amsterdam, The Netherlands
`a.ozgun@uva.nl`

Intentional modals, such as knowability [4], knowledge [12], belief [8, 15], and, of particular interest for this talk, *imagination* [5, 6, 9, 16], have recently received topic-sensitive treatment, which proceeds by taking seriously their intentionality – by focusing on the specific subject matter they are about. The central idea is that, much like linguistic expressions, these modals have topics, and the space of topics is conveniently structured by means of a parthood relation. This approach has been particularly useful in the development of logics for imagination, especially the notion of imagination as *reality-oriented mental simulation*—a concept that plays a crucial role in, e.g., counterfactual thought, pretense, contingency planning, and decision-making.

A prominent logic of imagination, proposed in [5], employs binary imagination operators, $I^\varphi \psi$, which represent an imagination or mental simulation operator and are read as: 'In an act of imagination starting with input φ, one imagines that ψ'. Semantically, this operator is modelled using Lewis-Stalnaker-style set-selection functions over possible worlds [13], combined with a (quasi-)mereology of topics [7, 10]. More precisely, according to this logic, $I^\varphi \psi$ is true at w when:

(**TC**) ψ is true at all worlds w' accessible via the set-selection function determined by φ.

(**AP**) ψ is fully on topic with respect to φ.

The truth-conditional component (TC) makes $I^\varphi \psi$ a variably strict quantifier over worlds, similar to a Lewis-Stalnaker-style conditional operator. The aboutness preservation component (AP) requires that the topic of the output of an imaginative episode be contained in the topic of the input. In other words, the topic of the imagined output can never go beyond the topic of the input that triggers the imaginative exercise.

In this talk, I will critically examine both TC and AP, identify some problems, and argue that there are reasons to weaken or revise both. I will propose weakening AP by introducing a topological closure or a more general, weaker algebraic operator (as developed in [14], joint work with A. J. Cotnoir). In light of this revision to AP, I will then reconsider TC, exploring how the two components of the new proposal align and how we might revise the truth-conditional component to achieve a more plausible formalization of imagination (based on ongoing joint work with Tianyi Chu and Tom Schoonen). Along the way, time permitting, I will discuss options borrowed from plausibility models for belief [2, 3], normality measures for epistemic operators [11], and logics of abstraction [1].

Disclosure of Interests. The author has no competing interests to declare that are relevant to the content of this article.

References

1. Baltag, A., Bezhanishvili, N., Ilin, J., Özgün, A.: Quotient dynamics: the logic of abstraction. In: Logic, Rationality, and Interaction (2017). https://api.semanticscholar.org/CorpusID:31373939
2. Baltag, A., Smets, S.: Conditional doxastic models: a qualitative approach to dynamic belief revision. Electron. Notes Theoret. Comput. Sci. **165**, 5–21 (2006). https://doi.org/https://doi.org/10.1016/j.entcs.2006.05.034. https://www.sciencedirect.com/science/article/pii/S1571066106005111
3. van Benthem, J.: Dynamic logic for belief revision. J. Appl. Non-Class. Logics **17**(2), 129–155 (2007). https://doi.org/10.3166/jancl.17.129-155
4. Berto, F., Hawke, P.: Knowability relative to information. Mind **130**(517), 1–33 (2018). https://doi.org/10.1093/mind/fzy045
5. Berto, F.: Aboutness in imagination. Philos. Stud. **175**(8), 1871–1886 (2018). https://doi.org/10.1007/s11098-017-0937-y
6. Berto, F.: Taming the runabout imagination ticket. Synthese **198**(Suppl 8), 2029–2043 (2018). https://doi.org/10.1007/s11229-018-1751-6
7. Berto, F., Hawke, P., Özgün, A.: Topics of Thought. The Logic of Knowledge, Belief, Imagination. Oxford University Press, Oxford (2022)
8. Berto, F.: Simple hyperintensional belief revision. Erkenntnis **84**(3), 559–575 (2019). https://doi.org/10.1007/s10670-018-9971-1
9. Canavotto, I., Berto, F., Giordani, A.: Voluntary imagination: a fine-grained analysis. Rev. Symb. Log. **15**(2), 362–387 (2022). https://doi.org/10.1017/S175502032000039
10. Fine, K.: Analytic implication. Notre Dame J. Formal Log. **27**(2), 169–179 (1986). https://doi.org/10.1305/ndjfl/1093636609
11. Goodman, J., Salow, B.: Epistemology normalized. Philos. Rev. **132**(1), 89–145 (2023). https://doi.org/10.1215/00318108-10123787
12. Hawke, P., Özgün, A., Berto, F.: The fundamental problem of logical omniscience. J. Philos. Log. **49**(4), 727–766 (2020). https://doi.org/10.1007/s10992-019-09536-6
13. Lewis, D.K.: Counterfactuals. Blackwell (1973)
14. Özgün, A., Cotnoir, A.J.: Imagination, mereotopology, and topic expansion. Rev. Symb. Log. **18**(1), 28–51 (2025). https://doi.org/10.1017/S1755020324000236
15. Özgün, A., Berto, F.: Dynamic hyperintensional belief revision. Rev. Symb. Log. **14**(3), 766–811 (2021). https://doi.org/10.1017/S1755020319000686
16. Özgün, A., Schoonen, T.: The logical development of pretense imagination. Erkenntnis **89**(6), 2121–2147 (2024). https://doi.org/10.1007/s10670-021-00476-9

On Verifying Graph Neural Networks with Logic

François Schwarzentruber

ENS de Lyon, CNRS, Université Claude Bernard Lyon 1, Inria, LIP, UMR 5668, 69342 Lyon, France
francois.schwarzentruber@ens-lyon.fr
https://perso.ens-lyon.fr/francois.schwarzentruber/

Abstract. Graph neural networks are algorithms used for classifying graphs or vertices (i.e. pointed graphs). We present a methodology for verifying GNNs based on logic. We then present several correspondence between GNNs and modal logic with counting and linear inequalities from the literature. When the activation function is truncated ReLU or when GNNs are quantized, verification is PSPACE-complete.

Keywords: Graph neural networks · Verification · Modal logic · Satisfiability problem

1 Introduction

Today, we are surrounded by programs built with neural network models that are considered as black boxes and that are difficult to trust. That is why it is important to formally verify them, see for instance [5] and [6].

While many ML models take data whose structure is fixed beforehand (vectors, bitmap image, etc.), graph neural networks (GNNs) [11] offer the possibility to handle data with arbitrary structure and size e.g. molecules, music scores, social networks (see [12]). A GNN takes as an input a graph (for graph classification) or a pointed graph, i.e. a graph with a designated vertex (for vertex classification).

GNNs need to be verified as well, especially in critical applications. That is why, researchers study GNNs by means of formal methods in order to understand their complexity [4], expressivity [7] and to develop verification methods. In this paper, we focus on the connections between GNNs and logic for verification.

2 Graph Neural Networks

We consider labelled directed graphs $G = (V, E, \ell)$ where V is a non-empty finite set of vertices, $E \subseteq V \times V$ is a set of edges and $\ell : V \to \mathbb{R}^d$ is a labelling function, i.e. each vertex u is labelled with a vector $\ell(u)$ containing d real numbers. We denote by $E(u)$

the set of successors of u. Formally, $E(u) = \{v \in V | (u, v) \in E\}$. We encode a standard Kripke structure by a labelled graph $G = (V, E, \ell_0)$ with $\ell_0 : V \to \{0, 1\}^d$.

A GNN N is usually defined as a tuple of parameters (see [1, 2, 8]). To make it more concrete, we present it as an algorithm (parametrized by the weights computed during some learning process), see figure:GNN_algorithm. It takes as an input a pointed graph made up of a labelled graph $G = (V, E, \ell_0)$ and a vertex u. It outputs yes/no.

$$
\begin{aligned}
&\textbf{main function } N((V, E, \ell_0), u) \\
&\quad \ell_1 := layer_1(V, E, \ell_0) \\
&\quad \ell_2 := layer_2(V, E, \ell_1) \\
&\quad \vdots \\
&\quad \ell_L := layer_L(V, E, \ell_{L-1}) \\
&\quad \textbf{return } \text{yes } \textbf{if } w^t \ell_L(u) + b \geq 0 \textbf{ else } \text{no}
\end{aligned}
$$

$$
\begin{aligned}
&\textbf{function } layer_i(V, E, \ell) \\
&\quad \ell' := \text{new labelling } V \to \mathbb{R}^d \\
&\quad \textbf{for } \text{vertices } u \in V \textbf{ faire} \\
&\quad\quad \ell'[u] := \alpha(A_i \times \ell[u] + B_i \times \sum \{\!\{\ell[v] \mid v \in E(u)\}\!\} + b_i) \\
&\quad \textbf{return } \ell'
\end{aligned}
$$

Fig. 1.   A graph neural network N presented as an algorithm. The main function is N. It computes a sequence of labellings $\ell_1, \ell_2, \ldots, \ell_L$ via functions $layer_i$. Learnt weights are w, b, A_i, B_i, b_i.

A GNN computes a sequence of labellings $\ell_1, \ell_2, \ldots, \ell_L$ via the application of so-called layers. For avoiding cumbersome notations, we suppose that all these labellings assign a vector of dimension d to each vertex (while in generality the dimension d may be different for each layer). In each layer $layer_i$, the function $\vec{\alpha} : \mathbb{R}^d \to \mathbb{R}^d$ is the point-wise application of an activation function $\alpha : \mathbb{R} \to \mathbb{R}$. The activation function α could be for instance $ReLU : x \mapsto \max(0, x)$ or $truncReLU : x \mapsto \max(0, \min(x, 1))$. Then $A_i \in \mathbb{R}^{d \times d}$ and $B_i \in \mathbb{R}^{d \times d}$ are matrices of weights, $\times$ is the standard matrix-vector multiplication, $b_i \in \mathbb{R}^d$ is a bias vector, $\{\!\{.\}\!\}$ is the multiset notation, and $\sum$ is the summation operation of all vectors in the multiset.

The ending linear inequality $w^t \ell_L(u) + b \geq 0$ where $w \in \mathbb{R}^d$ and $b \in \mathbb{R}$ are weights is used to classify the vertex u. In the sequel, the set of pointed labelled graphs positively classified by a GNN N is denoted by $[[N]] := \{(G, u) | N(G, u) \text{returnsyes}\}$.

Verification Tasks

We consider formulas in some logic evaluated on pointed graphs (for instance, modal logic or graded modal logic). Given a formula φ, we write $[[\varphi]]$ for the set of pointed graphs (G, u) satisfying φ. Let us list some verification tasks.

1. Satisfiability problem of a GNN: Given a GNN N, is there an input $(G, E, \updownarrow)$ that is positively classified by N? $([[N]] \neq \varnothing)$
2. Given a GNN N, given a specification formula φ, are the inputs positively classified by N exactly the inputs satisfying φ? $([[N]] = [[\varphi]])$
3. Given a GNN N, given a specification formula φ, are the inputs positively classified by N satisfying φ? $([[N]] \subseteq [[\varphi]])$
4. Given a GNN N, given a specification formula φ, are the inputs satisfying φ classified positively by N? $([[\varphi]] \subseteq [[N]])$
5. Given a GNN N, given a specification formula φ, does there exist an input satisfying φ and classified positively by N? $([[\varphi]] \cap [[N]] \neq \varnothing)$

4 Methodology

The verification tasks can be solved as follows. We design a "low-level" logical language $\mathcal{L}$ that is expressive enough to capture the computation of a GNN, and the specification language. Specifically, we aim for the existence of a function tr such that:

- for all GNNs N, $tr(N)$ is a $\mathcal{L}$-formula such that $[[N]] = [[tr(N)]]$;
- for all specification formulas φ, $tr(\varphi)$ is a $\mathcal{L}$-formula such that $[[\varphi]] = [[tr(\varphi)]]$.

To solve some verification task, the idea is to reduce it to the satisfiability problem of $\mathcal{L}$. For instance, the verification task 5 can be solved as follows:

i. transform N into a $\mathcal{L}$-formula $tr(N)$; transform φ into a $\mathcal{L}$-formula $tr(\varphi)$;
ii. check that $tr(N) \wedge tr(\varphi)$ is $\mathcal{L}$-satisfiable.

The seminal work [1] (also restated in [3]) shows that any formula in graded modal logic (GML) can be represented by a GNN. However, the reverse direction, that is, a GNN N can be transformed into a GML-formula is true only if N is expressible in first-order logic. In fact, GML is not a good candidate for the "low-level" language $\mathcal{L}$.

In modal logic $K^{\#}$ [8, 9] we reason about linear inequalities on expressions $\#\varphi$ which counts the number of successors satisfying φ. There is a correspondence between GNNs with truncReLU and *integer* weights and formulas in $K^{\#}$. Thus, $K^{\#}$ is a good candidate for $\mathcal{L}$; furthermore the satisfiability problem against $K^{\#}$-formulas is PSPACE-complete [8]. In [2], they present a logic similar to $K^{\#}$ but with a first-order syntax. Interestingly, they give a correspondence between GNNs with truncReLU and *rational* weights, and formulas in $K^{\#}$. They also prove that the satisfiability problem of a GNN with ReLU is decidable, but the exactly complexity is open.

In [10], they propose a logic $\mathcal{L}$ that directly mimics the computation of a GNNs. They consider *quantized* GNNs where numbers are represented by a fixed number of bits (e.g. 64 bits). They proved that $\mathcal{L}$ is PSPACE-complete, regardless of the activation function being used, and provide a tableau method implemented in Python[1].

---

[1] https://github.com/francoisschwarzentruber/ijcai2025-verifquantgnn.

5 Variants

*Directed vs undirected.*Some GNNs take undirected graphs as an input. As shown in [2], PSPACE-completeness with truncReLU also holds for undirected graphs. However, decidability with ReLU is still open for undirected graphs.

Global readout. Some GNNs have a mechanism to globally read all vertices in a layer. In [2], they show that the satisfiability problem for GNNs with truncReLU and global readout is undecidable.

Acknowledgments. I thank the co-authors of works on which this extended abstract is based on: Pierre Nunn that helped to define logic $K^{\#}$ in [9], and Marco Sälzer and Nicolas Troquard than have joined the adventure in [8, 10]. I also thank the organizers of LORI-25 for their invitation.

Disclosure of Interests. The author has no competing interests to declare that are relevant to the content of this article.

References

1. Barceló, P., Kostylev, E.V., Monet, M., Pérez, J., Reutter, J.L., Silva, J.P.: The logical expressiveness of graph neural networks. In: 8th International Conference on Learning Representations, ICLR 2020, Addis Ababa, Ethiopia, April 26–30, 2020. OpenReview.net (2020). https://openreview.net/forum?id=r1lZ7AEKvB
2. Benedikt, M., Lu, C., Motik, B., Tan, T.: Decidability of graph neural networks via logical characterizations. In: Bringmann, K., Grohe, M., Puppis, G., Svensson, O. (eds.) 51st International Colloquium on Automata, Languages, and Programming, ICALP 2024, July 8–12, 2024, Tallinn, Estonia. LIPIcs, vol. 297, pp. 127:1–127:20. Schloss Dagstuhl - Leibniz-Zentrum für Informatik (2024). https://doi.org/10.4230/LIPIcs.ICALP.2024.127
3. Grohe, M.: The logic of graph neural networks. In: 36th Annual ACM/IEEE Symposium on Logic in Computer Science, LICS 2021, Rome, Italy, June 29 – July 2, 2021, pp. 1–17. IEEE (2021). https://doi.org/10.1109/LICS52264.2021.9470677
4. Grohe, M.: The descriptive complexity of graph neural networks. TheoretiCS **3** (2024). https://doi.org/10.46298/theoretics.24.25
5. Huang, X., et al.: A survey of safety and trustworthiness of deep neural networks: Verification, testing, adversarial attack and defence, and interpretability. Comput. Sci. Rev. **37**, 100270 (2020). https://doi.org/10.1016/J.COSREV.2020.100270
6. Marques-Silva, J., Ignatiev, A.: Delivering trustworthy AI through formal XAI. In: Thirty-Sixth AAAI Conference on Artificial Intelligence, AAAI 2022, pp. 12342–12350. AAAI Press (2022). https://doi.org/10.1609/aaai.v36i11.21499
7. Morris, C., et al.: Weisfeiler and leman go neural: higher-order graph neural networks. In: The Thirty-Third AAAI Conference on Artificial Intelligence, AAAI 2019. Honolulu, Hawaii, USA, January 27 – February 1, 2019, pp. 4602–4609. AAAI Press (2019). https://doi.org/10.1609/aaai.v33i01.33014602
8. Nunn, P., Sälzer, M., Schwarzentruber, F., Troquard, N.: A logic for reasoning about aggregate-combine graph neural networks. In: Proceedings of the 33rd International

Joint Conference on Artificial Intelligence, IJCAI 2024, Jeju, South Korea, 3–9 August 2024. ijcai.org (2024)

9. Nunn, P., Schwarzentruber, F.: A modal logic for explaining some graph neural networks. CoRR **abs/2307.05150** (2023). https://doi.org/10.48550/arXiv.2307.05150

10. Sälzer, M., Schwarzentruber, F., Troquard, N.: Verifying quantized graph neural networks is pspace-complete. In: Proceedings of the 34rd International Joint Conference on Artificial Intelligence, IJCAI 2025, Montréal, Canada, 16–22 August 2025 (2025), (to appear)

11. Scarselli, F., Gori, M., Tsoi, A.C., Hagenbuchner, M., Monfardini, G.: The graph neural network model. IEEE Trans. Neural Netw. **20**(1), 61–80 (2009). https://doi.org/10.1109/TNN.2008.2005605

12. Zhou, J., et al.: Graph neural networks: a review of methods and applications. AI Open **1**, 57–81 (2020). https://doi.org/10.1016/J.AIOPEN.2021.01.001

Logic in Machine Morality

Marija Slavkovik

Department of Information Science and Media Studies, University of Bergen, Norway
marija.slavkovik@uib.no

Abstract. This extended abstract summaries the main points of my recent work in machine ethics. The intention with this discussion is to argue that in the field of machine ethics, and the broader field of AI alignment, logic reasoning is a necessary tool to accomplish artificial moral agents. While much attention at present is on sub-symbolic artificial intelligence approaches, the future puts new challenges and makes new demands on logic based methods.

Keywords: machine ethics · AI alignment · moral reasoning

1 Introduction

The issue that will be discussed here is what is the future of logic reasoning in the context of the field of machine ethics. Let us start with machine ethics. Machine ethics as an interdisciplinary field of study is concerned 'is concerned with the behavior of machines towards human users and other machines" [3].

When we make decisions and pursue goals in an environment shared with others, we do not only do what is most efficient, but also pay consideration to other people. When an autonomous system or an artificial intelligence makes decisions or pursues goals on our behalf, we would expect the same. Otherwise, the impact or using these systems on society and our individual lives, might not be worth the benefits.

In the past few years, the artificial intelligence (AI) community has had more affinity for the *AI alignment* field [11], which is concerned with ensuring that the operations with AI and of an AI agent are aligned with specific human designated values, including but not limited to safety. AI alignment underscores the need for, not only that the task is done, using AI, but that there is relevance in *how* it is done. The relationship between the fields of AI alignment and machine ethics is not clear cut [24], but the interest in both is well documented [14, 18, 22], which in turn, can be taken to demonstrate the pressing need for morally behaving computation, agents, autonomous systems.

2 Challenges

Earlier approaches in machine ethics were dominated by rule based approaches that used logic specification and reasoning [6]. Most recently, the interest of the technology developers has been turned towards so called "agentic AI" approaches[2] These predominantly non-academic AI communities [1, 12, 17] reference AI systems that involve deep learning and foundational models and that are developed for handling open-ended tasks that extend beyond their initial training data. One of the core approaches is reinforcement learning [21] where learning from human feedback [19], without maintaining a strong connection with moral philosophy or formal ethics, is the dominating approach [23]. Is there a need for logic reasoning in modern machine ethics? What follows are arguments that answer that question in the affirmative.

3 The Right Thing to Do

Making a moral decision is about choosing the right thing to do. To make this choice an agent needs to make a moral judgment of what the right thing to do is. This can be based on some value or principle, such as for example "do not do harm" or grounded in some moral theory, such as for example consequentalist theory, deontic theory or virtue ethics. Machine learning has proven very good at classification problems. Can an AI system machine extrapolate patterns of what is the right thing to do?

Delphi is an example of an AI system that morally evaluates natural language sentences [13]. However, systems based on large language or foundational models have many limitations. One is of course their accuracy and anoter their lack of ability to reason, see for example [15]. When it comes to moral evaluation, that choices are not made in isolation and not all options are available at the same time. Consider for example the choice to feed the poor and buy medicine for your family member. Both are the right thing to do, but you can afford one.

The future appears to lie somewhere in the intersection of the scalability of machine learning and the precision of logic reasoning. Having a moral evaluation specified as a logic statement would allow using existing automated reasoning methods.

In her masters thesis, Bjørkas [7] considers using GPT-4 to encode moral evaluations (normative statements) expressed in natural language (English) into first order (FOL) Horn logic propositions. She finds that this encoding system is able to encode 51% of the statements accurately to FOL in terms of syntactic properties and semantic faithfulness. The more interesting question she asks is what kind of mistakes the LLM system does and how easy it is to automatically detect the wrong encoding. Errors that occur can be as the following that is mentioned in the thesis:

Norm: "it's fine to want good things for yourself"

FOL encoding: $\forall x \forall y (GoodThings(y) \land WantsFor(x, y) \rightarrow evaluation(GOOD))$

---

[2] See e.g. https://www.mckinsey.com/capabilities/mckinsey-digital/our-insights/why-agents-are-the-next-frontier-of-generative-ai and https://hbr.org/2024/12/what-is-agentic-ai-and-how-will-it-change-work.

The encoding lacks a variable in the predicate *WantsFor(x,y)*, which should be *WantsFor(x,y,x)*.

A big open problem is how should moral evaluations be formally specified. The Delphi system uses a scale of goodness and badness. Labels such as these are not sufficient to make meaningful moral decisions. It has been argued that ethical reasons for decisions are necessary [2]. This implies the need for an argumentation structure to support claims of moral rightness [10, 16].

4 Moral Proxy

It has been anecdotally argued that the biggest problem of machine ethics is that we do not have a firm grasp on human ethics. What does it mean for a machine to do the right thing? Or to put it more plainly: where should the information of what is the right thing to do come from? While there are some aspects of the moral behavior of machines that should be governed by well identified regulations and rights, others should reflect the morality ideals of the users on whose behalf the machines act. It has been argued that the ethics of machines should be determined by social choice [5, 8].

Doing what is best for the collective and society is indeed in line with most of moral behavior ideas. On one hand, social choice does not absolve us from the hard moral reasoning problems we encounter [4]. However, smart technology penetrates many aspects of our lives and certain behavior of this technology should be supported by everyone, not just a majority. Consider for instance the option to constrain a smartphone from recording pictures of children on the beach. While such a design would positively contribute to unwanted sharing of some images, it would negatively affect the joy of grandparents. While social choice can identify the most representative option in the quantitative sense, logic reasoning can be used to construct qualitative compromises. We give two examples of such approaches.

Liao et al. [16] consider the option of building an argumentation framework to simulate negotiation between different stakeholders with different opinions on what should a smart machine do. The limitations of this approach lie in the scalability of the requirement to elicit and specify in logic the positions of the stakeholders.

Ozaki et al. [20] consider resolving some of the possible conflicts between the positions of different stakeholders. Their approach is based on a set of postulates that guarantee maximal retention of information from the stakeholders: no one's position is overturned. For example, when presented with two requirements: a) police should be notified when shoplifting is detected, and b) guardians should be notified when a child is shoplifting, the computed retained requirement is c) police should be notified when shoplifting is detected by an adult and d) guardians should be notified when shoplifting is detected by a child. The limitation of their work is in the expressivity of the used language (propositional Horn logic) and, same as [16] requirement to elicit and specify the positions of stakeholders.

5 Summary and Future work

The requirement of having autonomous systems and intelligent artificial agents that are not only efficient but well adjusted and of moral behavior is here to stay. Although much can be accomplished with machine learning and related approaches, particularly on machine ethics, reasoning is required. We discussed here two aspects of machine ethics where logic reasoning has been identified as quintessential: the specification of the moral qualities of an option or action and the resolving of conflicting positions by stakeholders. In addition to these there is also the requirement of verifying that indeed the behavior of an artificial moral agent is adequate [9]. While we as a society have evolved many societal and psychological mechanism to handle the immoral behavior of others, when machines are participants in society more formal guarantees should be required. As this field of verification of moral (or aligned) behavior develops, more need for logic specification and reasoning will present itself.

Disclosure of Interests. The author has no competing interests.

References

1. Acharya, D.B., Kuppan, K., Divya, B.: Agentic AI: autonomous intelligence for complex goals–a comprehensive survey. IEEE Access **13**, 18912–18936 (2025). https://doi.org/10.1109/ACCESS.2025.3532853
2. Alcaraz, B., Knoks, A., Streit, D.: Estimating weights of reasons using metaheuristics: A hybrid approach to machine ethics. In: et al., S.D. (ed.) Proceedings of the Seventh AAAI/ACM Conference on AI, Ethics, and Society (AIES-2024), pp. 27–38. ACM Press (2024)
3. Anderson, M., Anderson, S.L.: The status of machine ethics: a report from the aaai symposium. Minds Mach. **17**(1), 1–10 (2007). https://doi.org/10.1007/s11023-007-9053-7
4. Baum, K., Slavkovik, M.: Aggregation problems in machine ethics and ai alignment. In: Proceedings of the 2025 AAAI/ACM Conference on AI, Ethics, and Society, October 20–22, 2025, Madrid, Spain (2025), fortcoming
5. Baum, S.: Social choice ethics in artificial intelligence. AI and Soc. **35**, 165–176 (2020). https://doi.org/10.1007/s00146-017-0760-1
6. Bello, P., Malle, B.F.: Computational approaches to morality. In: Sun, R. (ed.) Cambridge Handbook of Computational Cognitive Sciences, pp. 1037–1063. Cambridge University Press, Cambridge (2023)
7. Bjørkas, E.: Automatic Encoding From Natural Language to First-Order Horn Clauses. Master's thesis, University of Bergen, Bergen, Norway (2024). https://bora.uib.no/bora-xmlui/handle/11250/3139356
8. Conitzer, V., et al.: Position: social choice should guide ai alignment in dealing with diverse human feedback. In: Proceedings of the 41st International Conference on Machine Learning, pp. 9346–9360 (2024)
9. Dennis, L.A., Fisher, M.: Specifying agent ethics (blue sky ideas). CoRR **abs/2403.16100** (2024). https://doi.org/10.48550/ARXIV.2403.16100

10. Dignum, V., Michael, L., Nieves, J.C., Slavkovik, M., Suarez, J., Theodorou, A.: Contesting black box ai decisions. In: Das, S., Nowe, A., Vorobeychik, E., Shah, N., Turrini, P. (eds.) Proceedings of the 19th International Conference on Autonomous Agents and Multiagent Systems, AAMAS'25, Detroit, USA, May 19–23, 2025, p. tbd. International Foundation for Autonomous Agents and Multiagent Systems (2025)

11. Gabriel, I., Keeling, G.: A matter of principle? AI alignment as the fair treatment of claims. Philos. Stud. **182**, 1951–1973 (2025). https://doi.org/10.1007/s11098-025-02300-4

12. Gulli, A., et al.: Agents companion. Technical report, Google (2025). https://www.kaggle.com/whitepaper-agent-companion,white paper,self-published

13. Jiang, L., et al.: Towards machine ethics and norms. ArXiv **abs/2110.07574** (2021). https://api.semanticscholar.org/CorpusID:238857096

14. Ji, J., et al.: AI alignment: a comprehensive survey (2025). https://arxiv.org/abs/2310.19852

15. Kucharavy, A.: Fundamental limitations of generative LLMs. In: Kucharavy, A., Plancherel, O., Mulder, V., Mermoud, A., Lenders, V. (eds) Large Language Models in Cybersecurity, pp. 55–64. Springer, Cham (2024). https://doi.org/10.1007/978-3-031-54827-7_5

16. Liao, B., Pardo, P., Slavkovik, M., van der Torre, L.: The jiminy advisor: moral agreements among stakeholders based on norms and argumentation. J. Artif. Intell. Res. **77**, 737–792 (2023). https://doi.org/10.1613/jair.1.14368

17. Mukherjee, A., Chang, H.H.: Agentic AI: autonomy, accountability, and the algorithmic society (2025). https://arxiv.org/abs/2502.00289

18. Nallur, V.: Landscape of machine implemented ethics. Sci. Eng. Ethics (2020). https://doi.org/10.1007/s11948-020-00236-y

19. Ouyang, L., et al.: Training language models to follow instructions with human feedback. In: Advances in Neural Information Processing Systems, vol. 35, pp. 27730–27744 (2022)

20. Ozaki, A., Rehman, A., Slavkovik, M.: Finding middle grounds for incoherent horn expressions: the moral machine case. Auton. Agents Multi-Agent Syst. **38**(2), 50 (2024). https://doi.org/10.1007/s10458-024-09681-6

21. Sutton, R.S., Barto, A.G.: Reinforcement Learning: An Introduction. A Bradford Book, Cambridge (2018)

22. Tolmeijer, S., Kneer, M., Sarasua, C., Christen, M., Bernstein, A.: Implementations in machine ethics: A survey. ACM Comput. Surv. **53**(6), 1–38 (2021). https://doi.org/10.1145/3419633

23. Vishwanath, A., Dennis, L.A., Slavkovik, M.: Reinforcement learning and machine ethics:a systematic review (2024). https://arxiv.org/abs/2407.02425

24. Vishwanath, A., Slavkovik, M.: Machine ethics or AI alignment? In: Jenssen, R., Bach, K. (eds.) Proceedings of the Symposium of the Norwegian AI Society 2025, Tromsø, Norway, June 17–18, 2025. CEUR Workshop Proceedings, vol. 3975, pp. 81–90. CEUR-WS.org (2025). https://ceur-ws.org/Vol-3975/paper9.pdf

On the Research Progress of Mathematical Fuzzy Logics

Hongjun Zhou(iD)

Shaanxi Normal University, Xi'an, 710119 China
`hjzhou@snnu.edu.cn`

Abstract. In this talk we will provide first an overview on the centenary developments of mathematical fuzzy logics from several perspectives including three development stages, three research levels, three axiomatic methods, three logical languages and some applications. Then we list some new research directions on general algebraic studies of mathematical fuzzy logics such as weakly implicative logics, semilinear logics, subresiduated logics, quantum B-algebras and importation algebras. Finally, we will report some our works on solving and applications of the laws of importation and migrativity between fuzzy logical connectives towards establishing new types of mathematical fuzzy logics.

Keywords: Mathematical fuzzy logic · Subresiduated logic · Importation law · Migrativity

The term "fuzzy logic" was used after the emergence of "fuzzy set" [1], referring to all logics that accept degrees of truth, i.e., its truth tables accommodate truth values beyond the classical "true" (represented as 1) and "false" (represented as 0). Typically, fuzzy logic specifically refers to multivalued logic with truth values in the unit interval [0, 1]. However, from the perspective of lattice-valued fuzzy sets [2], any multivalued logic with truth values structured as a bounded lattice can be termed fuzzy logic [3]. Thus, fuzzy logic may also broadly encompass multivalued logic in general.

In terms of scope, fuzzy logic can be divided into broad and narrow definitions. Broadly, it refers to all symbolic logical calculus systems addressing the nature of fuzzy phenomena, based on the representation of fuzzy sets and formalized at the levels of attributes, relations, and operations (e.g., assignment, modification, extraction, extension, combination, decomposition, derivation, and evolution). Narrowly, it denotes a semantically complete formal logical calculus system that processes graded truth values over the unit interval [0, 1] using classical logical methods, also known as mathematical fuzzy logic [4, 5].

Mathematical fuzzy logic is a branch of mathematics that employs traditional methods of mathematical logic to study formal logical systems whose algebraic semantics include truth values beyond just 0 and 1. While it extends the bivalence principle of classical mathematical logic, it retains the principle of truth-functionality, meaning that the truth value of a compound formula is entirely determined by the truth values of its subformulas. Like classical mathematical logic, it is a system of inference rules and principles that provides methods for deriving conclusions from given premises which may involve uncertain or fuzzy information. Such logical reasoning is developed from both syntactic and semantic perspectives, ultimately unified by completeness theorems [6–8].

A key distinction from classical mathematical logic is the existence of diverse logical systems within fuzzy logic, such as Łukasiewicz logic, Gödel logic, BL-logic, and MTL-logic, along with their combinations and extensions (including predicate extensions). Unlike other non-classical logics (e.g., intuitionistic logic, residuated lattice logic, substructural logic), mathematical fuzzy logic not only focuses on algebraic completeness (relative to semantic algebras) and linear completeness (relative to linearly ordered algebras) but also prioritizes proving standard completeness with respect to algebraic semantics on the unit interval [0, 1]. In this talk we will provide an overview on the centenary developments of mathematical fuzzy logics from several perspectives including three development stages, three research levels, three axiomatic methods, three logical languages and some applications [9, 10]. We conclude from the overview that mathematical fuzzy logics are a branch of mathematical logic that are rooted in real numbers.

The well-established formal systems of mathematical fuzzy logic as reviewed above are usually based on the residuation law between conjunction and implication connectives [7, 8, 11, 12]. In recent years, there has been a growing trend of more general algebraic study towards mathematical fuzzy logic by choosing different primitive connectives or other basic laws, such as Cintula's weakly implicative logics and semilinear logics [13] starting with a weak implication, subresiduated logics [14–17] based on the subresiduation law between conjunction and implication connectives, Rump-Yang's quantum B-algebras [18, 19] by the exchange principle of implication connective, and Gupta-Jayaram's importation algebras [20] based on the importation law between conjunctive and implication connectives.

The importation algebras in [20] are not compatible well with ordered structures, and on the other hand, the solving of importation equation involving fuzzy conjunction and implications has been a longstanding open problem [21–24]. We will report some our works on solving and applications of the laws of importation and migrativity between fuzzy logical connectives towards establishing new types of mathematical fuzzy logics [25, 26, 28–33].

Acknowledgments. This work was supported by National Natural Science Foundation of China (Grant No. 12171292) and Science Fund for Distinguished Young Scholars of Shaanxi Province (Grant No. 2024JC-JCQN-01).

Disclosure of Interests. The author has no competing interests.

References

1. Zadeh, L. A.: Fuzzy sets. Inform. Control **8**(3), 338–353 (1965)
2. Goguen, J. A.: L-fuzzy sets. J. Math. Anal. Appl. **18**(1), 145–174 (1967)
3. Rosser, J. B., Turquette, A. R.: Many-valued Logics. North-Holland, Amsterdam (1952)
4. Hájek, P.: What is mathematical fuzzy logic. Fuzzy Sets Syst. **157**, 597–603 (2006)

5. Gottwald, S.: Mathematical fuzzy logic as a tool for the treatment of vague information. Inform. Sci. **172**, 41–71 (2005)
6. Gottwald, S.: A Treatise on Many-Valued Logic. Resrach Studies Press, Baldok (2001)
7. Hájek, P.: Metamathematics of Fuzzy Logic. Kluwer, Dordrecht (1998)
8. Cintula, P., et al., Handbook of Mathematical Fuzzy Logic, Vols. 1–3. College Publications, London (2015)
9. Bělohlávek, R., Dauben, J. W., Klir, J. J.: Fuzzy Logic and Mathematics: A Historical Perspective. Oxford University Press, New York (2017)
10. Zhou, H.: Probabilistic Quantitative Logic and its Applications. Chinese edn. Scicence Press, Beijing (2015)
11. Metcalfe, G., Paoli, F., Tsinakis, C.: Residuated Structures in Algebra and Logic. American Mathematical Society, Providence (2023)
12. Galatos, N., Přenosil, A.: Complemented MacNeille completions and algebras of fractions. J. Alg. **623**, 288–357 (2023)
13. Cintula, P., Noguera C.: Logic and Implication: An Introduction to the General Algebraic Study of Non-classical Logics. Springer, Cham (2021). https://doi.org/10.1007/978-3-030-85675-5
14. Epstein, G., Horn, A.: Logics which are characterized by subresiduated lattices. Z. Math. Logik Grundlagen Math. **22**, 199–210 (1976)
15. Celani, S., San Martín, H. J.: On the variety of strong subresiduated lattices. Math. Logic Quart. **69** 207–220 (2023)
16. Cornejo, J. M., San Martín, H. J., Sígal, V. A.: Subresiduated lattice ordered commutative monoids. Fuzzy Sets Syst. **463**, 108447 (2023)
17. Cintula, P., Metcalfe, G., Tokuda, N.: Algebraic semantics for one-variable lattice-valued logics (2022). https://arxiv.org/pdf/2209.08566
18. Rump, W., Yang, Y.: Non-commutative logical algebras and algebraic quantales. Ann. Pure Appl. Logic **165**, 759–785 (2014)
19. Rump, W.: L-algebras and three main non-classical logics. Ann. Pure Appl. Logic **173**, 103121 (2022)
20. Gupta, V. K., Jayaram, B.: Importation lattices. Fuzzy Sets Syst. **405**, 1–17 (2021)
21. Mesiar R., Klement, E. P.: Open problems posed at the eighth international conference on fuzzy set theory and applications. Kybernetika **42**, 225–235 (2006)
22. Baczyński M., Jayaram, B.: Fuzzy Implications. Springer, Berlin (2008)
23. Baczyński, M., Beliakov, G., Bustince, H., Pradera, A.: Advances in Fuzzy Implication Functions. Springer, Berlin (2013). https://doi.org/10.1007/978-3-642-35677-3
24. Baczyński, M., Jayaram, B., Massanet, S., Torrens, J.: Fuzzy implications: past, present, and future. In: Kacprzyk, J., Pedrycz, W. (eds.) Springer Handbook of Computational Intelligence, pp. 183–202. Springer, Heidelberg (2015). https://doi.org/10.1007/978-3-662-43505-2_12
25. Massanet, S., Ruiz-Aguilera, D., Torrens, J.: Characterization of a class of fuzzy implication functions satisfying the law of importation with respect to a fixed uninorm-Part I, Part II. IEEE Trans. Fuzzy Syst. **26** (4), 1983–2003 (2018)
26. Li, W., Qin, F.: Characterization of a class of fuzzy implications satisfying the law of importation with respect to uninorms with continuous underlying operators. IEEE Trans. Fuzzy Syst. **30**(5), 1343–1356 (2022)

27. Zhou, H., Song, Y.: Characterization of a class of fuzzy implication solutions to the law of importation. Fuzzy Sets Syst. **441**, 58–82 (2022)
28. Zhou, H.: Characterizations and applications of fuzzy implications generated by a pair of generators of t-norms and the usual addition of real numbers. IEEE Trans. Fuzzy Syst. **30**(6), 1952–1966 (2022)
29. Zhou, H.: Characterizations of fuzzy implications generated by continuous multiplicative generators of t-norms. IEEE Trans. Fuzzy Syst. **29**(10), 2988–3002 (2021)
30. Pan, D., Zhou, H., Yan, X.: Characterizations for the migrativity of continuous t-conorms over fuzzy implications. Fuzzy Sets Syst. **456**, 173–C196 (2023)
31. Zhou, H., Chang, Q., Baczyński, M.: Characterizations on migrativity of continuous triangular conorms over N-ordinal sum implications. Inform. Sci. **637**, 118926 (2023)
32. Yan, X., Zhou, H.: Migrativity properties of general grouping (overlap) functions with regard to null-norms. Comput. Appl. Math. **43**, 251 (2024)
33. Zhou, H.: Ordinal sum combinations of continuous t-norms and their related ordered algebraic structures. Fuzzy Sets Syst. **519**, 109521 (2025)

Contents

Intentionally Anonymous Public Announcements

Thomas Ågotnes[1,2(✉)], Rustam Galimullin[1], Ken Satoh[3], and Satoshi Tojo[4]

[1] University of Bergen, Bergen, Norway
`thomas.agotnes@uib.no`
[2] Shanxi University, Taiyuan, China
[3] Center for Juris-Informatics, Tokyo, Japan
[4] Asia University, Tokyo, Japan

Abstract. We formalise the notion of an *intentionally anonymous public announcement* in the tradition of public announcement logic. An anonymous announcement can be seen as in-between a public announcement from "the outside" (an announcement of φ) and a public announcement by one of the agents a (an announcement of $K_a\varphi$): we get more information than just φ, but not (necessarily) about exactly who made it. In this paper we assume that it is common knowledge that the announcer *intended* to be anonymous. Like in the Russian Cards puzzle, with that assumption, anonymous announcements in fact reveal more information than without. We introduce an operator for intentionally anonymous announcements, and show that in several ways it all boils down to the notion of a "safe" announcement (again, similarly to Russian Cards). We model safety via a fixed-point operator that is similar to common knowledge. Main formal results include comparisons of expressivity and axiomatic completeness for a language expressing safety.

Keywords: Public announcement logic · Dynamic epistemic logic · Modal logic · Expressive power · Completeness · Anonymity · Privacy · Security

1 Introduction

Taken at face value, the title of this paper seems to be an oxymoron. Indeed, if "public announcement" is taken literally, as in an agent *saying* something in front of everyone else, it will not be anonymous. However, anonymous public communication is almost ubiquitous in our day-to-day lives. Think of posts on social media and message boards done under a username instead of a real name of the poster. Or an anonymous letter to an editor of a news outlet. Other examples include anonymous emails, transactions on a blockchain, whistle blower reports, and even cultural artifacts created anonymously under an alias, like Elena Ferrante and MF DOOM.

The type of anonymity we are interested in here focuses on *action anonymity*, i.e., the inability of an attacker to identify who performed a given action (also

V. Goranko et al. (Eds.): LORI 2025, LNCS 16010, pp. 1–17, 2026.
https://doi.org/10.1007/978-981-95-2481-5_1

sometimes referred to as *unlinkability* in the literature [22]). This is in contrast to *data anonymity*, i.e., the inability of an attacker to know the identity of a subject in an anonymised database, e.g., in medical records (see, e.g., [9]). One of the standard requirements of both types of anonymity is that they satisfy k-anonymity [25], which intuitively means that a data record or an action cannot be distinguished from at least $k - 1$ other records or actions. It is clear that in the case of public communication by someone in a group of agents, we should have at least 3-anonymity. Indeed, if a public announcement is so specific that it could be made only by two agents (and these two agents know that), then the non-announcing agent would be able to deduce the identity of the announcer. In the literature, such a scenario is called "background knowledge attack" [17].

In this paper we formalise anonymous public announcements inspired by *public announcement logic* (PAL) [23], an extension of multi-agent *epistemic logic* with constructs of the form $[\varphi!]\psi$ intuitively meaning that after φ is truthfully announced, ψ is true. In PAL the announced formula φ does not have to actually be known by any agent in the system – the identity of the announcer is left out of the picture. If the announcer indeed is one of the agents a in the system, the announcement in fact contains *more* information: in that case it would be modelled by the announcement $K_a\varphi!$. In this paper we formalise *anonymous* public announcements, conceptually somewhere in-between $\varphi!$ and $K_a\varphi!$ – we get *more* information than just φ but *less* than $K_a\varphi$ for a specific agent a. We make an additional crucial assumption: that it is common knowledge that the anonymous announcer *intended* to be anonymous, i.e., to not reveal her identity. Similarly to the Russian Cards puzzle [7], we shall see that assumption actually means that the announcement reveals *more* information. To this end, we introduce and study an intentionally anonymous public announcement operator $[\varphi\ddagger]$, such that $[\varphi\ddagger]\psi$ intuitively means that after φ is anonymously announced by some agent, no matter who, and it is common knowledge that the announcement as intended to be anonymous, ψ is true.

Reasoning about anonymity based on (variants of) epistemic logic has been studied in [11,26], with k-anonymity being discussed in [11]. Building on the runs-and-systems approach of [11], further extensions focus on privacy and onymity [27,28] and electronic voting [18]. A knowledge-based approach to data anonymity was presented in [15]. Other logical approaches to anonymity and privacy include [6,14,24,32]. The themes of anonymity and privacy are also related to the research on secrets in multi-agent systems (see, e.g., [12,20,21,31]). None of these approaches model anonymous *announcements*. Finally, an approach to anonymity based on *dynamic epistemic logic* (DEL) [8] was presented in [29], where the authors consider scenarios of secret communication between agents in a system. Such secret communication is captured by private announcements, as special type of *action models* [5]. In this setting, verifying whether secret communication remains secret boils down to model checking an epistemic formula in the resulting updated model, i.e. the model that is obtained after an application of a dynamic operator. While our approach is also DEL-based, there are two major differences. First, we model anonymous *public* communication, as opposed to *private* communication. Second, we tackle the crucial issue of *intentional* anonymity, where it is common knowledge that only *safe* announcements are made.

We start out by introducing the machinery of EL and DEL in Sect. 2. In Sect. 3, we formalise intentional anonymous announcements as well as introducing a *safety* modality, which is a fixed-point modality somewhat similar to common knowledge. In Sect. 4 we look at expressivity and show that, in several senses, safety is fundamental, and in particular that the logic with the intentionally anonymous announcement operators can be reduced to epistemic logic with safety (but not to EL, PAL or, in general, DEL *without* the safety modality). In Sect. 5 we give a sound an complete axiomatisation of the latter. We end with a discussion in Sect. 6.

2 Background

We give a brief review of relevant background concepts and refer the reader to [8] for further details. Let N be a finite set of *agents*, and P be a countable set of *propositional variables*. All logics in this paper are interpreted on epistemic models.

Definition 1. *An* (epistemic) model *is a triple $M = (S, \sim, V)$, where S is a non-empty set of states, $\sim: N \to 2^{S \times S}$ gives an equivalence relation for each $i \in N$, and $V : P \to 2^S$ is the valuation function. For $s \in S$, a pair M, s is called a* pointed (epistemic) model.

Definition 2 (Action Models). *Let $\mathcal{L}$ be a language defined over the signature $\langle N, P \rangle$. An* action model *is a triple $\mathsf{M} = (\mathsf{S}, \sim, \mathsf{pre})$, where S is a non-empty set of states, $\sim: N \to 2^{\mathsf{S} \times \mathsf{S}}$ gives an equivalence relation for each $i \in N$, and $\mathsf{pre} : \mathsf{S} \to \mathcal{L}$ is the precondition function. For $\mathsf{s} \in \mathsf{S}$, we will call a pair M, s a* pointed action model.

Definition 3. *Languages of* epistemic logic *($\mathcal{L}_K$)*, public announcement logic *($\mathcal{L}_!$)*, and action model logic *($\mathcal{L}_\otimes$)* are defined by the following BNFs:*

$$\mathcal{L}_K \quad \varphi ::= p \mid \neg\varphi \mid (\varphi \wedge \varphi) \mid K_i\varphi$$
$$\mathcal{L}_! \quad \varphi ::= p \mid \neg\varphi \mid (\varphi \wedge \varphi) \mid K_i\varphi \mid [\varphi!]\varphi$$
$$\mathcal{L}_\otimes \quad \varphi ::= p \mid \neg\varphi \mid (\varphi \wedge \varphi) \mid K_i\varphi \mid [\pi]\varphi \quad \pi ::= (\mathsf{M}, \mathsf{s}) \mid \pi \cup \pi$$

where $p \in P$, $i \in N$, and (M, s) is a pointed action model with a finite set of states S, and such that for all $\mathsf{s} \in \mathsf{S}$, the precondition $\mathsf{pre}(\mathsf{s})$ is some $\varphi \in \mathcal{L}_\otimes$ that was constructed in a previous stage of the inductively defined hierarchy. We write $\hat{K}_i\varphi$ for $\neg K_i \neg\varphi$, and when $G \subseteq N$ we write $E_G\varphi$ for $\bigwedge_{i \in G} K_i\varphi$.

The semantics is now defined as follows.

Definition 4. *Let $M, s = (S, \sim, V)$ be a model, $p \in P$, $i \in N$, and (M, s) be an action model.*

$$M, s \models p \qquad \textit{iff } s \in V(p)$$
$$M, s \models \neg\varphi \qquad \textit{iff } M, s \not\models \varphi$$
$$M, s \models \varphi \wedge \psi \qquad \textit{iff } M, s \models \varphi \textit{ and } M, s \models \psi$$
$$M, s \models K_i\varphi \qquad \textit{iff } M, t \models \varphi \textit{ for all } t \in S \textit{ such that } s \sim_i t$$
$$M, s \models [\psi!]\varphi \qquad \textit{iff } M, s \models \psi \textit{ implies } M^{\psi!}, s \models \varphi$$
$$M, s \models [\mathsf{M}, \mathsf{s}]\varphi \quad \textit{iff } M, s \models \mathsf{pre(s)} \textit{ implies } (M \otimes \mathsf{M}, (s, \mathsf{s})) \models \varphi$$
$$M, s \models [\pi \cup \rho]\varphi \textit{ iff } M, s \models [\pi]\varphi \textit{ and } M, s \models [\rho]\varphi$$

We write $[\![\varphi]\!]_M$ for the set $\{s \in S \mid M, s \models \varphi\}$. The updated model $M^{\varphi!}$ is $(S^{\varphi!}, \sim^{\varphi!}, V^{\varphi!})$, where $S^{\varphi!} = [\![\varphi]\!]_M$, $\sim_i^{\varphi!} = \sim_i \cap (S^{\varphi!} \times S^{\varphi!})$ for all $i \in N$, and $V^{\varphi!}(p) = V(p) \cap [\![\varphi]\!]_M$ for all $p \in P$. The updated model $M \otimes \mathsf{M}$ is $(S', \sim', V')$, where $S' = \{(s, \mathsf{s}) \mid s \in S, \mathsf{s} \in \mathsf{S}, M, s \models \mathsf{pre(s)}\}$, $(s, \mathsf{s}) \sim_i' (t, \mathsf{t})$ iff $s \sim_i t$ and $\mathsf{s} \sim_i \mathsf{t}$, and $(s, \mathsf{s}) \in V'(p)$ iff $s \in V(p)$. We call a formula φ valid, or a validity, if for all M, s it holds that $M, s \models \varphi$.

Definition 5. *Let $M^1 = (S^1, \sim^1, V^1)$ and $M^2 = (S^2, \sim^2, V^2)$ be two epistemic models. We say that M^1 and M^2 are* bisimilar *(denoted $M^1 \leftrightarroweq M^2$) if there is a non-empty relation $Z \subseteq S^1 \times S^2$, called a* bisimulation, *such that for all sZt:*

Atoms *for all $p \in P$: $s \in V^1(p)$ if and only if $t \in V^2(p)$,*
Forth *for all $i \in N$ and $u \in S^1$ such that $s \sim_i^1 u$, there is a $v \in S^2$ such that $t \sim_i^2 v$ and uZv,*
Back *for all $i \in N$ and $v \in S^2$ such that $t \sim_i^2 v$, there is a $u \in S^1$ such that $s \sim_i^1 u$ and uZv.*

We say that M^1, s and M^2, t are bisimilar and denote this by $M^1, s \leftrightarroweq M^2, t$ if there is a bisimulation linking states s and t.

It is a standard result that $M^1, s \leftrightarroweq M^2, t$ implies $M^1, s \models \varphi$ if and only if $M^2, t \models \varphi$ for $\varphi \in \mathcal{L}_K \cup \mathcal{L}_! \cup \mathcal{L}_\otimes$ (see, e.g., [8, Chapter 5]).

Definition 6. *Let $\mathcal{L}_1$ and $\mathcal{L}_2$ be two languages defined over the same class of models, and let $\varphi \in \mathcal{L}_1$ and $\psi \in \mathcal{L}_2$. We say that φ and ψ are* equivalent, *when for all M, s: $M, s \models \varphi$ if and only if $M, s \models \psi$. If for every $\varphi \in \mathcal{L}_1$ there is an equivalent $\psi \in \mathcal{L}_2$, we write $\mathcal{L}_1 \preccurlyeq \mathcal{L}_2$ and say that $\mathcal{L}_2$ is at least as expressive as $\mathcal{L}_1$. We write $\mathcal{L}_1 \prec \mathcal{L}_2$ iff $\mathcal{L}_1 \preccurlyeq \mathcal{L}_2$ and $\mathcal{L}_2 \not\preccurlyeq \mathcal{L}_1$, and we say that $\mathcal{L}_2$ is strictly more expressive than $\mathcal{L}_1$. Finally, if $\mathcal{L}_1 \preccurlyeq \mathcal{L}_2$ and $\mathcal{L}_2 \preccurlyeq \mathcal{L}_1$, we say that $\mathcal{L}_1$ and $\mathcal{L}_2$ are* equally expressive *and write $\mathcal{L}_1 \approx \mathcal{L}_2$.*

It is well known that $\mathcal{L}_! \approx \mathcal{L}_\otimes \approx \mathcal{L}_K$ [8, Chapter 8].

3 Intentionally Anonymous Public Announcements

An anonymous public announcement of φ (made by an agent in the system) is clearly not identical to a standard public announcement of φ, as in the latter case an announced formula may not even be known to any of the agents. Thus a necessary precondition for an (anonymous or not) announcement made by some agent a would be that a knows φ, i.e. that $K_a\varphi$. However, this is not sufficient to guarantee anonymity: for anonymity we also need someone *else* to know φ – in fact, as discussed in the introduction, we need at least *3-anonymity*. We are modeling announcements that are *intentionally* anonymous in the sense that it is common knowledge that an agent will only make the announcement if she knows that it is *safe*. We will discuss in detail what "safe" means, but the intuition is that it is safe for an agent to make an announcement if somebody else could have made it. One subtlety is that the potential announcement by that "somebody else" also should be safe, so the definition has a recursive (or even circular!) flavour. To make our intuitions more precise, consider model M in Fig. 1.

- In state s, all three agents a, b and c know p
- In state s, a considers it possible that we are in state t, where it would not be safe for her to announce p since then b would know that it was her. Thus it is not safe for a to announce p in s.
- In state s, c knows that it is not safe for a to announce p in s, and thus it is not safe for c either: if she announces p in s, b would know that it was her since b knows that it is not safe for a to announce p in s.
- It is thus not safe for b to announce p in s either.
- It is not safe for a or b to announce p in t (only two agents know p).
- p cannot be announced by anyone in u.

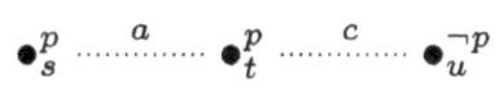

$$\bullet^p_s \quad\cdots\cdots\overset{a}{\cdots\cdots}\quad \bullet^p_t \quad\cdots\cdots\overset{c}{\cdots\cdots}\quad \bullet^{\neg p}_u$$

Fig. 1. Three-agent epistemic model with the agent b's relation being the identity.

Thus, p cannot be safely announced in this model. To see when it is actually safe for agents to make anonymous announcements, we need to provide precise definitions of what we mean by safety and intentionally anonymous announcements.

As we have seen, it is not enough that three agents know φ to ensure that φ can be safely announced by any of them; those three agents also need to know that φ is safe, i.e., that three agents know φ and that φ is safe ... and so on. This clearly has the flavour of *common knowledge*. However, the existence of a group of three different agents having common knowledge of φ, $M, s \models \bigvee_{a \neq b \neq c} C_{\{a,b,c\}}\varphi$, is sufficient but not necessary for φ to be safe in s. A weaker condition would also be sufficient: we don't need the three agents to know that *the same* three agents can safely announce φ. It might be, for example, that a considers it possible that a, b and d safely can announce φ.

In order to define that weaker condition, let us recall the fixpoint definition of common knowledge (see, e.g., [10]). It is well known that $C_G\varphi$ is the greatest fixpoint of $E_G(\varphi \wedge x)$ w.r.t. a variable x not occurring in φ, or, more accurately, given an epistemic model M, $[\![C_G\varphi]\!]_M$ is the greatest fixpoint of the function $f(S) = [\![E_G(\varphi \wedge x)]\!]^M_{V[x=S]}$ where $[\![\psi]\!]^M_{V[x=S]}$ is the extension of ψ in the model M where the valuation function has been changed so that the extension of the variable x is S. The greatest fixed-point of monotonic functions is equal to the union of all post-fixed points, so we have that:

$$[\![C_G\varphi]\!]_M = \bigcup\left\{S : S \subseteq [\![E_G(\varphi \wedge x)]\!]^M_{[x=S]}\right\}.$$

We now define a similar, weaker, notion of common knowledge in order to capture safety. Let us introduce a safety operator: $\blacktriangle\varphi$ intuitively means that φ is safe in the current state. We let, where $N^3 = \{\{a,b,c\} \mid a,b,c \in N; a \neq b, a \neq c, b \neq c\}$:

$$M,s \models \blacktriangle\varphi \text{ iff } s \in \bigcup\left\{S : S \subseteq [\![\bigvee_{G \in N^3} E_G(\varphi \wedge x)]\!]^M_{[x=S]}\right\}.$$

In other words, $\blacktriangle\varphi$ is the greatest fixpoint of $\bigvee_{G \in N^3} E_G(\varphi \wedge x)$. Thus, in particular we have that

$$\models \blacktriangle\varphi \leftrightarrow \bigvee_{G \in N^3} E_G(\varphi \wedge \blacktriangle\varphi),$$

and, similarly to the iterative definition of common knowledge, we have the following.

Lemma 1. $M,s \models \blacktriangle\varphi$ iff $M,s \models \varphi$ and $M,s \models \bigvee_{G_1 \in N^3} E_{G_1}\varphi$ and $M,s \models \bigvee_{G_1 \in N^3} E_{G_1}(\varphi \wedge \bigvee_{G_2 \in N^3} E_{G_2}\varphi)$ and $\cdots$.

While $\blacktriangle\varphi$ means that φ can safely be announced, $K_a\blacktriangle\varphi$ means that φ can safely be announced *by* a. It is easy to see that: $M,s \models K_a\blacktriangle\varphi$ iff $M,s \models \varphi$ and $M,s \models \bigvee_{G_1 \in N^3, a \in G_1} E_{G_1}\varphi$ and $M,s \models \bigvee_{G_1 \in N^3, a \in G_1} E_{G_1}(\varphi \wedge \bigvee_{G_2 \in N^3} E_{G_2}\varphi)$ and $\cdots$.

Finally we can introduce an operator for intentionally anonymous announcements, $[\varphi\ddagger]$, with the corresponding model update defined as follows.

Definition 7. *The update of epistemic model $M = (S, \sim, V)$ by the intentionally anonymous announcement of φ is the epistemic model $M^{\varphi\ddagger} = (S', \sim', V')$ where:*

- $S' = \{(s,a) : s \in S, a \in N, M, s \models K_a\blacktriangle\varphi\}$
- $(s,a) \sim'_c (t,b)$ *iff* $s \sim_c t$ *and* $a = c$ *iff* $b = c$
- $V'(s,a) = V(s)$

Intuitively, in the updated model it is common knowledge that someone has (truthfully) intentionally anonymously announced φ. A state (s,a) corresponds to that someone being a, in state s of the original model. The precondition is that a knows that φ is safe. An agent can only discern between a situation (s,a) where a made the announcement and a situation (t,b) where a different agent b did, if she could already discern between s and t before the announcement or she is exactly one of a or b. We then let:

$$M,s \models [\varphi\ddagger]\psi \Leftrightarrow \forall a \in N, \left(M,s \models K_a\blacktriangle\varphi \Rightarrow M^{\varphi\ddagger}, (s,a) \models \psi\right)$$

- after the intentionally anonymous announcement of φ, ψ is true no matter who the announcer was.

Thus, we have introduced two new operators: $\blacktriangle$ and $[\varphi\ddagger]$. Do we want both in the formal language? Let's consider all three combinations, which will be useful later[1].

$$
\begin{aligned}
\mathcal{L}_\ddagger \quad &\varphi ::= p \mid \neg\varphi \mid \varphi \wedge \varphi \mid K_i\varphi \mid [\varphi\ddagger]\varphi \\
\mathcal{L}_\blacktriangle \quad &\varphi ::= p \mid \neg\varphi \mid \varphi \wedge \varphi \mid K_i\varphi \mid \blacktriangle\varphi \\
\mathcal{L}_{\ddagger\blacktriangle} \quad &\varphi ::= p \mid \neg\varphi \mid \varphi \wedge \varphi \mid K_i\varphi \mid [\varphi\ddagger]\varphi \mid \blacktriangle\varphi
\end{aligned}
$$

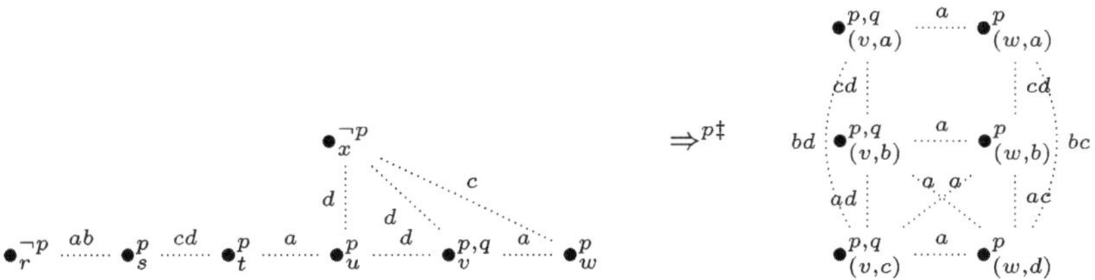

Fig. 2. Four-agent epistemic model M (left), with updated model $M^{p\ddagger}$ (right).

Let's look at some examples. In the model in Fig. 2:

- $M, s \models \neg\blacktriangle p$. p can't be safely announced in s: only two agents know p.
- $M, t \models \neg\blacktriangle p$. While all four agents know p in t, they do not all know that three agents know p: c and d consider it possible that only two agents know p. In other words, even though they know p, they do not know that it can be safely announced.
- $M, u \models \neg\blacktriangle p$. Three agents, a, b and c, know p. They also know that at least three agents know p. But they still do not know that p is safe: they do not know that at least three agents know that at least three agents know p.
- $M, v \models \blacktriangle p$ and $M, w \models \blacktriangle p$. p can be safely announced in v and in w. In fact, $\{w, v\}$ is the greatest fixed-point of $\bigvee_{G \in N^3} E_G(\varphi \wedge x)$.
- $M, v \models K_a\blacktriangle p$. p can be safely announced by a in v.
- Even though a knows *that* p is safe in v, she does not know *why*. She considers it possible that we indeed are in v in which it would be safe for a, b and c to announce p, or in w in which it would be safe for a, b and d to announce p, but she doesn't know which it is. We can say that she knows *de dicto* that p is safe, but not *de re*.
- $M, v \models [p\ddagger]K_cK_dq$. After an intentionally anonymous announcement of p in v, c knows that d has learned that q is true.
- In every state in the updated model, *only the announcer knows who the announcer was.*

The latter point is no coincidence: safe announcements are indeed always safe.

---

[1] $\mathcal{L}_\ddagger$ is well defined without $\blacktriangle$ in the syntax, even though $\blacktriangle$ is used in the semantic condition for $[\varphi\ddagger]$. That semantic condition could of course be written without explicitly mentioning $\blacktriangle$.

Lemma 2. *In any update by an intentionally anonymous announcement, it is common knowledge that no-one except the announcer knows who the announcer is. Formally, let M be a model, let $M^{\varphi\ddagger} = (S', \sim', V')$ be the updated model after the intentionally anonymous announcement of φ, and let $(s,a) \in S'$. Then for any agent $i \neq a$, there is a state $(s', a') \in S'$ such that $(s,a) \sim'_i (s', a')$ and $a \neq a'$.*

Proof. Let M be a model. Let (s,a) be a state in $M^{\varphi\ddagger}$ and let $d \neq a$. By definition, $M, s \models K_a \blacktriangle \varphi$, so there are b, c such that $\{a, b, c\} \in N^3$ and $M, s \models K_b \blacktriangle \varphi \wedge K_c \blacktriangle \varphi$. I.e., (s,b) and (s,c) are states in $M^{\varphi\ddagger}$. Either $b \neq d$ or $c \neq d$ (or both); assume the former. Since all of a, b and d are different, and $\sim_d$ is reflexive, we have that $(s,a) \sim'_d (s,b)$. $\qquad\square$

Thus, $K_a \blacktriangle \varphi$ is a *sufficient* condition for a safe anonymous announcement: after it is announced, no one, apart from a herself, will know who the announcer was. It is also *necessary*, in the sense that a must know that at least two other agents know φ, and must know that those two agents know that at least two other agents know φ, and so on, i.e., it is the case that $K_a \blacktriangle \varphi$. That justifies the definition of the model update from an intentionally anonymous announcement: that $K_a \blacktriangle \varphi$ holds for *some* a is the information that is revealed.

The reader might have observed that the updated model $M^{p\ddagger}$ in Fig. 2 is bisimilar to the submodel of M it is projected on, i.e., the submodel consisting of states v and w – exactly the submodel where $K_a \blacktriangle p$ holds for some a. This is, in fact, always the case. In the following lemma the model update $M^{\blacktriangle\varphi!}$ is defined like for PAL (Sect. 2).

Lemma 3. *For any φ, ψ, M, s: $M, s \models [\varphi\ddagger]\psi$ iff $\left(M, s \models \blacktriangle\varphi \Rightarrow M^{\blacktriangle\varphi!}, s \models \psi\right)$.*

Proof. We first show that $M^{\varphi\ddagger}$ and $M^{\blacktriangle\varphi!}$ are bisimilar. Let Z be such that $(s,a)Zs'$ iff $s = s'$, and let $(s,a)Zs$. *Atoms* is straightforward. *Forth* is as well: if $(s,a) \sim_c (t,b)$, then $s \sim_c t$ by definition. For *Back*, let $s \sim_c t$. First, consider that $c \neq a$. Since t is in $M^{\blacktriangle\varphi!}$, $M, t \models \blacktriangle\varphi$. Thus, there are different $b, d, e \in N$ such that $M, t \models K_b \blacktriangle \varphi \wedge K_d \blacktriangle \varphi \wedge K_e \blacktriangle \varphi$. At least one of b, d, e is different from both a and c. Say, b. Thus $(s,a) \sim_c (t,b)$. Second, consider that $c = a$. Since $M, s \models K_c \blacktriangle$ and $s \sim_c t$, we also have that $M, t \models K_c \blacktriangle$. Thus (t,c) is in $M^{\varphi\ddagger}$, and we have that $(s,a) \sim_c (t,c)$.

Note that satisfaction of $\blacktriangle\varphi$ is invariant under bisimulation (this follows, e.g., from Lemma 1). Then we have that $M, s \models [\varphi\ddagger]\psi$ iff for all a, $M, s \models K_a \blacktriangle \varphi$ implies $M^{\varphi\ddagger}, (s,a) \models \psi$ iff for all a, $M, s \models K_a \blacktriangle \varphi$ implies $M^{\blacktriangle\varphi!}, s \models \psi$. We argue that this is equivalent to $M, s \models \blacktriangle\varphi$ implies $M^{\blacktriangle\varphi!}, s \models \psi$. For the implication from left to right, if $M, s \models \blacktriangle\varphi$ then $M, s \models K_a \blacktriangle \varphi$ for some a by the fixed-point property of $\blacktriangle$ (actually that holds for three different $a \in N$). From $M, s \models K_a \blacktriangle \varphi \Rightarrow M^{\blacktriangle\varphi!}, s \models \psi$ for all a, we get that $M^{\blacktriangle\varphi!}, s \models \psi$. For the other direction, let $a \in N$. If $M, s \models K_a \blacktriangle \varphi$, then $M, s \models \blacktriangle\varphi$ by reflexivity, so $M^{\blacktriangle\varphi!}, s \models \psi$. $\qquad\square$

As an immediate corollary of Lemma 3 we have that if we allow public announcement operators in the language:

$$\models [\varphi\ddagger]\psi \leftrightarrow [\blacktriangle\varphi!]\psi.$$

So, safe anonymous announcements are exactly public announcements of safety. Thus, we get a "simpler", logically equivalent semantics. The existence of the original semantics and the fact that it corresponds exactly to this alternative semantics is of course still crucial: it is what ensures that safety is safe (Lemma 2)[2]. It also means that PAL extended with $\blacktriangle$ can express intentionally anonymous announcements. However, as we shall see when we now move on to comparing expressive power of the languages introduced above, even less is needed.

4   Expressive Power

There is a close relationship between intentionally anonymous announcements and action models. Indeed, consider the language $\mathcal{L}_{\otimes\blacktriangle}$ of action model logic extended with the safety modality. The semantics of $\varphi\ddagger$ given above is equivalent to a certain class of action models of $\mathcal{L}_{\otimes\blacktriangle}$: *anonymous action models*.

Definition 8. *The* anonymous action model *for N and formula φ is the action model* $\mathsf{M}^{\mathsf{N}}_{\varphi} = (\mathsf{S}, \sim, \mathsf{pre})$ *where* $\mathsf{S} = N$, $\mathsf{a} \sim_{\mathsf{c}} \mathsf{b}$ *iff* $(\mathsf{a} = \mathsf{c} \Leftrightarrow \mathsf{b} = \mathsf{c})$, *and* $\mathsf{pre}(\mathsf{a}) = K_a\blacktriangle\varphi$.

An intentionally anonymous announcement now corresponds to the *union* of events $(\mathsf{M}^{\mathsf{N}}_{\varphi}, \mathsf{a})$ for all agents a. The proof of the following is straightforward.

Lemma 4. *For any pointed epistemic model M, s and formulas φ and ψ, $M, s \models [\varphi\ddagger]\psi$ iff $M, s \models \left[\bigcup_{i\in\mathsf{N}}(\mathsf{M}^{\mathsf{N}}_{\varphi}, i)\right]\psi$.*

Consider now the three languages with safety and/or intentional anonymous announcements: $\mathcal{L}_{\ddagger\blacktriangle}, \mathcal{L}_{\blacktriangle}$ and $\mathcal{L}_{\ddagger}$. The first is similar to that of action model logic with common knowledge $\mathcal{L}_{\otimes C}$. In the same sense, $\mathcal{L}_{\blacktriangle}$ is similar to epistemic logic with common knowledge $\mathcal{L}_{KC}$. $\mathcal{L}_{\ddagger}$ is in-between: it would correspond to $\mathcal{L}_{\otimes C}$ but where common knowledge appears only in the preconditions of action models. It is known that $\mathcal{L}_{\otimes C}$ is strictly more expressive than $\mathcal{L}_{KC}$ [8, Chapter 8]; in particular expressions of the form $[M, s]C_G\varphi$ cannot always be reduced to an $\mathcal{L}_{KC}$ formula. Nevertheless, we shall now see that, perhaps contrary to intuition,

---

[2] We used invariance under standard bisimulation as a technical tool to establish the correspondence. Of course, when we say that "only the announcer knows who the announcer is" and so on, we implicitly mean that agents can potentially discern between states (s, a) and (s, b), even though, as shown above, they are bisimilar. An extended notion of bisimulation could be introduced to take into account this assumption, but it is in any case not picked up by the logical language. It is of course also possible to think about ways this distinction *could* be picked up by the language, e.g., by special atoms, but that is complicated by the possibility of iterated announcements leading to states like $((s, a), b)$ and so on.

all three languages $\mathcal{L}_{\ddagger\blacktriangle}, \mathcal{L}_{\blacktriangle}$ *and* $\mathcal{L}_{\ddagger}$ *are actually equally expressive.* Let us start with the "intermediate" language $\mathcal{L}_{\ddagger}$. By Lemma 4, every $\varphi\ddagger$ has an equivalent action model with preconditions $K_a\blacktriangle\varphi$. Because preconditions contain the $\blacktriangle$ modality, we cannot translate formulas of $\mathcal{L}_{\otimes\blacktriangle}$ into $\mathcal{L}_K$ since we do not have a reduction axiom for $\blacktriangle\varphi$. However, we *can* reduce formulas of $\mathcal{L}_{\otimes\blacktriangle}$ to formulas of $\mathcal{L}_{\blacktriangle}$.

Lemma 5. $\mathcal{L}_{\ddagger} \preccurlyeq \mathcal{L}_{\blacktriangle}$.

Proof. Consider the language $\mathcal{L}_{\otimes\blacktriangle}^{-}$, which is like $\mathcal{L}_{\otimes\blacktriangle}$ but with the restriction that $\blacktriangle$ is not allowed in the scope of a $[\pi]$ modality. It is easy to see that the reduction axioms for $\mathcal{L}_{\otimes}$ [8, Chapter 6] are still valid for this extended language. Take a $\varphi \in \mathcal{L}_{\ddagger}$, and let ψ be the corresponding $\mathcal{L}_{\otimes\blacktriangle}$ formula by replacing every $[\varphi\ddagger]$ with the corresponding action model expression $[\pi]$ (Lemma 4). Now we recursively reduce the number of occurrences of $[\pi]$ modalities in ψ by starting with an *outermost* one, i.e., one that is not in the scope of any other. Use the corresponding reduction axiom of $\mathcal{L}_{\otimes}$, and repeat until all $[\pi]$ modalities are gone. Every time we use a reduction axiom we might introduce a new subformula $K_a\blacktriangle\psi'$, but never in the scope of a $[\pi]$ modality since we started with the outermost one. Thus, every time we use a reduction axiom the result is in $\mathcal{L}_{\otimes\blacktriangle}^{-}$, and when we are done the result is in pure epistemic logic extended with $\blacktriangle$, i.e. in $\mathcal{L}_{\blacktriangle}$. $\qquad\square$

Thus, $\mathcal{L}_{\ddagger}$ can be "reduced" to $\mathcal{L}_{\blacktriangle}$ (even though the latter is not a sublanguage). The second "surprise" is that the $\blacktriangle$ operator can actually be expressed by $[\varphi\ddagger]$.

Lemma 6. *For any M, s and any $\varphi \in \mathcal{L}_{\ddagger\blacktriangle}$, $M, s \models \blacktriangle\varphi$ iff $M, s \models \neg[\varphi\ddagger]\bot$.*

Proof. $M, s \models \neg[\varphi\ddagger]\bot$ iff there is an $a \in N$ such that $M, s \models K_a\blacktriangle\varphi$ and $M^{\varphi\ddagger}, (s, a) \not\models \bot$ iff there is an $a \in N$ such that $M, s \models K_a\blacktriangle\varphi$ iff (by reflexivity in one direction and the fixed-point definition of $\blacktriangle$ in the other) $M, s \models \blacktriangle\varphi$. $\square$

Corollary 1. $\mathcal{L}_{\ddagger\blacktriangle} \approx \mathcal{L}_{\ddagger} \approx \mathcal{L}_{\blacktriangle}$.

Thus, yet again we see that dynamic announcement operators can actually be expressed in a purely static language: $\mathcal{L}_{\blacktriangle}$. We thus move on to axiomatising safety.

5   Axiomatisation of Safety

Observe that $\blacktriangle$ does not distribute over implication: $\not\models \blacktriangle(\varphi \rightarrow \psi) \rightarrow (\blacktriangle\varphi \rightarrow \blacktriangle\psi)$. In particular, $\bigvee_{G \in N^3} E_G(\varphi \rightarrow \psi)$ and $\bigvee_{G \in N^3} E_G\varphi$ does not imply $\bigvee_{G \in N^3} E_G\psi$ – it might not be the same G in the first two cases. For similar reasons the conjunctive closure axiom, $(\blacktriangle\varphi \wedge \blacktriangle\psi) \rightarrow \blacktriangle(\varphi \wedge \psi)$ does not hold. This is similar to *somebody knows* [2], but unlike most other group knowledge operators which are normal modalities. The other direction of conjunctive closure, *monotonicity,* $\blacktriangle(\varphi \wedge \psi) \rightarrow (\blacktriangle\varphi \wedge \blacktriangle\psi)$, does hold.

The axiomatic system $\mathbf{S}_\blacktriangle$ for $\mathcal{L}_\blacktriangle$ is shown in Table 1. The first two parts form a standard axiomatisation of propositional logic and the individual knowledge operators. The Monotonicity rule combines the monotonicity axiom and the replacement of equivalents rule standard in weak modal logics. We note that necessitation for $\blacktriangle$, from φ derive $\blacktriangle\varphi$, follows (we have $\blacktriangle\top$ from the Induction rule). The Mix axiom (or *fixed-point axiom*) says that $\blacktriangle\varphi$ is indeed a fixed-point of $\bigvee_{G \in N^3} E_G(\varphi \wedge x)$. Finally, the Induction rule give us a way to derive $\blacktriangle\varphi$. Mix and the Induction rule can be seen as adaptions of similar axioms/rules for common knowledge, see, e.g., [8, 10].

We will use the following shorthand:

$$\blacktriangle_n\varphi = \varphi \wedge \bigvee_{G_1 \in N^3} E_{G_1}(\varphi \wedge \bigvee_{G_2 \in N^3} E_{G_2}(\varphi \wedge \bigvee_{G_3 \in N^3} E_{G_3}(\varphi \wedge \cdots \wedge \bigvee_{G_n \in N^3} E_{G_n}\varphi))).$$

Thus, $\blacktriangle_0 = \varphi$, $\blacktriangle_1 = \varphi \wedge \bigvee_{G_1 \in N^3} E_{G_1}\varphi$, and so on, and we have that $M, s \models \blacktriangle\varphi$ iff $M, s \models \blacktriangle_n\varphi$ for all n.

Table 1. The proof system $\mathbf{S}_\blacktriangle$.

all instances of propositional tautologies	Prop
From $\varphi \rightarrow \psi$ and φ, derive ψ	Modus ponens
$K_a(\varphi \rightarrow \psi) \rightarrow (K_a\varphi \rightarrow K_a\psi)$	Distribution
$K_a\varphi \rightarrow \varphi$	Truth
$\neg K_a\varphi \rightarrow K_a \neg K_a\varphi$	Negative introspection
From φ, derive $K_a\varphi$	Necessitation
$\blacktriangle\varphi \rightarrow \bigvee_{G \in N^3} E_G(\varphi \wedge \blacktriangle\varphi)$	Mix
From $\varphi \rightarrow \bigvee_{G \in N^3} E_G\varphi$, derive $\varphi \rightarrow \blacktriangle\varphi$	Induction rule
From $\varphi \rightarrow \psi$, derive $\blacktriangle\varphi \rightarrow \blacktriangle\psi$	Monotonicity rule

It is straightforward to see that the Mix axiom is valid and that the Induction and Monotonicity rules preserve validity, and thus that the following holds.

Lemma 7. $\mathbf{S}_\blacktriangle$ *is sound.*

We now prove that $\mathbf{S}_\blacktriangle$ is also complete. Similarly to epistemic logic with common knowledge, the logic is not *compact*. For example, $\{\blacktriangle_n\varphi : n \geq 0\} \cup \{\neg\blacktriangle\varphi\}$ is not satisfiable, but any finite subset of it is. Thus, we have to settle for *weak* rather than *strong* completeness. To that end, we adapt the standard technique for common knowledge of defining a finite canonical model based on the finite syntactic closure of some formula. This is complicated by the fact that $\blacktriangle$ is not normal: in the axiomatisation we have the (weaker) monotonicity rule instead of the K axiom for $\blacktriangle$. The consequence, aside from the fact that we cannot rely on distribution over implication like in proofs for common knowledge, is that there is no normal relational semantics for $\blacktriangle$ (like "φ is true on all states reachable by a G-path" for $C_G\varphi$). We now define an alternative definition of the semantics and show that it is equivalent, which we will use in the completeness proof.

Given a model M, a *group assignment function* for M is a function $f : S \to \mathcal{P}(N)$ such that $f(s) \in N^3$ or $f(s) = \emptyset$ for each s, i.e., assigning a group of three agents to some of the states, such that for all $s, t \in S$ and $i \in f(s)$, if $s \sim_i t$ then $f(t) \neq \emptyset$. The idea is that f works as an assignment of groups of three agents to certain states, such that if we follow the accessibility relations for those agents, f again assigns a group of three agents to those states, and so on. An f-*consistent path* in M is a finite sequence of states $s_0 s_1 s_2 \cdots s_m$ such that for all $0 \leq i < m$, $s_i \sim_a s_{i+1}$ for some $a \in f(s_i)$. We say that φ is true on a path if it is true in every state on that path.

Lemma 8. $M, s \models \blacktriangle\varphi$ *iff there is a group assignment function* f *such that* $f(s) \neq \emptyset$ *and* φ *is true on every* f-*consistent path starting in* s.

Proof. Left-to-right: let $M, s \models \blacktriangle\varphi$. Define f as follows: for any state t, $f(t) = G$ for some $G \in N^3$ such that $M, t \models E_G \blacktriangle\varphi$ if such a G exists, $f(t) = \emptyset$ if not. If there are several such G chose any of them. We must show that f is indeed a group assignment function. Let $i \in f(s')$ and $s' \sim_i t$; we must show that $f(t) \neq \emptyset$. Since $i \in f(s')$, there is a $G \in N^3$ such that $M, s' \models E_G \blacktriangle\varphi$ and $i \in G$. Thus $M, s \models K_i \blacktriangle\varphi$ and $M, t \models \blacktriangle\varphi$. By Mix, $M, t \models E_{G'} \blacktriangle\varphi$ for some G', so $f(t) \neq \emptyset$. Finally, $f(s) \neq \emptyset$ from $M, s \models \blacktriangle\varphi$ and Mix.

We proceed by induction on the length of any f-consistent path starting in s. The base case is immediate. Consider an f-consistent path $s_0 \cdots s_n s_{n+1}$ where $s_0 = s$. By the induction hypothesis $s_0, \cdots, s_n$ are φ-states. Since $s_n \sim_i s_{n+1}$ for some $i \in f(s_n)$ and $M, s_n \models E_{f(s_n)} \blacktriangle\varphi$, we thus also have $M, s_{n+1} \models \varphi$.

Right-to-left: by contraposition. Assume that $M, s \not\models \blacktriangle\varphi$, i.e., that $M, s \not\models \blacktriangle_n\varphi$ for some n. Let f be a group assignment function such that $f(s) \neq \emptyset$. We must show that there is a f-consistent path starting in s where φ is not true. We have that $M, s \models \neg\varphi \vee \bigwedge_{G_1 \in N^3} \bigvee_{i_1 \in G_1} \hat{K}_{i_1}(\neg\varphi \vee \bigwedge_{G_2 \in N^3} \bigvee_{i_2 \in G_2} \hat{K}_{i_2}(\neg\varphi \vee \cdots \vee \bigwedge_{G_n \in N^3} \bigvee_{i_n \in G_n} \hat{K}_{i_n}\neg\varphi))$. Thus, there is a $k \leq n$ such that

$$M, s \models \bigwedge_{G_1 \in N^3} \bigvee_{i_1 \in G_1} \hat{K}_{i_1} \bigwedge_{G_2 \in N^3} \bigvee_{i_2 \in G_2} \hat{K}_{i_2} \cdots \bigwedge_{G_k \in N^3} \bigvee_{i_k \in G_n} \hat{K}_{i_k}\neg\varphi.$$

Hence there is a path $s_0 s_1 \cdots s_{k+1}$ where:

$s_0 = s$, $s_0 \sim_{i_1} s_1$ for some $i_1 \in f(s_0)$, $s_1 \sim_{i_2} s_2$ for some $i_2 \in f(s_1)$, $\cdots$, $s_k \sim_{i_{k+1}} s_{k+1}$ for some $i_{k+1} \in f(s_k)$,
such that $M, s_{k+1} \models \neg\varphi$. This is an f-consistent path. $\qquad\square$

We now proceed with defining the finite canonical model, and proving a truth lemma. We adapt the standard proof for common knowledge (see [8, Chapter 7][3]).

---

[3] There are also other complications, in addition to the lack of normality, including that the correspondent to the *induction axiom* for common knowledge, $\blacktriangle(\varphi \to \bigvee_{G \in N^3} E_G\varphi) \to (\varphi \to \blacktriangle\varphi)$, is not valid. We instead use a variant of the induction *rule* for common knowledge, used in, e.g., [10].

Definition 9. *The closure $cl(\gamma)$ of a formula γ is the smallest set that contains all subformulas of γ, is closed under single negations, and that contains $E_G \blacktriangle \varphi$ for all $G \subseteq N$ whenever it contains $\blacktriangle \varphi$.*

The closure of any formula is finite. As usual we say that a set of formulas Γ is maximal consistent in $cl(\gamma)$ iff $\Gamma \subseteq cl(\gamma)$, Γ is consistent, and there is no consistent $\Gamma' \subseteq cl(\gamma)$ such that $\Gamma \subset \Gamma'$. When Δ is a finite set of formulas we write $\underline{\Delta}$ for $\bigwedge_{\delta \in \Delta} \delta$.

The canonical model is defined as follows.

Definition 10. *The canonical model for some formula γ is $M^\gamma = (S^\gamma, \sim^\gamma, V^\gamma)$ where: S^γ is the set of all sets of formulas maximal consistent in $cl(\gamma)$, $\Gamma \sim^\gamma_i \Delta$ iff $\{K_i \varphi : K_i \varphi \in \Gamma\} = \{K_i \varphi : K_i \varphi \in \Delta\}$, and $V^\gamma(p) = \{\Gamma \in S^\gamma : p \in \Gamma\}$.*

This definition of the canonical model is identical to the one used for common knowledge in, e.g., [8], except that the closure is wider. Many of the properties of that model, like deductive closure of maximal consistent sets in $cl(\gamma)$, carry over, and we only need to focus on the following property for the $\blacktriangle$ modality. A φ-path in the canonical model is a path $\Gamma_0 \Gamma_1 \cdots$ where $\varphi \in \Gamma_i$ for every i.

Lemma 9. *For any γ and $\Gamma \in S^\gamma$ and $\blacktriangle \varphi \in cl(\gamma)$, $\blacktriangle \varphi \in \Gamma$ iff there is a group assignment function f such that $f(\Gamma) \neq \emptyset$ and every f-consistent path from Γ in M^γ is a φ-path.*

Proof. Left-to-right: let $\blacktriangle \varphi \in \Gamma$. Define f as follows: for any $\Delta \in S^\gamma$, $f(\Delta) = G$ if $E_G \blacktriangle \varphi \in \Delta$ for some $G \in N^3$ and $f(\Delta) = \emptyset$ otherwise. If there are several such G, chose one. Consider a Δ such that $f(\Delta) \neq \emptyset$ and $\Delta \sim^\gamma_i \Delta'$ for some $i \in f(\Delta)$. We must show that $f(\Delta') \neq \emptyset$. Since $K_i \blacktriangle \varphi \in \Delta$, $K_i \blacktriangle \varphi \in \Delta'$ by definition of $\sim^\gamma_i$, and thus $\blacktriangle \varphi \in \Delta'$ by the Truth axiom, and $E_{G'} \blacktriangle \varphi \in \Delta'$ for some G' by Mix and the fact that $E_{G'} \blacktriangle \varphi \in cl(\gamma)$. Finally, $f(\Gamma) \neq \emptyset$ follows from Mix and definition of the closure.

We must show that every f-consistent path from Γ is a φ-path. The proof is by induction on the length of the path. For the base case, we have that $\varphi \in \Gamma$ from the Mix and Truth axioms. For the induction step, consider an f-consistent path $\Gamma_0 \cdots \Gamma_n \Gamma_{n+1}$ where $\varphi \in \Gamma_j$ for all $j \leq n$. From f-consistency we have that $\Gamma_n \sim^\gamma_i \Gamma_{n+1}$ for some $i \in G$ for some $G \in N^3$ such that $E_G \blacktriangle \varphi \in \Gamma_n$. It follows from the definition of $\sim^\gamma_i$ that $K_i \blacktriangle \varphi \in \Gamma_{n+1}$, and thus that $\blacktriangle \varphi \in \Gamma_{n+1}$ from the Truth axiom, and thus that $\varphi \in \Gamma_{n+1}$ from the Mix axiom and Truth axiom again.

Right-to-left: let f be such that $f(\Gamma) \neq \emptyset$ and every f-consistent path is a φ-path. Let $S_{f,\varphi}$ be the set of all sets Δ maximal consistent in $cl(\gamma)$ such that $f(\Delta) \neq \emptyset$ and every f-consistent path from Δ is a φ-path. Let $\chi = \bigvee_{\Delta \in S_{f,\varphi}} \underline{\Delta}$. We first show that $\vdash \chi \rightarrow \bigvee_{G \in N^3} E_G \chi$ (*).

Assume, towards a contradiction, that $\chi \wedge \neg \bigvee_{G \in N^3} E_G \chi$ is consistent. Then $\underline{\Delta} \wedge \neg \bigvee_{G \in N^3} E_G \chi$ is consistent for some $\Delta \in S_{f,\varphi}$. Then, for any $G \in N^3$ there must be a $i_G \in G$ such that $\underline{\Delta} \wedge \hat{K}_{i_G} \neg \chi$ is consistent. $\underline{\Delta} \wedge \hat{K}_{i_G} \bigvee_{\Theta \in S^\gamma \setminus S_{f,\varphi}} \underline{\Theta}$ is consistent for each i_G, and by modal reasoning for individual knowledge,

$\underline{\Delta} \wedge \bigvee_{\Theta \in S^\gamma \setminus S_{f,\varphi}} \hat{K}_{i_G} \underline{\Theta}$ is consistent for each i_G. Thus, for every $G \in N^3$ there is an $i_G \in G$ and a $\Theta_{i_G} \in S^\gamma \setminus S_{f,\varphi}$ such that $\underline{\Delta} \wedge \hat{K}_{i_G} \Theta_{i_G}$ is consistent. In particular, since $\Delta \in S_{f,\varphi}$ and thus $f(\Delta) \neq \emptyset$, there is an $i_G \in G = f(\Delta)$ and a $\Theta_{i_G} \in S^\gamma \setminus S_{f,\varphi}$ such that $\underline{\Delta} \wedge \hat{K}_{i_G} \Theta_{i_G}$ is consistent. It follows (by a standard property of the canonical model [8, Item 4 of Lemma 7.14]) that $\Delta \sim^\gamma_{i_G} \Theta_{i_G}$. Since $i_G \in f(\Delta)$, $f(\Theta_{i_G}) \neq \emptyset$ and since Θ_{i_G} is not in $S_{f,\varphi}$ that means that there is a an f-consistent path from Θ_{i_G} which is not a φ-path. It follows that there is an f-consistent path from Δ that is not a φ-path. But $\Delta \in S_{f,\varphi}$, which leads to a contradiction. Thus, we have shown (*).

From (*) and the induction rule, we get that $\vdash \chi \to \blacktriangle\chi$. Since Γ is one of the disjuncts in χ we have that $\vdash \underline{\Gamma} \to \chi$. Thus, $\vdash \underline{\Gamma} \to \blacktriangle\chi$. Since $\varphi \in \bigcap_{\Delta \in S_{f,\varphi}} \Delta$, we also have that $\vdash \chi \to \varphi$. By Monotonicity, $\vdash \blacktriangle\chi \to \blacktriangle\varphi$, and thus $\vdash \underline{\Gamma} \to \blacktriangle\varphi$. Since $\blacktriangle\varphi \in cl(\gamma)$, we have that $\blacktriangle\varphi \in \Gamma$. □

Lemma 10 (Truth). *For any γ, $\varphi \in cl(\gamma)$ and $\Gamma \in S^\gamma$, $\varphi \in \Gamma$ iff $M^\gamma, \Gamma \models \varphi$.*

Proof. The proof is by structural induction on φ. We only show the induction step for $\varphi = \blacktriangle\psi$, the other cases are straightforward and/or standard. We have that $M^\gamma, \Gamma \models \blacktriangle\psi$ iff (by Lemma 8) there is an f such that $f(\Gamma) \neq \emptyset$ and ψ is true on every f-consistent path starting in Γ iff (by the ind. hyp.) there is an f such that $f(\Gamma) \neq \emptyset$ and every f-consistent path starting in Γ is a ψ-path iff (by Lemma 9) $\blacktriangle\psi \in \Gamma$. □

We immediately get the following.

Theorem 1 (Completeness). $S\blacktriangle$ *is complete.*

6 Discussion

Intentionally anonymous announcements $\varphi\ddagger$ capture exactly the announcements that are guaranteed to ensure anonymity. It all boils down to *safety*: intentional anonymous announcements $\varphi\ddagger$ are exactly public announcements of safety $\blacktriangle\varphi!$. Furthermore, the $\varphi\ddagger$ operators can be expressed in epistemic logic extended with only the safety operator $\blacktriangle$. We gave a complete axiomatisation of the latter.

There are still many open problems. While the expressivity results "reduce" the languages with $\varphi\ddagger$ to equivalent languages we have axiomatisations for, they don't directly give us "native" axiomatisations of those languages themselves.

A conceptually related work is van Ditmarsch's analysis of the *Russian Cards Problem* [7] which also models *safe announcements* – albeit with a different notion of safety, namely that a secret "card deal", instead of the identity of the announcer, is not revealed. A safe announcement of φ by a is captured by $[K_a\varphi \wedge [K_a\varphi]C\,cignorant!]$, where C is common knowledge among all agents and *cignorant* is the safety or secrecy condition – corresponding to only the announcer knowing the identity of the announcer in our case. That is again equivalent to the sequence $[K_a\varphi!][C\,cignorant!]$, which means that φ is safe to announce for a if $C\,cignorant$ holds after the announcement of $K_a\varphi$. Our $[\varphi\ddagger]$ operator also satisfies that definition of safety (Lemma 2) (although it cannot be expressed in the logical language).

Safety can be strengthened by using some variant of *distributed knowledge* [13,30], so that announcements are safe even when non-announcing agents share their knowledge to deduce the identity of the announcer. Another avenue of further research is to extend the presented formalisms with *quantification over anonymous announcements* like in [1,3]. This will allow us to express properties like "there is a safe anonymous announcement, such that φ holds afterwards". Also of interest for future work, is to model *self-referential* intentionally anonymous announcements of the form "after this very announcement, no-one will know", as studied recently in [4,16][4]. Finally, in this paper we have focussed on *knowledge* of agents. In the future, we would also like to explore anonymous announcements also in the setting of non-S5 modal logics that can capture, for example, *beliefs*, potentially false ones, of agents about the identity of the announcer. That would also enable relaxing the strict definition of the safety modality, e.g., similarly to the approximation of common knowledge in [19].

Acknowledgments. We thank Hans van Ditmarsch and the LORI reviewers for detailed and constructive comments which has helped us improve the manuscript significantly.

Disclosure of Interests. This work was supported by ROIS-DS-JOINT (047RP2024)

References

1. Ågotnes, T., Balbiani, P., van Ditmarsch, H., Seban, P.: Group announcement logic. J. Appl. Log. **8**(1), 62–81 (2010)
2. Ågotnes, T., Wáng, Y.N.: Somebody knows. In: Bienvenu, M., Lakemeyer, G., Erdem, E. (eds.) Proceedings of the 18th KR, pp. 2–11 (2021). https://doi.org/10.24963/kr.2021/1
3. Balbiani, P., Baltag, A., Ditmarsch, H., Herzig, A., Hoshi, T., Lima, T.: Knowable as known after an announcement. Rev. Symb. Logic **1**(3), 305–334 (2008)
4. Baltag, A., Bezhanishvili, N., Fernández-Duque, D.: The topology of surprise. In: Kern-Isberner, G., Lakemeyer, G., Meyer, T. (eds.) Proceedings of the 19th KR (2022). https://doi.org/10.24963/kr.2022/4
5. Baltag, A., Moss, L.S.: Logics for epistemic programs. Synthese **139**(2), 165–224 (2004). https://doi.org/10.1023/B:SYNT.0000024912.56773.5E
6. Barthe, G., Gaboardi, M., Arias, E.J.G., Hsu, J., Kunz, C., Strub, P.: Proving differential privacy in hoare logic. In: Proceedings of the 27th CSF, pp. 411–424. IEEE Computer Society (2014). https://doi.org/10.1109/CSF.2014.36
7. van Ditmarsch, H.: The Russian cards problem. Studia logica **75**, 31–62 (2003). https://doi.org/10.1023/A:1026168632319
8. van Ditmarsch, H., van der Hoek, W., Kooi, B.: Dynamic Epistemic Logic, Synthese Library, vol. 337. Springer (2007)

---

[4] This comment also applies to the Russian Cards problem.

9. Domingo-Ferrer, J., Sánchez, D., Soria-Comas, J.: Database anonymization: privacy models, data utility, and microaggregation-based inter-model connections. In: Synthesis Lectures on Information Security, Privacy, & Trust, Morgan & Claypool Publishers (2016). https://doi.org/10.2200/S00690ED1V01Y201512SPT015

10. Fagin, R., Halpern, J.Y., Moses, Y., Vardi, M.Y.: Reasoning About Knowledge. The MIT Press, Cambridge (1995)

11. Halpern, J.Y., O'Neill, K.R.: Anonymity and information hiding in multiagent systems. J. Comput. Secur. **13**(3), 483–512 (2005). https://doi.org/10.3233/JCS-2005-13305

12. Halpern, J.Y., O'Neill, K.R.: Secrecy in multiagent systems. ACM Trans. Inform. Syst. Secur. **12**(1), 5:1–5:47 (2008). https://doi.org/10.1145/1410234.1410239

13. van der Hoek, W., van Linder, B., Meyer, J.J.: Group knowledge is not always distributed (neither is it always implicit). Math. Soc. Sci. **38**(2), 215–240 (1999). https://doi.org/10.1016/S0165-4896(99)00013-X

14. Hughes, D.J.D., Shmatikov, V.: Information hiding, anonymity and privacy: a modular approach. J. Comput. Secur. **12**(1), 3–36 (2004). https://doi.org/10.3233/JCS-2004-12102

15. Jiang, J., Naumov, P.: De re/de dicto distinction: a logicians' perspective on data anonymity. J. Cybersecurity **11**(1) (2025). https://doi.org/10.1093/CYBSEC/TYAE025

16. Li, Y., Ren, J., Ågotnes, T.: The surprise exam in full modal fixed-point logic. In: Ågotnes, T., Doder, D. (eds.) Logic and Argumentation, pp. 185–200. Springer, Singapore (2025)

17. Machanavajjhala, A., Kifer, D., Gehrke, J., Venkitasubramaniam, M.: L-diversity: privacy beyond k-anonymity. ACM Trans. Knowl. Discov. Data **1**(1), 3 (2007). https://doi.org/10.1145/1217299.1217302

18. Mano, K., Kawabe, Y., Sakurada, H., Tsukada, Y.: Role interchange for anonymity and privacy of voting. J. Log. Comput. **20**(6), 1251–1288 (2010). https://doi.org/10.1093/LOGCOM/EXQ013

19. Monderer, D., Samet, D.: Approximating common knowledge with common beliefs. Games Econom. Behav. **1**(2), 170–190 (1989)

20. More, S.M., Naumov, P.: Hypergraphs of multiparty secrets. Ann. Math. Artif. Intell. **62**(1–2), 79–101 (2011). https://doi.org/10.1007/S10472-011-9252-Z

21. More, S.M., Naumov, P.: Logic of secrets in collaboration networks. Ann. Pure Appl. Logic **162**(12), 959–969 (2011). https://doi.org/10.1016/J.APAL.2011.06.001

22. Pfitzmann, A., Hansen, M.: A terminology for talking about privacy by data minimization: anonymity, unlinkability, undetectability, unobservability, pseudonymity, and identity management (2010). http://dud.inf.tu-dresden.de/literatur/Anon_Terminology_v0.34.pdf, v0.34

23. Plaza, J.: Logics of public communications. In: Proceedings of the 4th ISMIS, pp. 201–216. Oak Ridge National Laboratory (1989)

24. Schneider, S.A., Sidiropoulos, A.: CSP and anonymity. In: Bertino, E., Kurth, H., Martella, G., Montolivo, E. (eds.) Proceedings of the 4th ESORICS. LNCS, vol. 1146, pp. 198–218. Springer (1996). https://doi.org/10.1007/3-540-61770-1_38

25. Sweeney, L.: k-anonymity: a model for protecting privacy. Internat. J. Uncertain. Fuzziness Knowl. Based Syst. **10**(5), 557–570 (2002). https://doi.org/10.1142/S0218488502001648

26. Syverson, P.F., Stubblebine, S.G.: Group principals and the formalization of anonymity. In: Wing, J.M., Woodcock, J., Davies, J. (eds.) FM 1999. LNCS, vol. 1708, pp. 814–833. Springer, Heidelberg (1999). https://doi.org/10.1007/3-540-48119-2_45
27. Tsukada, Y., Mano, K., Sakurada, H., Kawabe, Y.: Anonymity, privacy, onymity, and identity: a modal logic approach. In: Proceedings of the 12th CSE, pp. 42–51. IEEE Computer Society (2009). https://doi.org/10.1109/CSE.2009.251
28. Tsukada, Y., Sakurada, H., Mano, K., Manabe, Y.: On compositional reasoning about anonymity and privacy in epistemic logic. Ann. Math. Artif. Intell. **78**(2), 101–129 (2016). https://doi.org/10.1007/s10472-016-9516-8
29. van Eijck, J., Orzan, S.: Epistemic verification of anonymity. In: Proceedings of the 2nd VODCA. Electronic Notes in Theoretical Computer Science, vol. 168, pp. 159–174 (2007). https://doi.org/10.1016/j.entcs.2006.08.026
30. Wáng, Y.N., Ågotnes, T.: Public announcement logic with distributed knowledge: expressivity, completeness and complexity. Synthese **190**(1), 135–162 (2013). https://doi.org/10.1007/s11229-012-0243-3
31. Xiong, Z., Ågotnes, T.: The logic of secrets and the interpolation rule. Ann. Math. Artif. Intell. **91**(4), 375–407 (2023). https://doi.org/10.1007/S10472-022-09815-0
32. Ye, X., Li, Z., Li, Y.: Capture inference attacks for k-anonymity with privacy inference logic. In: Kotagiri, R., Krishna, P.R., Mohania, M., Nantajeewarawat, E. (eds.) DASFAA 2007. LNCS, vol. 4443, pp. 676–687. Springer, Heidelberg (2007). https://doi.org/10.1007/978-3-540-71703-4_57

Next-Time Coalition Logic

Thomas Ågotnes[1,2] and Fengkui Ju[3(✉)]

[1] Department of Information Science and Media Studies, University of Bergen, Bergen, Norway
[2] School of Philosophy, Shanxi University, Taiyuan, China
[3] School of Philosophy, Beijing Normal University, Beijing, China
fengkui.ju@bnu.edu.cn

Abstract. Coalition Logic (CL) can be seen as the next-time fragment of Alternating-time Temporal Logic (ATL). Modalities in ATL are combinations of strategy quantifiers and tense modalities "glued" together. In ATL* this requirement is relaxed, and the two types of modalities can be mixed freely, resulting in more expressive logics. That comes at a cost: for example, as far as we know, no complete axiomatisation of ATL* exists. In this paper we study the next-time fragment of ATL*: CL*. Like in ATL and unlike in CL, we must now consider strategies rather than just actions, and like in ATL* but unlike in ATL, it matters whether we use full-memory or just memoryless strategies. The question of whether CL* actually is more expressive than CL hinges on that choice. We characterise the relative expressive power of the two variants of CL* compared to CL, and give a sound and complete axiomatisation of CL* with full-memory strategies.

Keywords: Coalition Logic · Alternating-time Temporal Logic · Expressivity · Completeness

1 Introduction

Several logical frameworks have been developed to reason about the *ability* of agents and in particular of *coalitions*: Pauly's Coalition Logic (CL) [16], Alur et al.'s Alternating-time Temporal Logic (ATL) [4], Seeing-to-it-that (STIT) logics by Belnap and others [5], van Benthem's forcing modalities [20], as well as Chatterjee et al.'s Strategy Logics [7]. CL is in a sense one of the "purest" coalitional ability logics: it extends propositional logic with (only) modalities of the form $\langle\!\langle C \rangle\!\rangle$ where C is a group of agents, a coalition, such that $\langle\!\langle C \rangle\!\rangle \varphi$ means that C has the ability to make φ true no matter what the other agents do. From a logical perspective it has neighbourhood semantics, but it can also be seen as directly capturing so-called α-effectivity in strategic form games.

CL can be seen as the next-time fragment of ATL, the latter which has combinations of $\langle\!\langle C \rangle\!\rangle$ with tense connectives, such as $\langle\!\langle C \rangle\!\rangle X\varphi$ and $\langle\!\langle C \rangle\!\rangle G\varphi$ – C has the ability to make φ true in the next state and in all future states

© The Author(s), under exclusive license to Springer Nature Switzerland AG 2026
V. Goranko et al. (Eds.): LORI 2025, LNCS 16010, pp. 18–33, 2026.
https://doi.org/10.1007/978-981-95-2481-5_2

respectively. ATL can again be seen as generalising the A and E path quantifiers of Computation-Tree Logic (CTL) [8] by replacing them with $\langle\!\langle C \rangle\!\rangle$.

In both CTL and ATL the path/coalition modalities $E/A/\langle\!\langle C \rangle\!\rangle$ and the tense connectives $X/G/F/U$ are "glued" together: in ATL a $\langle\!\langle C \rangle\!\rangle$ is always immediately followed by a tense connective and the latter are always preceded by a $\langle\!\langle C \rangle\!\rangle$. In CTL* and ATL* that restriction is relaxed, and expressions such as $\langle\!\langle C \rangle\!\rangle G(sent \rightarrow F arrive)$ are allowed, intuitively meaning: *C has a strategy to guarantee whenever a message is sent, it will arrive later.* While complete axiomatisations of CTL [6], CTL* [18] and ATL [13] exist, a complete axiomatisation of ATL* remains an open problem.

In this paper we consider CL^*: the next-time fragment of ATL*, or, equivalently, the extension of CL with explicit next-time tense connectives X that can mix freely with $\langle\!\langle C \rangle\!\rangle$ (the latter which no longer implicitly refer to the next state). Some examples:

- $\langle\!\langle C \rangle\!\rangle XXp$: C can ensure that p is true in the state after next;
- $\langle\!\langle C \rangle\!\rangle X(\neg p \wedge Xp)$: C can ensure that p becomes false and then immediately becomes true;
- $\langle\!\langle C \rangle\!\rangle (Xp \vee XXp)$: C can ensure that p becomes true either in the next state or in the state after that.

While the semantics of $\langle\!\langle C \rangle\!\rangle$ in CL is defined by quantifying over *joint actions*, in CL* we must, like in ATL, consider joint *strategies*. Two types of strategies are often considered in ATL-like logics: *memoryless* strategies, which map each state to an action for each agent, and *full-memory* (or *history-based*) strategies which map any finite history to an action for each agent. The former can be seen as a special case of the latter, always mapping histories ending with the same state to the same action. It is well known that for ATL, the choice between the two types doesn't matter, we get the same satisfaction relation in both cases [19], but for ATL* it *does*. The latter is true also for CL*. Here we want to mention that CL* with full-memory strategies can be seen as generalising the next-time fragment of CTL*, whose completeness is shown by Ju, Grilletti and Goranko [14].

In this paper we study the expressive power of the two variants of CL* corresponding to the two types of strategies, and their relationships to CL. We shall see that in a certain sense the memoryless semantics is more fundamental than the full-memory semantics, perhaps contrary to intuition. We also give a sound and complete axiomatisation for the case of full-memory strategies.

The remainder of the paper is organised as follows. In the next section we give a quick overview of relevant definitions and results from CL and ATL/ATL* that will be useful in the following. In Sect. 3 we formally define the language and semantics of CL*. Section 4 looks at expressive power, while the mentioned axiomatisation is given and studied in Sect. 5. We conclude in Sect. 6.

2   Background

2.1   Coalition Logic

Given a non-empty set P of primitive propositions and a finite non-empty set N of agents, the language of Coalition Logic, $\mathcal{L}_{CL}$, is defined by the following grammar:

$$\varphi ::= p \mid \neg\varphi \mid \varphi \wedge \varphi \mid [\![C]\!]\varphi$$

where $p \in P$ and $C \subseteq N$. We use the usual derived propositional connectives. We write $\overline{C}$ for $N \setminus C$.

As mentioned in the introduction, the intended meaning of a formula of the form $[\![C]\!]\varphi$ is that the group C (henceforth called a "coalition") has the ability to make φ true, in the sense that they can choose a joint action such that no matter what the other agents do, φ will necessarily be the case. Formally, Pauly [16] gives two equivalent semantics for the language: interpretation in a certain class of neighbourhood models with so-called playable[1] effectivity functions, and interpretation in game structures where there is a normal game form associated with every state. The latter structures and interpretation are again equivalent to the interpretation in *concurrent game structures* used in ATL (and ATL*) [4], and CL thus forms the next-time fragment of ATL [9]. In this paper we will use concurrent game structures.

A *concurrent game structure* (*CGS*, or sometimes just *model*) is a tuple $M = (S, V, ACT, d, \delta)$ where

- S is a nonempty set of states.
- $V : P \to 2^S$ is a valuation function.
- ACT is a nonempty set of actions.
- $d_i(s) \subseteq ACT$ is a nonempty set of actions for agent $i \in N$ in state $s \in S$. We let $D(s) = \times_{i \in N} d_i(s)$ be the set of *full joint actions* in s. When $C \subseteq N$, $D_C(s) = \times_{i \in C} d_i(s)$ is the set of C-*actions* in s. If $\mathbf{a} \in D_C(s)$, a_i $(i \in C)$ denotes the action taken from $d_i(s)$. When $C_1 \cap C_2 = \emptyset$ and $\mathbf{a} \in D_{C_1}(s)$ and $\mathbf{b} \in D_{C_2}(s)$, then $(\mathbf{a}, \mathbf{b})$ denotes the $C_1 \cup C_2$ action $\mathbf{c}$ with $c_i = a_i$ for $i \in C_1$ and $c_i = b_i$ for $i \in C_2$.
- δ is a *transition function*, mapping each state $s \in S$ and full joint action $\mathbf{a} \in D(s)$ to a state $\delta(s, \mathbf{a}) \in S$.

The interpretation of a formula in a state of a CGS is defined as follows:

$$\begin{aligned}
M, s &\models p &&\Leftrightarrow s \in V(p) \\
M, s &\models \neg\varphi &&\Leftrightarrow M, s \not\models \varphi \\
M, s &\models \varphi \wedge \psi &&\Leftrightarrow M, s \models \varphi \text{ and } M, s \models \psi \\
M, s &\models [\![C]\!]\psi &&\Leftrightarrow \exists \mathbf{a}_C \in D_C(s) \forall \mathbf{a}_{\overline{C}} \in D_{\overline{C}}(s), \ M, \delta(s, (\mathbf{a}_C, \mathbf{a}_{\overline{C}})) \models \psi
\end{aligned}$$

A formula φ is valid, $\models \varphi$, iff $M, s \models \varphi$ for all M and s in M. The axiomatic system in Fig. 1 is sound and complete [16]; in particular, a formula is valid iff it is derivable.

---

[1] There is a subtlety here: [11] showed that it is actually not the case that all playable effectivity functions correspond to games, but that only so-called *truly* playable ones do, but for the interpretation of the CL language it makes no difference.

instances of propositional tautologies $(Prop)$
$\neg \langle\!\langle C \rangle\!\rangle \bot$ $(\bot)$
$\langle\!\langle C \rangle\!\rangle \top$ $(\top)$
$\neg \langle\!\langle \emptyset \rangle\!\rangle \neg\varphi \to \langle\!\langle N \rangle\!\rangle \varphi$ (N)
$\langle\!\langle C \rangle\!\rangle (\varphi \wedge \psi) \to \langle\!\langle C \rangle\!\rangle \psi$ (M)
$(\langle\!\langle C_1 \rangle\!\rangle \varphi_1 \wedge \langle\!\langle C_2 \rangle\!\rangle \varphi_2) \to \langle\!\langle C_1 \cup C_2 \rangle\!\rangle (\varphi_1 \wedge \varphi_2)$, where $C_1 \cap C_2 = \emptyset$ (S)
From $\varphi \leftrightarrow \psi$, derive $\langle\!\langle C \rangle\!\rangle \varphi \leftrightarrow \langle\!\langle C \rangle\!\rangle \psi$ (E)
From φ and $\varphi \to \psi$, derive ψ (MP)

Fig. 1. Axiomatisation of CL.

2.2 Alternating-Time Temporal Logic and ATL*

The language of alternating-time temporal logic (ATL) is defined as follows.

$$\varphi ::= p \mid \neg\varphi \mid \varphi \wedge \varphi \mid \langle\!\langle C \rangle\!\rangle X\varphi \mid \langle\!\langle C \rangle\!\rangle G\varphi \mid \langle\!\langle C \rangle\!\rangle (\varphi U \varphi)$$

where $p \in P$ and $C \subseteq N$. ATL adds combinations of $\langle\!\langle C \rangle\!\rangle$, which also here intuitively means that C has the ability to make something come about, with temporal modifiers X (next), G (always in the future) and U (until). CL can be seen as the next-time fragment of ATL with an implicit X after each $\langle\!\langle C \rangle\!\rangle$.

ATL is interpreted in CGSs as follows. Since the satisfaction of, e.g., formulas of the form $\langle\!\langle C \rangle\!\rangle G\varphi$ does not only depend on the actions chosen by C in the current state (like for CL) but also on actions chosen in other states, we need the notion of a strategy. Given a CGS, a *memoryless strategy* for agent $i \in N$ is a function $f_i : S \to ACT$ such that $f_i(s) \in d_i(s)$. A *full-memory strategy* for agent i is a function $f_i : S^+ \to ACT$ that maps any finite sequence of states to an action $f_i(s_0 \cdots s_k) \in d_i(s_k)$ for i. A joint (memoryless or full-memory) strategy for a coalition $C \subseteq N$ consists of one (memoryless or full-memory) strategy for each member of C. Note that we can view a memoryless strategy as a full-memory strategy where $f_i(s_0 \cdots c_k) = f_i(s_0' \cdots c_l')$ whenever $c_k = c_l'$. Finally, $out(s_0, f_C)$ is the set of paths starting in s_0 compatible with joint strategy f_C for C, i.e., all infinite sequences $\lambda = s_0 s_1 \cdots$ where for each $i \geq 0$, there is a full joint action $\mathbf{a} \in D(s_i)$ where for all $j \in C$, $a_j = f_j(s_0 \cdots s_i)$ and $s_{i+1} = \delta(s_i, \mathbf{a})$. The interpretation is defined as follows, where only featured clauses are presented:

$$M, s \models \langle\!\langle C \rangle\!\rangle X\psi \;\Leftrightarrow\; \exists f_C \forall \lambda \in out(s, f_C), M, \lambda[1] \models \psi$$
$$M, s \models \langle\!\langle C \rangle\!\rangle G\psi \;\Leftrightarrow\; \exists f_C \forall \lambda \in out(s, f_C) \forall i \geq 0, M, \lambda[i] \models \psi$$
$$M, s \models \langle\!\langle C \rangle\!\rangle \varphi U\psi \;\Leftrightarrow\; \exists f_C \forall \lambda \in out(s, f_C) \exists i \geq 0,$$
$$M, \lambda[i] \models \psi \text{ and } \forall j < i, M, \lambda[j] \models \varphi$$

where $\lambda[i] = s_i$ when $\lambda = s_0 s_1 \cdots$. It is easy to see that the interpretation of $\langle\!\langle C \rangle\!\rangle X$ is equivalent to the interpretation of $\langle\!\langle C \rangle\!\rangle$ in CL. In this definition f_C is a joint strategy for C, but the definition doesn't say which kind of strategy – full-memory or memoryless. The reason is that it does not matter: it turns out that the semantics is equivalent in both cases [19].

As we have seen from the definition, in ATL the strategy quantifiers $\langle\!\langle C \rangle\!\rangle$ and temporal modifiers X, G and U are "glued" together. In ATL* [4] this restriction is relaxed. The language is defined as follows, where φ is a *state formula* and γ is a *path formula*:

$$\varphi ::= p \mid \neg\varphi \mid \varphi \wedge \varphi \mid \langle\!\langle C \rangle\!\rangle \gamma \qquad\qquad \gamma ::= \varphi \mid \neg\gamma \mid \gamma \wedge \gamma \mid X\gamma \mid G\gamma \mid \gamma U \gamma$$

State formulas are interpreted in states, and path formulas in paths, of a CGS as follows, where only featured clauses are presented:

$$M, s \models \langle\!\langle C \rangle\!\rangle \gamma \Leftrightarrow \exists f_C \forall \lambda \in out(s, f_C), M, \lambda \models \gamma$$
$$M, \lambda \models \varphi \quad\ \Leftrightarrow M, \lambda[0] \models \varphi$$
$$M, \lambda \models X\gamma \ \Leftrightarrow M, \lambda[1, \infty] \models \gamma$$
$$M, \lambda \models G\gamma \ \Leftrightarrow M, \lambda[i, \infty] \models \gamma \text{ for all } i \geq 0$$
$$M, \lambda \models \gamma U \delta \ \Leftrightarrow \text{ for some } i \geq 0, M, \lambda[i, \infty] \models \delta \text{ and for all } j < i, M, \lambda[j, \infty] \models \gamma$$

Here, $[i, \infty]$ is the postfix of the path λ starting at $\lambda[i]$. Again, we deliberately left out what kind of strategy we quantify over, but now *it matters*: we get a different semantics for full-memory strategies than for memoryless strategies [19]. In the original publication on ATL [4], full-memory strategies are used.

3 CL*: Language and Semantics

The language of CL* is on the one hand an extension of the language of CL with (free occurrences of) X, on the other the restriction of ATL* where U and G are not allowed. It is formally defined as follows.

Given non-empty P and finite non-empty N, the *state formulas* φ and *path formulas* γ are defined by mutual recursion:

$$\varphi ::= p \mid \neg\varphi \mid \varphi \wedge \varphi \mid \langle\!\langle C \rangle\!\rangle \gamma \qquad\qquad \gamma ::= \varphi \mid \neg\gamma \mid \gamma \wedge \gamma \mid X\gamma$$

where $p \in P, C \subseteq N$. $\mathcal{L}_{CL^*}$ is the set of all state formulas. Note that while we use φ for state formulas and γ for path formulas in the BNF grammar defining the language, in the following we will not always strictly follow the same convention, and often, e.g., use symbols φ, ψ to stand for arbitrary path formulas as well as state formulas – which will be clear from context.

The standard coalition logic language $\mathcal{L}_{CL}$ is the fragment where every occurrence of $\langle\!\langle C \rangle\!\rangle$ is followed by X, and every X is immediately preceeded by a $\langle\!\langle C \rangle\!\rangle$[2].

The interpretation of the language in a state of a concurrent game structure as follows is given by the definition for ATL* in Sect. 2 restricted to the $\mathcal{L}_{CL^*}$ language:

$$M, s \models p \quad\quad\ \Leftrightarrow s \in V(p)$$
$$M, s \models \neg\varphi \quad\ \Leftrightarrow M, s \not\models \varphi$$
$$M, s \models \varphi \wedge \psi \Leftrightarrow M, s \models \varphi \text{ and } M, s \models \psi$$
$$M, s \models \langle\!\langle C \rangle\!\rangle \gamma \Leftrightarrow \exists f_C \forall \lambda \in out(s, f_C), M, \lambda \models \gamma$$
$$M, \lambda \models \varphi \quad\quad\ \Leftrightarrow M, \lambda[0] \models \varphi \text{ when } \varphi \text{ is a state formula}$$
$$M, \lambda \models \neg\varphi \quad\ \Leftrightarrow M, \lambda \not\models \varphi$$
$$M, \lambda \models \varphi \wedge \psi \Leftrightarrow M, \lambda \models \varphi \text{ and } M, \lambda \models \psi$$
$$M, \lambda \models X\gamma \quad\ \Leftrightarrow M, \lambda[1, \infty] \models \gamma$$

---

[2] In $\mathcal{L}_{CL}$ the X is implicit, $\langle\!\langle C \rangle\!\rangle$ corresponding to $\langle\!\langle C \rangle\!\rangle X$ in $\mathcal{L}_{CL^*}$. In $\mathcal{L}_{CL^*}$, $\langle\!\langle C \rangle\!\rangle \varphi$ is actually equivalent to φ, when φ is a state formula.

Like for ATL* it matters whether we quantify over all full-memory strategies, or only memoryless strategies (we will show this formally in the next section). We thus write $\models_L$ when the quantification over strategies is restricted to memoryless strategies, and $\models_F$ when we also allow full-memory strategies. It is easy to check that for $\mathcal{L}_{CL}$ we always have that $M, s \models_F \varphi$ iff $M, s \models_L \varphi$ and in that case we often write $M, s \models \varphi$. We use $CL_L^\star$ and $CL_F^\star$ to refer to the resulting logics, with memoryless strategies and with full-memory strategies, respectively.

4   Expressivity

We now have three ways to talk about the same models: standard CL, and CL* with memoryless strategies and with full-memory strategies. Let's compare the expressive power. We start with comparing the former and the latter.

4.1   *CL* vs. $CL_F^\star$

It is not too hard to see that we have the following:

$$\models_F (\!(C)\!) X(p \wedge Xq) \leftrightarrow (\!(C)\!) X(p \wedge (\!(C)\!) Xq) \tag{1}$$

Note that the right hand side is a $\mathcal{L}_{CL}$ formula. This equivalence is thus like a "reduction axiom", reducing a $\mathcal{L}_{CL^\star}$ formula to an equivalent $\mathcal{L}_{CL}$ formula. A natural question then is whether we can do a similar reduction for *any* formula, in the style of Public Announcement Logic [17]. That in fact turns out to be the case, although the details are a little more complicated. We are going to prove the following.

Theorem 1. *For every $\mathcal{L}_{CL^\star}$ formula φ there a $\mathcal{L}_{CL}$ formula φ' such that for any M, s,*

$$M, s \models_F \varphi \Leftrightarrow M, s \models_F \varphi'$$

To prove Theorem 1 we first need some preliminary results and definitions. The following is straightforward.

Lemma 2. *For any model M, path λ and path formulas γ and γ':*

1. $M, \lambda \models_F \neg X\gamma$ iff $M, \lambda \models_F X\neg\gamma$
2. $M, \lambda \models_F X\gamma \wedge X\gamma'$ iff $M, \lambda \models_F X(\gamma \wedge \gamma')$

We say that a state formula $\varphi \in \mathcal{L}_{CL^\star}$ is *clean* if $\varphi \in \mathcal{L}_{CL}$. We say that a path formula is clean if every state formula it contains is clean. For every path formula φ, recursively define $||\varphi||$, called the *path-depth* of φ, as follows: $||p|| = ||(\!(A)\!)\psi|| = 0$; $||\neg\psi|| = ||\psi||$; $||\psi \wedge \chi|| = \max(||\psi||, ||\chi||)$; $||X\psi|| = ||\psi|| + 1$.

Lemma 3. *Every clean path formula γ is equivalent to a path formula of the form*

$$\gamma' = (\beta_1 \wedge X\chi_1) \vee (\beta_2 \wedge X\chi_2) \vee \cdots \vee (\beta_m \wedge X\chi_m)$$

where m is a natural number, all $\beta_i \in \mathcal{L}_{CL}$, all χ_i are clean path formulas, and $||\gamma'|| = ||\gamma||$.

Proof. Let γ be a clean path formula. A *general literal* is either a propositional literal, or of the form $X\gamma'$ or $\neg X\gamma'$, or of the form $\langle\!\langle C\rangle\!\rangle\gamma'$ or $\neg\langle\!\langle C\rangle\!\rangle\gamma'$. We can write γ on disjunctive normal form:

$$\gamma' = (\alpha_1^1 \wedge \cdots \wedge \alpha_{k_1}^1) \vee \cdots \vee (\alpha_1^m \wedge \cdots \wedge \alpha_{k_m}^m)$$

where each α_i^j is a clean general literal. From Lemma 2.1 we can replace every $\neg X\gamma'$ with $X\neg\gamma'$, and by Lemma 2.2 we can replace all the α_i^j that starts with a X in a clause with a single one: $X\gamma_1 \wedge X\gamma_2 \wedge \cdots \wedge X\gamma_l$ is replaced by $X(\gamma_1 \wedge \cdots \wedge \gamma_l)$. Clearly, $\neg X\gamma$ is clean iff $X\neg\gamma$ is clean, and $X(\gamma_1 \wedge \gamma_2)$ is clean iff both $X\gamma_1$ and $X\gamma_2$ are clean. Thus, every clause can be rewritten as $(\beta_i \wedge X\chi_i)$ where β_i is a conjunction of propositional literals and all χ_i are clean. It is easy to see that $||\gamma'|| = ||\gamma||$.

Lemma 4. *Every formula $\langle\!\langle C\rangle\!\rangle\gamma'$ where γ' is as in Lemma 3 is equivalent to a formula of the form*

$$\varphi' = (\beta_1' \wedge \langle\!\langle C\rangle\!\rangle X\chi_1') \vee \cdots \vee (\beta_k' \wedge \langle\!\langle C\rangle\!\rangle X\chi_k')$$

where k is a natural number, each β_i' and χ_i' are clean, and $||\chi_i'|| < ||\gamma'||$.

Proof. We show that $\langle\!\langle C\rangle\!\rangle\gamma'$ is equivalent to:

$$\bigvee_{Y \subseteq \{1,\ldots,m\}} \left(\bigwedge_{i \in Y} \beta_i \wedge \langle\!\langle C\rangle\!\rangle X \bigvee_{i \in Y} \chi_i \right) \tag{2}$$

For the implication towards the right, let $M, s \models_F \langle\!\langle C\rangle\!\rangle\gamma'$. There is a f_C such that for all $\lambda \in out(s, f_C), M, \lambda \models_F \gamma_1 \vee \cdots \vee \gamma_m$, where $\gamma_i = (\beta_i \wedge X\chi_i)$. Thus, for each $\lambda \in out(s, f_C), M, \lambda \models_F \gamma_i$ for some i.

Let $Y = \{1 \leq i \leq m : M, \lambda \models_F \gamma_i$ for some $\lambda \in out(s, f_C)\}$. Let $\lambda \in out(s, f_C)$. Let $i \in Y$. Note $M, \lambda' \models_F \beta_i$ for some $\lambda' \in out(s, f_C)\}$. Note β_i is a state formula and $\lambda'[0] = \lambda[0]$. Then $M, \lambda \models_F \beta_i$. Then $M, \lambda \models_F \bigwedge_{i \in Y} \beta_i$. Then $M, s \models_F \bigwedge_{i \in Y} \beta_i$. It is easy to see $M, \lambda \models_F X\chi_i$ for some $i \in Y$. Then $M, \lambda \models_F \bigvee_{i \in Y} X\chi_i$. Then $M, \lambda \models_F X \bigvee_{i \in Y} \chi_i$. Then $M, s \models_F \langle\!\langle C\rangle\!\rangle X \bigvee_{i \in Y} \chi_i$.

For the implication to the left, let $M, s \models_F (\bigwedge_{i \in Y} \beta_i \wedge \langle\!\langle C\rangle\!\rangle \bigvee_{i \in Y} \chi_i)$ for some $Y \subseteq \{1,\ldots,m\}$. Let f_C be such that for all $\lambda \in out(s, f_C), M, \lambda[1,\infty] \models_F \bigvee_{i \in Y} \chi_i$. Since $M, \lambda[0] \models_F \beta_i$ for all $i \in Y$, and $M, \lambda[1,\infty] \models_F \chi_i$ for some $i \in Y$, we have that $M, \lambda \models_F \beta_i \wedge X\chi_i$ for some $i \in Y$ and thus that $M, s \models_F \langle\!\langle C\rangle\!\rangle\gamma'$. That $\bigwedge_{i \in Y} \beta_i$ and $\bigvee_{i \in Y} \chi_i$ are clean follow from the fact that each β_i, χ_i are clean. Since $\chi_j' = \bigvee_{i \in Y} \chi_i$ and $X\chi_i$ occurs in γ' for each i, $||\chi_j'|| < ||\gamma'||$.

Lemma 5.
$$\models_F \langle\!\langle C\rangle\!\rangle X\chi \leftrightarrow \langle\!\langle C\rangle\!\rangle X\langle\!\langle C\rangle\!\rangle\chi$$

Proof. For the direction to the right, let $M, s \models_F \langle\!\langle C\rangle\!\rangle X\chi$. Let f_C be a full-memory strategy such that for all $\lambda \in out(s, f_C), M, \lambda[1,\infty] \models_F \chi$. Let $\lambda \in out(s, f_C)$ and $t = \lambda[1]$. We are going to show that $M, t \models_F \langle\!\langle C\rangle\!\rangle\chi$.

Let f'_C be defined as follows: for each $i \in C$, for each finite sequence τ starting from t, $f'_i(\tau) = f_i(s\tau)$. The definition of f'_C for other finite sequences doesn't matter.

Let $\lambda' \in out(t, f'_C)$. Then $s\lambda' \in out(s, f_C)$. Then $M, s\lambda'[1, \infty] \models_F \chi$, that is, $M, \lambda' \models_F \chi$. Then $M, t \models_F (\!(C)\!)\chi$.

For the direction to the left, let $M, s \models_F (\!(C)\!)X(\!(C)\!)\chi$, and let f_C be such that for all $\lambda \in out(s, f_C)$, $M, \lambda[1] \models_F (\!(C)\!)\chi$. Let σ be the C-action specified by f_C at s. Then for each successor t of σ at s, there is a full-memory strategy f^t_C such that for all $\lambda' \in out(t, f^t_C)$, $M, \lambda' \models_F \chi$.

Let f'_C be defined as follows: for each $i \in C$, (1) $f'_i(s) = f_i(s)$, and (2) for every finite sequence $ss_0 \cdots$, where s_0 is a successor of σ at s, $f'_i(ss_0 \cdots) = f^{s_0}_i(s_0 \cdots)$. The definition of f'_C for other finite sequences doesn't matter.

Let $\lambda \in out(s, f'_C)$. Let $t = \lambda[1]$. Note t is a successor of σ at s. It is easy to see $\lambda[1, \infty] \in out(t, f^t_C)$. Then $M, \lambda[1, \infty] \models_F \chi$. Then $M, \lambda \models_F X\chi$. Then $M, s \models_F (\!(C)\!)X\chi$.

We can finally prove Theorem 1.

Proof (of Theorem 1). Let φ be a $\mathcal{L}_{CL^*}$ formula. Let $(\!(C)\!)\gamma$ be a subformula of φ such that $(\!(C)\!)\gamma$ is not clean but γ is clean (if there are none we are done). We will show that we can replace it with an equivalent $\mathcal{L}_{CL}$ formula. The theorem follows: start with the "innermost" such formula, replace it with a $\mathcal{L}_{CL}$ formula, and continue "outwards".

First translate γ to γ', as in Lemma 3. $(\!(C)\!)\gamma$ is equivalent to $(\!(C)\!)\gamma'$, and $||\gamma'|| = ||\gamma||$. By Lemma 4, $(\!(C)\!)\gamma'$ is again equivalent to a formula of the form $(\beta'_1 \wedge (\!(C)\!)X\chi'_1) \vee (\beta'_2 \wedge (\!(C)\!)X\chi'_2) \vee \cdots (\beta'_k \wedge (\!(C)\!)X\chi'_k)$ where each β'_i, χ'_i are clean and $||\chi'_i|| < ||\gamma'|| = ||\gamma||$. Consider all $(\!(C)\!)X\chi'_i$ not in $\mathcal{L}_{CL}$. By Lemma 5, each $(\!(C)\!)X\chi'_i$ is equivalent to $(\!(C)\!)X(\!(C)\!)\chi'_i$, and we have that $||\chi'_i|| < ||\gamma||$. We replace each $(\!(C)\!)X\chi'_i$ with $(\!(C)\!)X(\!(C)\!)\chi'_i$. If all $(\!(C)\!)\chi'_i$ are clean we are done, if not repeat the process for $(\!(C)\!)\chi'_i$ in place of $(\!(C)\!)\gamma$. Each time we replace a formula $(\!(C)\!)\gamma$ where γ is not clean with possibly several $(\!(C)\!)\chi'$ where $||\chi'|| < ||\gamma||$, and eventually the process will end and we are left with a clean formula.

Thus, in the case of full-memory strategies, $\mathcal{L}_{CL^*}$ is no more expressive than $\mathcal{L}_{CL}$.

4.2 CL^*_F Vs. CL^*_L

Let us turn to memoryless strategies. There is a difference between using memoryless strategies and full-memory strategies, unlike the case for ATL. Consider the following formula and single-agent model:

$$\varphi = \langle\!\langle i \rangle\!\rangle (Xp \wedge XX\neg p) \qquad M:$$

We have that $M, q \models_F \varphi$, but $M, q \not\models_L \varphi$. We immediately get the following expressivity result (see [2] for a definition of bisimulation for CGSs).

Lemma 6. *With full-memory strategies, satisfaction of $\mathcal{L}_{CL^\star}$ formulas is invariant under bisimulation. With memoryless strategies, satisfaction of $\mathcal{L}_{CL^\star}$ formulas is not invariant under bisimulation.*

Proof. The first claim follows immediately from Theorem 1 and the fact that $\mathcal{L}_{CL}$ is invariant under bisimulation [2]. For the second, consider the *tree-unfolding* [2] M' of the model M above, in which we have $M', q \models_L \varphi$. The two models are bisimilar.

Thus, perhaps contrary to intuition[3], the $\mathcal{L}_{CL^\star}$ language interpreted using only memoryless strategies is *more expressive* than if it is interpreted using full-memory strategies – even though memoryless strategies *are* (a special case of) full-memory strategies.

Let us formalise these claims about expressive power. In general a modal language $\mathcal{L}_1$ is said to be *at least as expressive* as language $\mathcal{L}_2$, written $\mathcal{L}_1 \succeq \mathcal{L}_2$, interpreted over the same class of models, iff for every $\varphi_2 \in \mathcal{L}_2$ there is a $\varphi_1 \in \mathcal{L}_1$ such that $M, s \models \varphi_2$ iff $M, s \models \varphi_1$ for any model M and state s. $\mathcal{L}_1 \succ \mathcal{L}_2$ means that $\mathcal{L}_1 \succeq \mathcal{L}_2$ and $\mathcal{L}_2 \not\succeq \mathcal{L}_1$; $\mathcal{L}_1 \approx \mathcal{L}_2$ that $\mathcal{L}_1 \succeq \mathcal{L}_2$ and $\mathcal{L}_2 \succeq \mathcal{L}_1$. When comparing $\mathcal{L}_{CL^\star}$ with the two types of strategies, we have the *same* language but with two different interpretations, but we will abuse notation and write $\mathcal{L}_{CL^\star}^L \succeq \mathcal{L}_{CL^\star}^F$ in the case that for every $\varphi \in \mathcal{L}_{CL^\star}$ there is a $\varphi' \in \mathcal{L}_{CL^\star}$ such that $M, s \models_F \varphi$ iff $M, s \models_L \varphi'$ for any model M and state s. We thus have that

$$\mathcal{L}_{CL^\star}^L \succ \mathcal{L}_{CL^\star}^F \approx \mathcal{L}_{CL}$$

As an example, the formula

$$\langle\!\langle a \rangle\!\rangle X(p \wedge \langle\!\langle a \rangle\!\rangle X\neg p) \wedge \neg\langle\!\langle a \rangle\!\rangle (Xp \wedge XX\neg p) \tag{3}$$

is unsatisfiable with full-memory semantics but with memoryless semantics it is a sufficient condition for the property that a can choose to loop back to the same state.

The difference in expressive power is indeed closely related to "looping". In fact, on tree-like CGSs, i.e., models where there is a unique path to every state from the "root" state, the two types of strategies coincide [2] and we immediately have the following.

Lemma 7. *For any tree-like model M and $\varphi \in \mathcal{L}_{CL^\star}$, $M, s \models_L \varphi$ iff $M, s \models_F \varphi$.*

---

[3] However it is not untypical that there is no direct relationship between widening nor narrowing the scope of a modal quantifier and an increase or decrease in expressive power, see, e.g., [1].

From the discussion around (3) above, we immediately have $\models_F \langle\!\langle a \rangle\!\rangle X(p \wedge \langle\!\langle a \rangle\!\rangle X \neg p) \rightarrow \langle\!\langle a \rangle\!\rangle (Xp \wedge XX\neg p)$ but $\not\models_L \langle\!\langle a \rangle\!\rangle X(p \wedge \langle\!\langle a \rangle\!\rangle X \neg p) \rightarrow \langle\!\langle a \rangle\!\rangle (Xp \wedge XX\neg p)$. Together, Lemmas 6 and 7 show that full-memory CL* is equivalent to memoryless CL* restricted to tree-like models. Viewing each of these logics as sets of valid formulas, $CL_L^* = \{\varphi \in \mathcal{L}_{CL^*} : \models_L \varphi\}$ and $CL_F^* = \{\varphi \in \mathcal{L}_{CL^*} : \models_F \varphi\}$, we get:

Theorem 8. $CL_L^* \subset CL_F^*$

While full-memory CL* is no more expressive than CL, as we have seen it has validities that are not CL validities since they are not even CL formulas. In the next section we give a complete axiomatisation of all validities of full-memory CL*.

5   Completeness of Full-Memory CL*

We define an axiomatic system over the $\mathcal{L}_{CL^*}$ language, extending the system for $\mathcal{L}_{CL}$ in Fig. 1.

Definition 9 (S_F).   *Let S_F be the Hilbert-style system with the following axioms and rules, where $C, D \subseteq N$. A state-formula (path-formula) instance of a propositional tautology is the result of uniformly replacing one or more propositional letters in a propositional tautology with state- (path-) formulas.*

Axioms, where φ, ψ, χ range over path formulas:

1. *Axioms of CL:*
   (a) *state-formula instances of propositional tautologies*
   (b) $\langle\!\langle C \rangle\!\rangle X(\varphi \wedge \psi) \rightarrow \langle\!\langle C \rangle\!\rangle X\varphi$[4]
   (c) $\neg\langle\!\langle C \rangle\!\rangle X\bot$
   (d) $\langle\!\langle C \rangle\!\rangle X\top$
   (e) $(\langle\!\langle C \rangle\!\rangle X\varphi \wedge \langle\!\langle D \rangle\!\rangle X\psi) \rightarrow \langle\!\langle C \cup D \rangle\!\rangle X(\varphi \wedge \psi)$, *where* $C \cap D = \emptyset$
   (f) $\neg\langle\!\langle \emptyset \rangle\!\rangle X\neg\varphi \rightarrow \langle\!\langle N \rangle\!\rangle X\varphi$

2. *New axioms:*
   (a) $\langle\!\langle C \rangle\!\rangle (\neg X\varphi \leftrightarrow X\neg\varphi)$
   (b) $\langle\!\langle C \rangle\!\rangle ((X\varphi \wedge X\psi) \leftrightarrow X(\varphi \wedge \psi))$
   (c) $\langle\!\langle C \rangle\!\rangle (\beta \wedge \varphi) \leftrightarrow (\beta \wedge \langle\!\langle C \rangle\!\rangle \varphi)$, *where β is a state formula*
   (d) $\langle\!\langle C \rangle\!\rangle X\chi \leftrightarrow \langle\!\langle C \rangle\!\rangle X\langle\!\langle C \rangle\!\rangle \chi$
   (e) $\langle\!\langle C \rangle\!\rangle \varphi$, *where φ is a path-formula instance of a propositional tautology*
   M2 $\langle\!\langle C \rangle\!\rangle (\varphi \wedge \psi) \rightarrow \langle\!\langle C \rangle\!\rangle \varphi$

---

[4] We notice that this formula is derivable from axiom (2b) and rule E2. We treat it as an axiom to make the system easier to understand.

Rules:

1. *MP: From φ and $\varphi \to \psi$, derive ψ, where φ and ψ are state formulas.*
2. *E2: From $\langle\!\langle C \rangle\!\rangle (\varphi \leftrightarrow \psi)$, derive $\chi \leftrightarrow \chi[\varphi/\psi]$, where φ, ψ are path formulas, χ is a state formula, and $\chi[\varphi/\psi]$ is the result of replacing an occurrence of φ in χ by ψ.*

The first part of the axioms are the schemas for CL, but now φ and ψ range (also) over path formulas. The new axioms (2) take care of relations between path formulas, and the E2 rule will help replace equivalent path formulas inside a formula (here we can't use derivability of $\varphi \leftrightarrow \psi$ directly as the antecedent, since it is not necessarily a state formula). The E rule of CL is derivable using E2 as we show below (where it is called E1).

We write $\vdash \varphi$ to denote that φ is derivable in S_F. The proof of the following is straightforward: the CL axiom schemas are valid also for the extended language; the new axioms are valid (validity of (2d) was shown in Lemma 5); and the rules preserve validity.

Theorem 10 (Soundness). *For any (state formula) $\varphi \in \mathcal{L}_{CL^\star}$, if $\vdash \varphi$ then $\models_F \varphi$.*

The following, all of which can be easily shown, will be useful going forward. In particular, E3 says that we can replace provably equivalent *state* formulas anywhere inside a formula. This will be used frequently in the following, often without explicit mention.

Lemma 11.

M: $\vdash \langle\!\langle C \rangle\!\rangle \varphi \to \langle\!\langle C \rangle\!\rangle \varphi'$, where $\varphi \to \varphi'$ is a path-formula instance of a propositional tautology
(2f): $\vdash \langle\!\langle C \rangle\!\rangle (\varphi \leftrightarrow (\varphi \wedge X\top))$
(2g): $\vdash \langle\!\langle A \rangle\!\rangle ((X\varphi \vee X\psi) \leftrightarrow X(\varphi \vee \psi))$
E1: If $\vdash \varphi \leftrightarrow \psi$, then $\vdash \langle\!\langle C \rangle\!\rangle X\varphi \leftrightarrow \langle\!\langle C \rangle\!\rangle X\psi$, where φ and ψ are state formulas
E3: If $\vdash \varphi \leftrightarrow \psi$, then $\vdash \chi \leftrightarrow \chi[\varphi/\psi]$, where φ, ψ and χ are state formulas, and $\chi[\varphi/\psi]$ is like in rule E2

We move on to completeness: that we can derive all state formulas that are valid in the full-memory semantics. The strategy is as follows: we will replicate the proof of Theorem 1 syntactically – "translating" a $\mathcal{L}_{CL^\star}$ formula to a provably equivalent $\mathcal{L}_{CL}$ formula. The new axioms and the E2 rule will help us do exactly that. Note that the E2 rule lets us replace *path* formulas that are equivalent, in the sense that we can prove $\langle\!\langle C \rangle\!\rangle (\varphi \leftrightarrow \psi)$, anywhere inside a formula which we will also make frequent use of (often without comment) in the following.

Recall the notions of clean state/path formulas, path-depth, and of general literals, from Sect. 4. We start with a syntactic variant of Lemma 3.

Lemma 12. *Let $\langle\!\langle C \rangle\!\rangle \gamma$ be a state formula in $\mathcal{L}_{CL^*}$ such that $\langle\!\langle C \rangle\!\rangle \gamma$ is not clean, but γ is clean and not in $\mathcal{L}_{CL}$. Then, there is a γ' in $\mathcal{L}_{CL^*}$ such that $\vdash \langle\!\langle C \rangle\!\rangle \gamma \leftrightarrow \langle\!\langle C \rangle\!\rangle \gamma'$, where:*

$$\gamma' = (\beta_1 \wedge X\chi_1) \vee (\beta_2 \wedge X\chi_2) \vee \cdots \vee (\beta_m \wedge X\chi_m),$$

all $\beta_i \in \mathcal{L}_{CL}$, all χ_i are clean path formulas, and $||\gamma'| = ||\gamma||$.

Proof. By axiom (2e) and E2, we can transform γ to δ s.t. $\vdash \langle\!\langle C \rangle\!\rangle \gamma \leftrightarrow \langle\!\langle C \rangle\!\rangle \delta$, and:

$$\delta = (\alpha_1^1 \wedge \cdots \wedge \alpha_{k_1}^1) \vee \cdots \vee (\alpha_1^m \wedge \cdots \wedge \alpha_{k_m}^m),$$

where each α_i^j is a clean general literal. Note $||\gamma|| = ||\delta||$.

Note that $\vdash \langle\!\langle \emptyset \rangle\!\rangle (\varphi \leftrightarrow (\varphi \wedge \top))$, so by E2 we can assume that every $\alpha_1^i \wedge \cdots \wedge \alpha_{k_i}^i$ contains $\top$, which is in $\mathcal{L}_{CL}$. Note that some $\alpha_1^i \wedge \cdots \wedge \alpha_{k_i}^i$ contains a formula in the form $X\chi$ or $\neg X\chi$; otherwise γ would be in $\mathcal{L}_{CL}$. By (2f) (Lemma 11) and E2, we can assume that every $\alpha_1^i \wedge \cdots \wedge \alpha_{k_i}^i$ contains $X\top$. The two assumptions do not change $||\delta||$, but can avoid δ to be in the forms such as $X\chi \vee p$.

By axiom (2a) and E2, we can replace every α_i^j of the form $\neg X\psi$ in $\langle\!\langle C \rangle\!\rangle \delta$ with $X\neg\psi$. By axiom (2b) and E2, we can replace all the α_i^j that start with an X in a clause with a single one: $X\gamma_1 \wedge \cdots \wedge X\gamma_l$ is replaced by $X(\gamma_1 \wedge \cdots \wedge \gamma_l)$. Thus, we can write $\langle\!\langle C \rangle\!\rangle \delta$ as $\langle\!\langle C \rangle\!\rangle \gamma'$, where γ' is of the form

$$\gamma' = (\beta_1 \wedge X\chi_1) \vee (\beta_2 \wedge X\chi_2) \vee \cdots \vee (\beta_m \wedge X\chi_m),$$

where every β_i is in $\mathcal{L}_{CL}$.

As in the proof of Lemma 3, every χ_i is clean. It is easy to check that $||\gamma'| = ||\delta|| = ||\gamma||$.

We next prove a syntactic variant of Lemma 4.

Lemma 13. *Let $A \subseteq N$, $I = \{1, \ldots, m\}$ be a set of indices, $\beta_i \in \mathcal{L}_{CL}$ for each $i \in I$, and χ_i be a clean path formula for each $i \in I$. Then:*

$$\vdash \langle\!\langle A \rangle\!\rangle \bigvee_{k \in I} (\beta_k \wedge X\chi_k) \leftrightarrow \bigvee_{I' \subseteq I} \left(\bigwedge_{i \in I'} \beta_i \wedge \langle\!\langle A \rangle\!\rangle X \bigvee_{i \in I'} \chi_i \right)$$

and $|| \bigvee_{i \in I'} \chi_i|| < || \bigvee_{k \in I} (\beta_k \wedge X\chi_k)||$ for all $I' \subseteq I$.

Proof. We use $I_1, \ldots, I_{2^m}$ to respectively indicate the 2^m subsets of $\{1, \ldots, m\}$, and $\overline{I_1}, \ldots, \overline{I_{2^m}}$ to indicate their complement with respect to $\{1, \ldots, m\}$. The following equivalences are derivable (we use the E1, E2 and E3 rules without further comment):

1.

$$\langle\!\langle A \rangle\!\rangle \bigvee_{k \in I} (\beta_k \wedge X\chi_k)$$

2. $\updownarrow$ (propositional logic)

$$\bigwedge_{i\in I_1}\beta_i \wedge \bigwedge_{i\in\overline{I_1}}\neg\beta_i \wedge \langle\!\langle A\rangle\!\rangle \bigvee_{k\in I}(\beta_k \wedge X\chi_k) \vee \cdots \vee$$

$$\bigwedge_{i\in I_{2m}}\beta_i \wedge \bigwedge_{i\in\overline{I_{2m}}}\neg\beta_i \wedge \langle\!\langle A\rangle\!\rangle \bigvee_{k\in I}(\beta_k \wedge X\chi_k)$$

3. $\updownarrow$ (2c)

$$\langle\!\langle A\rangle\!\rangle\left(\bigwedge_{i\in I_1}\beta_i \wedge \bigwedge_{i\in\overline{I_1}}\neg\beta_i \wedge \bigvee_{k\in I}(\beta_k \wedge X\chi_k)\right) \vee \cdots \vee$$

$$\langle\!\langle A\rangle\!\rangle\left(\bigwedge_{i\in I_{2m}}\beta_i \wedge \bigwedge_{i'\in\overline{I_{2m}}}\neg\beta_i \wedge \bigvee_{k\in I}(\beta_k \wedge X\chi_k)\right)$$

4. $\updownarrow$ (2e)

$$\langle\!\langle A\rangle\!\rangle\left(\bigwedge_{i\in I_1}\beta_i \wedge \bigwedge_{i'\in\overline{I_1}}\neg\beta_i \wedge \bigvee_{i\in I_1}(\beta_i \wedge X\chi_i)\right) \vee \cdots \vee$$

$$\langle\!\langle A\rangle\!\rangle\left(\bigwedge_{i\in I_{2m}}\beta_i \wedge \bigwedge_{i\in\overline{I_{2m}}}\neg\beta_i \wedge \bigvee_{i\in I_{2m}}(\beta_i \wedge X\chi_i)\right)$$

5. $\updownarrow$ (2e)

$$\langle\!\langle A\rangle\!\rangle\left(\bigwedge_{i\in I_1}\beta_i \wedge \bigwedge_{i\in\overline{I_1}}\neg\beta_i \wedge \bigvee_{i\in I_1}X\chi_i\right) \vee \cdots \vee \langle\!\langle A\rangle\!\rangle\left(\bigwedge_{i\in I_{2m}}\beta_i \wedge \bigwedge_{i\in\overline{I_{2m}}}\neg\beta_i \wedge \bigvee_{i\in I_{2m}}X\chi_i\right)$$

6. $\updownarrow$ (2c)

$$\bigwedge_{i\in I_1}\beta_i \wedge \bigwedge_{i\in\overline{I_1}}\neg\beta_i \wedge \langle\!\langle A\rangle\!\rangle \bigvee_{i\in I_1}X\chi_i \vee \cdots \vee \bigwedge_{i\in I_{2m}}\beta_i \wedge \bigwedge_{i\in\overline{I_{2m}}}\neg\beta_i \wedge \langle\!\langle A\rangle\!\rangle \bigvee_{i\in I_{2m}}X\chi_i$$

7. $\updownarrow$ (*)

$$\bigwedge_{i\in I_1}\beta_i \wedge \langle\!\langle A\rangle\!\rangle \bigvee_{i\in I_1}X\chi_i \vee \cdots \vee \bigwedge_{i\in I_{2m}}\beta_i \wedge \langle\!\langle A\rangle\!\rangle \bigvee_{i\in I_{2m}}X\chi_i$$

8. $\updownarrow$ (2g)

$$\bigwedge_{i\in I_1}\beta_i \wedge \langle\!\langle A\rangle\!\rangle X \bigvee_{i\in I_1}\chi_i \vee \cdots \vee \bigwedge_{i\in I_{2m}}\beta_i \wedge \langle\!\langle A\rangle\!\rangle X \bigvee_{i\in I_{2m}}\chi_i$$

We now show that the equivalence in the step marked (*) from line 6 to 7 is derivable. The implication from the left to the right is straighforward. Consider the other direction. Fix $I_n \subseteq I$. Let $J = I - I_n$ and $|J| = l$. We use $J_1, \ldots, J_{2^l}$ to respectively indicate the 2^l subsets of J, and use $\overline{J_1}, \ldots, \overline{J_{2^l}}$ to respectively indicate their complement with respect to J. By propositional reasoning, $(\bigwedge_{i \in I_n} \beta_i \wedge [\![A]\!] \bigvee_{i \in I_n} X\chi_i) \leftrightarrow (\bigwedge_{i \in I_n} \beta_i \wedge \bigwedge_{x \in J_1} \beta_x \wedge \bigwedge_{y \in \overline{J_1}} \neg\beta_y \wedge [\![A]\!] \bigvee_{i \in I_n} X\chi_i \vee \cdots \vee \bigwedge_{i \in I_n} \beta_i \wedge \bigwedge_{x \in J_{2^l}} \beta_x \wedge \bigwedge_{y \in \overline{J_{2^l}}} \neg\beta_y \wedge [\![A]\!] \bigvee_{i \in I_n} X\chi_i)$, the right hand side which again implies, by using (M) (Lemma 11) $\bigwedge_{i \in I_n} \beta_i \wedge \bigwedge_{x \in J_1} \beta_x \wedge [\![A]\!] \bigvee_{i \in I_n \cup J_1} X\chi_i \vee \cdots \vee \bigwedge_{i \in I_n} \beta_i \wedge \bigwedge_{x \in J_{2^l}} \beta_x \wedge [\![A]\!] \bigvee_{i \in I_n \cup J_{2^l}} X\chi_i.$

It is easy to see that $|| \bigvee_{i \in I'} \chi_i || < || \bigvee_{k \in I} (\beta_k \wedge X\chi_k)||$ for all $I' \subseteq I$.

Lemma 14. *For every $\mathcal{L}_{CL^\star}$ state formula φ, there a $\mathcal{L}_{CL}$ formula φ' s.t. $\vdash \varphi \leftrightarrow \varphi'$.*

Proof. Let φ be a $\mathcal{L}_{CL^\star}$ state formula. Let $[\![C]\!]\gamma$ be a subformula of φ such that $[\![C]\!]\gamma$ is not clean but γ is clean (like in the proof of Theorem 1 if there are none we are done). We will show that it is provably equivalent to a $\mathcal{L}_{CL}$ formula. The result immediately follows: start with the "innermost" such formula, replace it with a $\mathcal{L}_{CL}$ formula using E3 (Lemma 11), and continue "outwards".

If γ is in $\mathcal{L}_{CL}$ it can be shown that $\vdash [\![C]\!]\gamma \leftrightarrow \gamma$, so assume γ is not in $\mathcal{L}_{CL}$. By Lemma 12, $\vdash [\![C]\!]\gamma \leftrightarrow [\![C]\!]\gamma'$, where:

$$\gamma' = (\beta_1 \wedge X\chi_1) \vee (\beta_2 \wedge X\chi_2) \vee \cdots \vee (\beta_m \wedge X\chi_m)$$

and all β_i are in $\mathcal{L}_{CL}$, all χ_i are clean path formulas, and $||\gamma'|| = ||\gamma||$. By Lemma 13,

$$\vdash [\![C]\!]\gamma' \leftrightarrow (\beta_1' \wedge [\![C]\!]X\chi_1') \vee (\beta_2' \wedge [\![C]\!]X\chi_2') \vee \cdots (\beta_k' \wedge [\![C]\!]X\chi_k')$$

where all β_i' are in $\mathcal{L}_{CL}$, and all χ_i' are clean such that $||\chi_i'|| < ||\gamma'||$.

We consider all $[\![C]\!]X\chi_i'$ not in $\mathcal{L}_{CL}$. By axiom (2d) and E3, we can replace each $[\![C]\!]X\chi_i'$ with $[\![C]\!]X[\![C]\!]\chi_i'$. Note that $||\chi_i'|| < ||\gamma'|| = ||\gamma||$. Repeat. Like in the proof of Theorem 1 eventually the process will end.

Theorem 15 (Completeness). *For any (state formula) $\varphi \in \mathcal{L}_{CL^\star}$, if $\models_F \varphi$ then $\vdash \varphi$.*

Proof. Let $\varphi \in \mathcal{L}_{CL^\star}$ s.t. $\models_F \varphi$. By Lemma 14, there is φ' in $\mathcal{L}_{CL}$ such that $\vdash \varphi \leftrightarrow \varphi'$. By the soundness of S_F, $\models_F \varphi \leftrightarrow \varphi'$. Then, $\models_F \varphi'$. By the completeness of CL (wrt. either type of strategy) and the fact that S_F is an extension of CL, $\vdash \varphi'$. Then, $\vdash \varphi$.

6 Discussion

For CL* *memory matters*: we get a different semantics if we allow full-memory strategies, than if we allow only memoryless strategies. The "only" here hints at a natural intuition, namely that the case with full-memory strategies is the general case and the memoryless case is a special case. As we have shown, in a precise sense the opposite can be argued to be the case: the logic with memoryless strategies is strictly more expressive. In a certain sense it is *all about memoryless strategies*: the logic of full-memory strategies is the logic of memoryless strategies restricted to tree-like models. From this perspective we can dispense with full-memory strategies.

We have given a complete axiomatisation of the full-memory case, effectively by translation to standard CL. The main open problem is a complete axiomatisation of the logic of memoryless strategies over all models. We note that the only of our axioms for the full-memory case that is not valid in the memoryless case, is (2d).

There are plenty of other opportunities for future work. Looking at other, limited, fragments of ATL* is one broad category. Another is generalising other extensions of CL in the same sense as for CL*, e.g., Epistemic Coalition Logic [21], Coalition Logic over stochastic games [15], Coalition Logic with resource bounds [3], or Coalition Logics with conditional operators [10,12].

Acknowledgments. We thank the LORI reviewers for their detailed and constructive comments which helped us improve the manuscript significantly.

Disclosure of Interests. The authors have no competing interests to declare that are relevant to the content of this article.

References

1. Ågotnes, T., Alechina, N., Galimullin, R.: Logics with group announcements and distributed knowledge: completeness and expressive power. J. Log. Lang. Inform. 1–26 (2022). https://doi.org/10.1007/s10849-022-09355-0
2. Ågotnes, T., Goranko, V., Jamroga, W.: Alternating-time temporal logics with irrevocable strategies. In: Samet, D. (ed.) Proceedings of the 11th Conference on Theoretical Aspects of Rationality and Knowledge (TARK XI), pp. 15–24. Presses Universitaires de Louvain/ACM DL, Brussels, Belgium (June 2007)
3. Alechina, N., Logan, B., Nguyen, H.N., Rakib, A.: Logic for coalitions with bounded resources. J. Log. Comput. **21**(6), 907–937 (2011)
4. Alur, R., Henzinger, T., Kupferman, O.: Alternating-time temporal logic. J. ACM **49**, 672–713 (2002)
5. Belnap, N., Perloff, M., Xu, M.: Facing the Future: Agents and Choices in Our Indeterminist World. Oxford University Press, Oxford (2001)
6. Ben-Ari, M., Manna, Z., Pnueli, A.: The temporal logic of branching time. In: Proceedings of the 8th ACM SIGPLAN-SIGACT Symposium on Principles of Programming Languages, pp. 164–176 (1981)
7. Chatterjee, K., Henzinger, T.A., Piterman, N.: Strategy logic. Inf. Comput. **208**(6), 677–693 (2010)
8. Clarke, E.M., Emerson, E.A.: Design and synthesis of synchronization skeletons using branching time temporal logic. In: Kozen, D. (eds.) Logics of Programs. Logic of Programs 1981. LNCS, vol. 131, pp. 52–71. Springer, Berlin, Heidelberg (1981). https://doi.org/10.1007/BFb0025774

9. Goranko, V.: Coalition games and alternating temporal logics. In: Theoretical Aspects Of Rationality And Knowledge: Proceedings of the 8 th Conference on Theoretical Aspects of Rationality and Knowledge, vol. 8, pp. 259–272 (2001)
10. Goranko, V., Enqvist, S.: Socially friendly and group protecting coalition logics. In: 17th International Conference on Autonomous Agents and Multiagent Systems (AAMAS 2018), Stockholm, Sweden, 10–15 July 2018, pp. 372–380. The International Foundation for Autonomous Agents and Multiagent Systems (2018)
11. Goranko, V., Jamroga, W., Turrini, P.: Strategic games and truly playable effectivity functions. Auton. Agent Multi-Agent Syst. **26**, 288–314 (2013)
12. Goranko, V., Ju, F.: A logic for conditional local strategic reasoning. J. Log. Lang. Inform. **31**(2), 167–188 (2022)
13. Goranko, V., Drimmelen, G.: Complete axiomatization and decidability of alternating-time temporal logic. Theor. Comput. Sci. **353**(1–3), 93–117 (2006)
14. Ju, F., Grilletti, G., Goranko, V.: A logic for temporal conditionals and a solution to the sea battle puzzle. In: Bezhanishvili, G., D'Agostino, G., Metcalfe, G., Studer, T. (eds.) Proceedings of the 12th International Conference on Advances in Modal Logic, pp. 407–426. College Publications (2018)
15. Naumov, P., Ros, K.: Strategic coalitions in stochastic games. J. Log. Comput. **31**(7), 1845–1867 (2021)
16. Pauly, M.: A modal logic for coalitional power in games. J. Log. Comput. **12**(1), 149–166 (2002)
17. Plaza, J.A.: Logics of public communications. In: Proceedings of ISMIS, pp. 201–216 (1989)
18. Reynolds, M.: An axiomatization of full computation tree logic. J. Symb. Log. **66**(3), 1011–1057 (2001)
19. Schobbens, P.Y.: Alternating-time logic with imperfect recall. Electron. Notes Theor. Comput. Sci. **85**(2) (2004)
20. Van Benthem, J.: Games in dynamic-epistemic logic. Bull. Econ. Res. **53**(4), 219–248 (2001)
21. Ågotnes, T., Alechina, N.: Coalition logic with individual, distributed and common knowledge. J. Log. Comput. **29**(7), 1041–1069 (2019)

How to Avoid Unexpected Exams

Alexandru Baltag and Sonja Smets[(✉)]

Institute for Logic, Language and Computation, Amsterdam, The Netherlands
`s.j.l.smets@uva.nl`

Abstract. We show how techniques from Belief Revision Theory and Dynamic Epistemic Logic can solve the Surprise Examination Paradox. Our solution is based on formalizing the role played by the attitude of the Student towards the Teacher as a source of information.

1 Introduction

The surprise exam paradox has intrigued philosophers and mathematicians for many years.[1] In a class with only one Student, the Teacher makes two announcements (just before the exam week): (1) *there will be an exam* in one of the five (working) days of next week; and (2) *the exam's day will be a surprise*, meaning that: even in the evening before the exam, the Student will still not be sure that the exam is tomorrow. *Is the Teacher lying, or is she telling the truth?*

Paradoxical Argumentation. Intuitively, the Student can prove (by backward induction, starting with Friday) that *the Teacher lied*: first, if the Teacher told the truth, then there will be an exam; but the exam cannot be on Friday (-since Friday is the last possible day, by Thursday evening the Student would know it, so the exam won't be a surprise, contrary to the Teacher's second statement); and once Friday is eliminated, the Student can repeat the same argument, until all days are eliminated, contradicting the Teacher's first statement (that there will be an exam). The only way out seems to be to conclude that the Teacher lied: *there will be no surprise exam.* As such, the Student *doesn't prepare for it,* being confident that the exam will either not be given at all, or else it will not be a surprise: he will somehow be sure of it the evening before, which should give him enough time to prepare in the last minute. But then, on (say) Tuesday, the exam comes, and it is a complete surprise! So, maybe the *Teacher wasn't lying* after all?! But then *what was the mistake* in the Student's reasoning?

Two Approaches: Epistemological Versus Self-referential. Most of the solutions proposed to the paradox are based on one of two approaches: the "epistemological" one, and the "self-referential" (or 'logical') one. According to the first, epistemological school [12,17,19,24,25], the Surprise Exam Paradox is similar to the so-called "Moore paradox", and the surprise announcement is akin

---

[1] See Chow [14] for an overview of the literature till 2011, Sorensen [26] for its philosophical analysis and Quine [22] for the Hangman version.

V. Goranko et al. (Eds.): LORI 2025, LNCS 16010, pp. 34–50, 2026.
https://doi.org/10.1007/978-981-95-2481-5_3

to a *Moore sentence* [20], or what Sorensen [24,26] calls an *epistemic 'blindspot'*: an unknowable yet consistent proposition. An example of a Moore sentence is "There will be an exam tomorrow but the Student doesn't know that". This proposition may well be true, but it cannot be known by the Student (today, i.e. at the same time when it is true). According to this interpretation, the surprise announcement is of this type, just looking more complicated because of the (irrelevant) addition of more possible days for the exam. This version is easy to solve, and no paradox ensues. We will review below J. Gerbrandy's formalization of this approach [18] using Public Announcement Logic (PAL) [21], as well as a limit-case epistemic concept: the notion of *infallible knowledge* (by a logically omniscient agent), understood as a "warranty of truth", providing absolute certainty of veracity, with no possibility of error, doubt or further revisions.

According to the second interpretation [13,16], the surprise announcement is *self-referential*: the Teacher announces that *the exam will be a surprise even after the Student hears (and presumably, believes) her announcement*. This school typically (but not unanimously) tries to avoid any reference to epistemic notions, interpreting the surprise announcement as saying that "there will be an exam next week, and moreover its date will not be deducible from *this very announcement* at any time prior to the exam"; hence, the name 'logical' approach.

Many authors consider the self-referential version to be genuinely paradoxical, but regard this as unremarkable: self-reference is known to be dangerous in Logic. According to this analysis, the Surprise Exam is similar to the Liar Paradox: a circular contradiction produced by an illicit form of self-reference, that should be expunged from a well-formed logical language.

Our Approach. In this paper, we propose that both schools are partially right: on the one hand, the puzzle is epistemological in nature, involving epistemic blindspots in an essential way; and moreover, there are various epistemic interpretations of the surprise announcement, all equally natural and interesting, and some going beyond simple logical deducibility: e.g. in terms of (lack of) correct belief. But on the other hand, the most natural reading of the story involves an announcement that is at the same time *both self-referential and blindspotting*. Nevertheless, we claim that *none of the versions of the puzzle are genuinely paradoxical, and they do not have a Liar-like structure.*[2]

Furthermore, we claim that a crucial (though often neglected) role is played by *the Student's attitude towards the Teacher as a source of information*. Our approach makes essential use of the tools of Dynamic Epistemic Logic (DEL) [2–7,11,15] for modeling belief change and knowledge updates. We formalize the various possible attitudes as different types of *(hard and soft) doxastic upgrades*.[3] Some of them capture 'positive' attitudes towards the source (when the Student is willing to believe the announcement if this is consistent with his knowledge). The 'paradox' only shows that *certain positive attitudes cannot be adopted*

---

[2] Mathematically, the reason is that the self-reference is not of the 'negative' type, and so (unlike the Liar Paradox) *it only involves a fixed point of a positive (thus monotonic) operator*.

[3] See van Benthem [10] for the distinction between soft and hard information upgrades.

(towards such a self-referential statement), on pain of contradiction. In particular, the *strongest* positive attitude (considering the Teacher as a *source of infallible knowledge*, in the above sense) proves to be untenable.

Our main question will then be: *is there any positive attitude that the Student can consistently adopt towards the Teacher, without falling into paradox?* Such an attitude would be a *(positive) solution* to the paradox. We show that the answer is *yes*, and that (in the most interesting version of the puzzle) *this solution is unique* and can be obtained as *the limit of an infinite iteration of upgrades.*[4]

2   Modelling Belief, Knowledge and Their Dynamics

In this section we provide formal tools for analyzing the Surprise Exam scenario. We use a *single-agent dynamic-epistemic-doxastic logic*, having only one (implicit) agent. In the Surprise puzzle, *the relevant agent will be the Student*. The Teacher only plays the role of *source* of information.

2.1   (Conditional) Belief and (Infallible) Knowledge

Our language is build from a given set of atomic sentences At, using standard Boolean connectives, as well as *knowledge* operators $K\varphi$ and *conditional belief* operators $B(\varphi|\psi)$. Here, $K\varphi$ means that the agent *infallibly knows* that φ, while the binary doxastic modality $B(\varphi|\psi)$ expresses the agent's belief that φ is true conditional on ψ being true. We use $B\varphi$ to denote *unconditional belief* that φ, defined as an abbreviation for $B(\varphi|\top)$. We will later add to this repertoire *dynamic modalities* $[T\varphi]\psi$, capturing various types of doxastic upgrades.

Plausibility Models provide the semantics for this logic and are used to encode exactly what in Belief Revision Theory is called *the "epistemic state" of an agent*: these models can represent the agent's beliefs, knowledge, belief-revision plans etc. i.e. *all that is available to the agent by pure introspection*. Given a set At of atomic sentences, a *plausibility model* $\mathbf{S} = (S, \leq, \|.\|)$ consists of a non-empty set of possible worlds (or states) S, equipped with a so-called plausibility relation, which is a total, converse-well-founded preorder[5]. For two given worlds s and w, we read $s \leq w$ as "world w is at least as plausible as world s", and write $s < w$ for strict plausibility (defined as $s \leq w \not\leq s$). We use $\|.\|$ to denote a valuation map which assigns to each atomic sentence p in At the set of worlds $\|p\| \subseteq S$ in which the sentence is true. The semantics will extend the valuation to an *interpretation* φ of all sentences φ in our language.

Infallible Knowledge. The agent *infallibly knows* φ (in any world) if φ is true in all the possible worlds of the model $\mathbf{S}$: formally, $\|K\varphi\|_{\mathbf{S}} = S$ iff $\|\varphi\|_{\mathbf{S}} = S$,

---

[4] See [6–8] for an investigation of upgrade behavior in the limit under infinite iteration.

[5] This means that $\leq$ is reflexive and transitive, as well as 'connected' or 'total' ($\forall s, t : s \leq t \vee t \leq s$), and there are no infinite strictly ascending chains $s_0 < s_1 < s_2 < \ldots$ of more and more plausible worlds.

and otherwise $\|K\varphi\|_{\mathbf{S}} = \emptyset$. Note that this is an "absolute" sense of knowledge, corresponding to the $S5$ concept used in Game Theory and Computer Science.[6]

Conditional Belief. Given a plausibility model, the sentence $B(\varphi|\psi)$ (φ is believed conditional on ψ) is true (in any world) if φ is true in all the most plausible worlds satisfying ψ; i.e. when $Max_{\leq}\|\psi\|_{\mathbf{S}} \subseteq \|\varphi\|_{\mathbf{S}}$, where $Max_{\leq}\|\psi\|_{\mathbf{S}} := \{w \in \|\psi\|_{\mathbf{S}} : s \leq w \text{ for all } s \in \|\psi\|_{\mathbf{S}}\}$ is the set of all maximal ψ-worlds. Conditional beliefs can be viewed as *'strategies', or "contingency plans" for belief change*: one can read $B(\varphi|\psi)$ as saying that 'in case I will find out that ψ was the case, I will believe that φ was the case'.

Example 1. Let us represent the initial situation of the puzzle, before the Teacher announces anything. We consider atomic sentences of the form e_i (with i ranging from 1 to 5), expressing that 'the *exam* takes places in the i-th day of the week'; and in addition, an atomic sentence n expressing that 'there will be *no* exam'. The valuation ensures that these sentences are mutually exclusive. The model will of course depend on our assumptions about the Student's knowledge and beliefs at the start of the story. One possible (and rather natural) assumption is that in which the Student starts by considering all 6 cases as *equally plausible*. This is based on the 'principle of indifference': he has no reason to prefer one or the other. The model for this situation is represented below.

Formally, $\mathbf{S} = (S, \leq, \|.\|)$, where $S = \{s_1, s_2, s_3, s_4, s_5, s_6\}$, with $\|e_i\| = \{s_i\}$ for $1 \leq i \leq 5$, $\|n\| = \{s_6\}$, and $\leq = S \times S$ is the universal relation (with all worlds equi-plausible). Here, the e_1-world s_1 represents the case in which the exam is on Monday, etc.; while in the n-world s_6 there is no exam.[7] Note that the sentences $K(\bigvee_{i=1}^{5} e_i \vee n)$, and $B(n|\neg \bigvee_{i=1}^{5} e_i)$ are both true in all possible worlds.

Example 2. Another possible initial situation is that the Student starts with the belief that there *will be an exam* (so the n-world is considered less plausible than the other cases), but still considers all working days to be equally plausible.

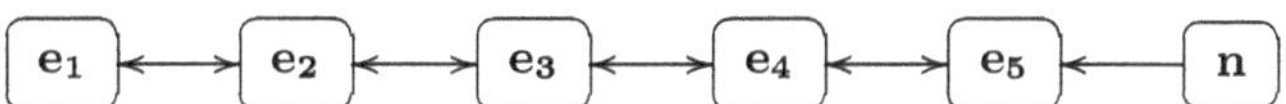

In this model, the Student knows exactly the same things as in Example 1; but he has *stronger beliefs*: $B(\bigvee_{i=1}^{5} e_i)$ now holds unconditionally (in all worlds).

---

[6] A notion of fallible or defeasible knowledge can also be defined in plausibility models, but will play no role in this paper.

[7] The names of the worlds are irrelevant, so we do not include them in the drawing. We also skip the reflexive loops and the arrows that can be obtained by transitivity. This is a standard convention when depicting preorders.

Generalization: 'Natural' Models. To generalize these examples, let a *'natural model' for the Surprise Story* be any model $\mathbf{S} = (S, \leq, \|.\|)$ based on the 6 possible worlds $S = \{s_1, s_2, s_3, s_4, s_5, s_6\}$, with $\|e_i\| = \{s_i\}$ for $1 \leq i \leq 5$ and $\|n\| = \{s_6\}$ (as in Example 1), but endowed with *any* (total) plausibility preorder. These are models in which the Student is *completely ignorant* concerning the exam's day, or even about whether or not there'll be an exam, but he is free to hold *any beliefs* about these issues.

2.2 'Dynamic' Attitudes Towards an Information Source

The above plausibility models were used to represent the static propositional attitudes of an agent: her beliefs, conditional beliefs and knowledge. Yet to fully represent our puzzling scenario we have to capture also the agent's possible *'dynamic' attitudes towards the reliability of a source of information.* These will dictate the agent's 'belief-revision policy': the way he will revise his beliefs when receiving any information φ from that source.

Belief Upgrades. A *doxastic upgrade* is essentially a sentence-indexed model transformer; more precisely, an upgrade T is an operation taking any sentence φ and any plausibility model $\mathbf{S}$, and returning a *new* model $T\varphi(\mathbf{S})$, having as a new set of worlds some *subset* $T\varphi(S) \subseteq S$, as valuation the restriction $\|.\| \cap T\varphi(S)$ of the old valuation to the subset $T\varphi(S)$, and as new plausibility relation some other total preorder $T\varphi(\leq)$ on $T\varphi(S)$. Such an upgrade $T\varphi$ induces a *partial map* on the set of worlds S of any model $\mathbf{S}$, a map which is also denoted by $T\varphi : S \rightharpoonup S$, and which is given by the *restriction of identity* to $T\varphi(S)$.[8]

Soft and Hard Upgrades. Following the terminology in [6], an upgrade $T\varphi$ is called 'soft' if, for every model $\mathbf{S}$, the map $T\varphi : S \rightarrow S$ is *total*; i.e. if and only if $T\varphi(S) = S$ for all $\mathbf{S}$. A soft upgrade *doesn't add anything to the agent's irrevocable knowledge*: it *only conveys "soft information"*, changing only the agent's beliefs or his belief-revision plans. In contrast, a 'hard' upgrade adds new knowledge, by shrinking the set of possible worlds to a *proper subset* $S' \subset S$.

Examples of Upgrades. Here are some examples of doxastic upgrades and their interpretations as dynamic attitudes towards a source of information.

- **Identity** id: a *neutral* attitude ("indifference"). The source is neither believed nor dis-believed, but simply ignored. The agent keeps all his old beliefs. Formally, $id\varphi(\mathbf{S}) = \mathbf{S}$, i.e. everything is left the same (same states, same plausibility order).
- **Update** $!$: *infallibility.* The source is infallibly known to be truthful. The source is *"known" (guaranteed) to be always truthful.* Formally, $!\varphi(\mathbf{S})$ is obtained by *deleting all the non-φ worlds* and *keeping the same plausibility order between the remaining worlds.*

---

[8] I.e., $T\varphi(s) = s$ iff $s \in T\varphi(S)$, and $T\varphi(s) = $ undefined, otherwise.

- **Radical upgrade** $\Uparrow$: *strong trust*. The source is considered (fallible, but) *highly reliable*, i.e. *strongly believed to be truthful*. Formally, in $\Uparrow \varphi(\mathbf{S})$, *all φ-worlds become "better" (strictly more plausible) than all $\neg\varphi$-worlds*, and *within the two zones, the old ordering remains*.
- **Conservative upgrade** $\uparrow$: the source is *trusted, but only 'barely'*. The source is *believed to be truthful*; but this belief can be easily given up later! Formally, in $\uparrow \varphi(\mathbf{S})$, *the "best" (most plausible) φ-worlds become better than all other worlds*, and *in rest the old order remains*.
- **Negative attitudes ('downgrades')** $!^-$, $\Uparrow^-$, $\uparrow^-$: the source is infallibly known to lie, or strongly distrusted, or just believed to lie. Formally, we apply the corresponding 'positive' upgrades $!$, $\Uparrow$, $\uparrow$ to the negation $\neg\varphi$.

Positive Attitudes: Willingness to Learn. An attitude T towards a source is said to be *positive* if we have that: $\|\varphi\|_{\mathbf{S}} \neq \emptyset$ implies $Max_{T\varphi(\leq)}T\varphi(S) \subseteq \|\varphi\|_{\mathbf{S}}$. Positivity expresses *(relative) trust* in the source: the agent is willing to learn from it, whenever this is possible. More precisely, it means that after $T\varphi$, the agent will *come to believe* that φ was the case (before the upgrade), *unless* he already *knew* (before the upgrade) that φ was false.[9]

Examples of Positive Attitudes. The update $!$, radical upgrade $\Uparrow$ and conservative upgrade $\uparrow$ are standard examples of positive attitudes. But there are many others: essentially, every belief-revision operator considered in the literature is an example.[10]

Dynamic Modalities. To express what will be true in a model after an upgrade, we add dynamic operators $[T\varphi]$ to our language, for an appropriate set of doxastic upgrades T. The sentence $[T\varphi]\psi$ says that ψ *will be true after the upgrade* $T\varphi$. Formally, $s \in \|[T\varphi]\psi\|_{\mathbf{S}}$ iff we have that $s \in T\varphi(S)$ implies $s \in \|\psi\|_{T\varphi(\mathbf{S})}$.

3 Solving the Surprise Examination Paradox

In this section we analyze various interpretations of the Surprise Exam story and use Dynamic Epistemic Logic to provide natural solutions. We start with the non-self-referential versions, as articulated by the 'epistemological' school.

3.1 The Non-self-Referential Interpretation

There are in fact two announcements involved in the story. The first is the sentence $exam = \bigvee_{i=1}^{5} e_i$ saying that "there will be an exam (in one of the 5 working days)". The second is the sentence "there will be a *surprise* exam", thought of as a *one-shot announcement*, that according to this interpretation does *not* involve any self-reference.

---

[9] The past-sense specification "was the case" is only needed here because φ might be a Moore sentence, that may have become false after being learnt.

[10] See e.g. [23] for a plethora of examples.

"Surprise' as Lack of Knowledge. Following Gerbrandy [18], we can formalize this as:

$$surprise_K \; = \; \bigvee_{i=1}^{5} \left(e_i \wedge [!(\bigwedge_{1 \leq j < i} \neg e_j] \neg K e_i) \right).$$

We call this $surprise_K$ since it interprets 'surprise' as meaning "the Student will *not know* the exam day in advance". Here, 'knowledge' is taken in the absolute, infallible sense (given by the $S5$ notion from the previous section). The use of updates $!\varphi$ in this sentence is meant to represent the information that will be gained by the Student after $i - 1$ days passed with no exam: he will learn the sentence $\bigwedge_{1 \leq j < i} \neg e_j$ (saying that the exam is not taking place in any of the days $j < i$). This is 'hard' information, hence the Student is entitled to use updates.

If we write $K(\varphi|\psi) := K(\psi \to \varphi)$ to express "conditional knowledge" (of φ given ψ), then the usual Reduction Laws for knowledge after an update in DEL (or rather PAL) [5,15] give us an equivalent formulation, that does not use dynamic modalities:

$$surprise_K \; = \; \bigvee_{1 \leq i \leq 5} \left(e_i \wedge \neg K(e_i| \bigwedge_{1 \leq j < i} \neg e_j) \right).$$

It is now easy to check that, in both the models of Example 1 and Example 2, the sentences *exam* and *suprise$_K$* are both *true in all worlds except for the n-world* (in which there is no exam).

Non-self-Referentiality. Note that the truth conditions for this sentence only depend on the situation (including the Students' knowledge and beliefs) *before* the sentence is announced: although the sentence refers to the Student's future (lack of) knowledge, the way this 'future' is expressed is by using an update that does *not* include Teacher's announcement. So we are talking about a *counterfactual future*: the way the Student's knowledge would have evolved by the ith-day if the Teacher did *not* make the announcement.

Alternative Reading: 'surprise' as lack of correct belief. The understanding of surprise as lack of absolutely certain, infallible 'knowledge' K may seem to set the bar too high for the Student: intuitively, if by the evening before the exam, the Student will correctly believe that the exam is tomorrow, then the exam won't be a surprise. This leads to an alternative, non-equivalent formulation of the Teacher's announcement, which we denote by $surprise_B$: it says that "just before the exam day, the Student will *not believe* that the exam is in that day".[11] Formally:

$$surprise_B \; = \; \bigvee_{i=1}^{5} \left(e_i \wedge [!(\bigwedge_{1 \leq j < i} \neg e_j] \neg B e_i) \right).$$

---

[11] There are other possible interpretations, in which we replace infallible knowledge K by *fallible or defeasible knowledge* (i.e. 'safe belief' $\Box$, rather than true belief B), cf. [6]. Their treatment is relegated to an extended journal version of this paper. Suffice it to say that $surprise_\Box$ turns out to behave similarly to $surprise_B$.

Using the Reduction Law for belief after an update [10], we can express this again without dynamic modalities, but using instead conditional belief operators:

$$surprise_B \;=\; \bigvee_{1 \leq i \leq 5} \left(e_i \wedge \neg B(e_i| \bigwedge_{1 \leq j < i} \neg e_j) \right).$$

Note the contrast with $surprise_K$: while in the model of Example 1 the sentence $surprise_B$ is still true in all worlds except for the n-world, this is *not* the case in Example 2. In that model, $surprise_B$ is only true in the first 4 worlds, but *false in both the e_5-world and the n-world*. Indeed, if the Student believes there'll be an exam, than he will be un-surprised if the exam is on Friday: by Thursday evening, the most plausible surviving world is the e_5-one, so at that moment he'll correctly believe that the exam is on Friday!

It is useful to provide a more direct way to assess the truth value of $surprise_K$ and $surprise_B$ in 'natural models' for the Surprise Story.

Lemma 1 (Characterization of *surprise* sentences). *Let* $\mathbf{S} = (S, \leq, \|.\|)$ *be any natural model for the Surprise Story (i.e., with* $S = \{s_1, s_2, s_3, s_4, s_5, s_6\}$, $\|e_i\| = \{s_i\}$ *for* $1 \leq i \leq 5$ *and* $\|n\| = \{s_6\}$, *and any plausibility preorder). Then*

1. $s_i \models surprise_K$ *iff* $s_i \models exam$ *(i.e., iff* $i \leq 5$*);*
2. $s_i \models surprise_B$ *iff* $s_i \models exam$ *and* $\exists j > i(s_i \leq s_j)$.

Proof. Easy verification, by unfolding the semantics of the two formulas.

Surprise Announcements as 'Moore Sentences'. It is easy to see that the sentence $surprise_K$ is an *epistemic 'blindspot'* (an epistemic Moore sentence): though it may be true, it *cannot be known* by the Student. Similarly, the sentence $surprise_B$ is a *doxastic 'blindspot'*: it is *unbelievable* by the Student.

Proposition 1. *Both formulas* $\neg K surprise_K$ *and* $\neg B surprise_B$ *are valid.*

Proof. For $surprise_K$, we show by backward induction (starting with the last disjunct) that all the disjuncts involved in $surprise_K$ are inconsistent. Note that $surprise_K$ entails $exam$, and thus $K surprise_K$ entails $K exam$ which entails $K(e_5| \neg \bigwedge_{1 \leq j < 5} e_j)$. But this contradicts the last disjunct in $surprise_K$. Given this, $K surprise_K$ is equivalent to knowledge of the disjunction of only the first 4 disjuncts. We repeat the same argument to similarly eliminate all the other disjuncts, concluding that $K surprise_K$ entails a contradiction, i.e. $\neg K surprise_K$ is valid. The proof of the validity of $\neg B surprise_B$ is similar.

How about knowing that the exam will *not* be a surprise? When does the Student know that the Teacher's statement is a lie?

Proposition 2. *Let* $\mathbf{S} = (S, \leq, \|.\|)$ *be any natural (initial) model for the Surprise Story. Then*

1. $K\neg surprise_K$ is false in **S** (at any/all states): at the start of the story, the Student cannot know that $surprise_K$ is false;
2. $K\neg surprise_B$ holds in **S** (at any/all states) iff $s_0 > s_1 > s_2 > s_3 > s_4 > s_5 > s_6$.

Proof. Both parts follow easily by noticing that $K\neg surprise_K$ is equivalent to requiring $s_i \models surprise_K$ for all i, and similarly that $K\,surprise_B$ is equivalent to requiring $s_i \models surprise_B$ for all i, and then applying the characterizations in Lemma 1.

Student's Attitude Towards the Teacher: Infallibility? What should be the Student's attitude towards the Teacher as a source of information? Here, there are many possible choices. Gerbrandy has the Student considering the Teacher to be an *infallible source*: the Teacher simply cannot lie. So he interprets the Teacher's announcement as **two successive updates** !*exam* and !*surprise*. Since in both models from Example 1 and Example 2, the sentence *exam* is false in the n-world, the first update eliminates the n-world, yielding:

In this updated model, both sentence $surprise_K$ and $surprise_B$ are false in the e_5-world (and true in all the others). So the second update (with either of the surprise sentences) eliminates Friday as a possibility, yielding:

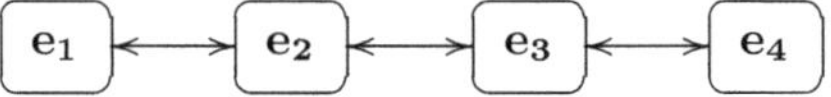

Gerbrandy's conclusion is: if the Teacher is an infallible source, then the *Student comes to know that there will be an exam, and that it cannot be on Friday*. But *this reasoning cannot be iterated*: none of the other days can be excluded (unless the Teacher *repeats* the announcement)! Furthermore, an infallible Teacher does not lie, so all her statements were true.

Other Possible Attitudes. We can of course alternatively postulate other attitudes, e.g. strong trust ⇑. In this case, the Student performs a radical upgrade ⇑ *exam*, followed by another upgrade ⇑ $surprise_K$, or alternatively ⇑ $surprise_B$ (depending on the interpretation). The first upgrade ⇑ *exam* produces (in both Example 1 and Example 2)

If we adopt the knowledge-interpretation and perform on this model the upgrade ⇑ $surprise_K$, *the model stays the same*: so the Student ends up believing (though not knowing) that there will be an exam, but he has no special belief or knowledge concerning the day. If the exam will indeed take place, then the Teacher didn't lie: her announcements were true. However, if we adopt instead the belief-interpretation and apply the upgrade $\Uparrow surprise_B$ on the last model above, this will yield

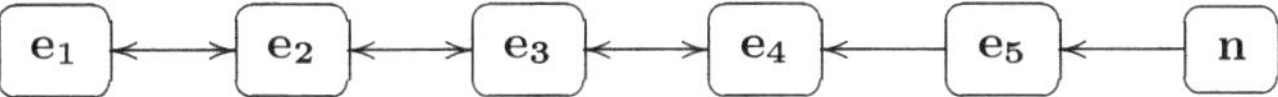

So in this case, the Student comes to believe (though he doesn't know) that there will be an exam in the first 4 days of the week. If this is indeed the case, the Teacher's statements were true; otherwise, she lied.

Other attitudes towards the Teacher produce different results, but in each case *the pattern is the same: the Teacher's announcements always have a definite truth-value* (depending on whether or not there is an exam, and in which day); and *the upgrades produce a well-defined output-model,* in which the Student has *definite and consistent beliefs* (that are justified by his attitude towards the Teacher as a source).

In all cases, there is no justification for iterating the second upgrade, so there is no backwards induction and the contradiction is blocked. This is due to the fact that $surprise_K$ and $surprise_B$ are Moore sentences, and so their truth-value is not stable under their own announcement: even if they were initially true, nothing guarantees that they stay true after the Student learns them. So, according to this analysis, the mistake in the 'paradoxical' reasoning by backward induction is that the Student is not entitled to iterate the upgrade.

3.2 The Self-referential Interpretation

Note that the above conclusion crucially depends on having a non-self-referential reading of the surprise announcement. For instance, according to Gerbrandy's solution (based on updates), e_5 (Friday) and n (no exam) are eliminated, so by Wednesday evening the Students will *know* that the exam is on Thursday: he will *not* be surprised. Despite this, the Teacher told the truth! Indeed, if accused of lying, the Teacher can later claim that he did *not* mean to say that exam will still be a surprise even after he announced that: he only meant that this was true before the announcement!

Many authors (and many Students) will regard such an answer by the Teacher as a form of "cheating". While the above formalizations of the sentence *surprise* seem natural at first sight, there is something profoundly odd about them. The teacher announced that the exam's date *will* be a surprise: this seemed to point to the *actual future,* as it will unfold after this announcement is made. However, the above formalization allows for the possibility that the announcement was meant to be true only *before* the announcement. But then in what sense can one still claim that the Teacher was truthful in her announcement about what 'will' happen?

Looking at the logical structure of our formal sentences $surprise_K$ and $surprise_B$, we can see that the best way to describe them in natural language is as *counterfactual statements* of the type: "the exam's date would have been a surprise, if I didn't make this very announcement". However, this is *not* what the

Teacher said. When referring to the future in an announcement, it is typically implicitly assumed that the speaker factors in her own announcing action: thus, she uses the word "will" to refer to what will actually happen after she makes the announcement. "It will be a surprise" means that it will be so, not that it would have been so in some other possible future.

Thus, to understand the Teacher's statement, we need to make explicit its *implicit self-referentiality*, reading it as "You will not know in advance the exam day (i.e. after hearing this very announcement)". Most authors who wrote about the paradox agree that this self-referential interpretation is the intended one. However, many of the same authors argue that this self-referential version leads to a genuine, Liar-like paradox, because of its inherent circularity. In contrast, other logicians such as Quine [22] argue that there is no real paradox: the student cannot even exclude the no-exam case, and thus the exam will be a surprise (regardless of the day).

Self-referential Announcements as Limits of Iterated Announcements.
One way to understand such a future-oriented self-referential announcement, of the form "φ holds (even after I'm telling you *this*)" is as the limit of an infinite iteration of simple (non-self-referential) announcements: "φ holds; and after learning that, φ will still hold; and after learning that, φ will still hold; etc.". Imagine that a smart Student wants to pre-empt the Teacher from adopting the 'cheating' counterfactual interpretation, and so after hearing the Teacher's *surprise* announcement, he presses her to reassert it, asking: "Now that I heard you, is the exam still going to be a surprise?". If the Teacher answers "yes", he presses again, with the same question, and she answers "yes" again; etc. In effect, we have an infinite sequence of iterated upgrades $T\varphi; T\varphi; T\varphi; \cdots$, where T represents the Student's attitude towards the Teacher. Starting with any model $\mathbf{S}$ for the initial situation (before any announcement is made) and putting $\mathbf{M} = Texam(\mathbf{S})$ (for the model after the Teacher announces that there will be an exam), this produces an infinite sequence of upgraded models $\mathbf{M}$, $T\varphi(\mathbf{M})$, $T^2\varphi(\mathbf{M}) = T\varphi(T\varphi(\mathbf{M}))$, ..., $T^n\varphi(\mathbf{M}), \ldots$ Intuitively, such an infinite iteration is equivalent to the future-oriented self-referential announcement with φ. If $T^n\varphi(\mathbf{M})$ stabilizes eventually on some fixed-point model, *we could thus denote that limit-model by $T^\infty\varphi(\mathbf{M})$, and take it to represent the output of revising with such a self-referential announcement.* If however, the infinite iteration keeps changing the model forever (as in the case of the Liar, according to the revision theory of truth), or if at some point the iterated upgrade cannot be performed (on pain of contradiction: the next model would be empty), then $T^\infty\varphi(\mathbf{M})$ does not exist. In this case, *attitude T cannot be consistently held towards a source that utters such a self-referential announcement.* This case could perhaps be regarded as a 'paradox', but *only if there are no other reasonable attitudes* that can be consistently adopted.

So the question is now: do $T^\infty surprise_K(\mathbf{M})$ or $T^\infty surprise_B(\mathbf{M})$ exist, for any interesting attitude T and any model $\mathbf{M} = Texam(\mathbf{S})$ (obtained by first upgrading the initial model $\mathbf{S}$ with the *exam* sentence)? Of course, the upgrades immediately stabilize if the Student adopts the *neutral attitude id*: we have

$id\varphi(\mathbf{M}) = id^\infty \varphi(\mathbf{M}) = \mathbf{M}$, for all φ and all $\mathbf{M}$. This is in a sense a 'solution' to the paradox, but a trivial one: an agent who ignores all incoming information never reaches any paradoxical conclusions, but never learns anything either.

We are thus lead to reformulate our question: *are there any positive attitudes that the Student can consistently hold towards a Teacher who utters the self-referential surprise sentence* (or equivalently, keeps answering "yes" when pressed to reassert the non-self-referential sentence)?

Infallibility Is No Longer Tenable. Here is a first negative result: the limit-models $!^\infty surprise_K(\mathbf{M})$ and $!^\infty surprise_B(\mathbf{M})$ *do not exist*, for any natural initial model $\mathbf{M} = !exam(\mathbf{S})$ for the Surprise story (such as the ones in Examples 1 and 2). The reason is that, after applying $!exam$ and four iterations of $!surprise$, all worlds but the e_1-world are eliminated, so the next update can no longer be performed (since the next model would be empty). This iterated elimination of all possibilities is a 'dynamic' formalization of the informal reasoning of the Student in the 'paradoxical' argument.[12] But the conclusion is *not* that the Teacher lied (as in the informal reasoning), but that *infallibility attitude* ! *cannot be consistently maintained*: a source (Teacher) uttering such a self-referential statement cannot be consistently considered as an infallible source of truth.

Tenable positive attitudes do exist. The good news is that *there exist soft positive attitudes that can be consistently adopted*. We start with the 'easy' case of $surprise_K$.

Proposition 3. *Let* $\mathbf{S} = (S, \leq, \|.\|)$ *be any 'natural model' for the Surprise Story. Let* $T \in \{\Uparrow, \uparrow\}$ *be either radical upgrade* $\Uparrow$ *('strong trust') or conservative upgrade* $\uparrow$ *('bare trust'), and put* $\mathbf{M} := Texam(\mathbf{S})$ *for the output of the first upgrade (with the exam announcement). Then* $T^\infty surprise_K(\mathbf{M})$ *exists, and moreover the sequence* $T^n surprise_K(\mathbf{M})$ *stabilizes in one-step, in fact it keeps the model* $\mathbf{M}$ *fixed:* $T^\infty surprise_K(\mathbf{M}) = \mathbf{M} = Texam(\mathbf{S})$. *In this fixed-point model, the Student* believes that there will be an exam and knows that the exam will not be a surprise: $Bexam \wedge K\neg surprise_K$ holds in $T^\infty surprise_K(\mathbf{M})$.

Proof. The upgrade $\Uparrow exam$ promotes all exam-worlds s_i (i.e. with $i \leq 5$), making them strictly more plausible than s_6 (and leaving everything else the same). Since $\Uparrow$ is soft, $\Uparrow exam$ preserves the set of states and the valuation, so $\mathbf{M} = \Uparrow exam(\mathbf{S})$ is still a natural model. By part 1 of Lemma 1, $surprise_K$ is true at any state s_i of $\mathbf{M}$ iff $i \leq 5$. Since all these worlds s_i with $i \leq 5$ are already strictly more plausible in $\mathbf{M}$ than s_6, the upgrade $\Uparrow surprise_B$ leaves $\mathbf{M}$ unchanged, so we reached the fixed point $\Uparrow surprise_B(\mathbf{M}) = \mathbf{M}$.

The argument for $\uparrow$ is similar, except that $\uparrow exam$ may only promote *some* exam-worlds s_i (namely, the most plausible ones with $i \leq 5$), making them strictly more plausible than all others (and leaving everything else unchanged). So, in the natural model $\mathbf{M}$, the most plausible worlds will all be of the form s_i with $i \leq 5$, and thus will satisfy $surprise_B$ (by part 1 of Lemma 1. This is enough to ensure that $\uparrow surprise_B$ leaves $\mathbf{M}$ unchanged.

---

[12] See [1] for interesting topological connections to Cantor derivative and the Cantor-Bendixson process of calculating the 'perfect core' of a set.

46   A. Baltag and S. Smets

Quine's (Unsatisfactory) Solution Regained. Part 1 of the above Proposition essentially matches Quine's solution [22]: the exam can be in any day (including Friday), and it will still be a surprise (since the Student cannot exclude the 'no exam' option). Indeed, if we start with any of the natural models **S** in Example 1 or Example 2, and take $T \in \{\Uparrow, \uparrow\}$, then the limit model $T^{\infty} surprise_K(\mathbf{M})$ is

So, if there will be an exam, the Teacher told the truth (regardless of the exam day)! Somehow, this still seems an unsatisfactory solution. It seems unfair to insist on the knowledge-interpretation $surprise_K$, when the upgrade is soft: in this case, the Teacher only provides 'soft' information, so *she can only change the Student's beliefs, not his knowledge.* So of course the exam day will remain a surprise (in the sense of knowledge).

So it seems to us that, when faced with a Teacher who only gives soft information, it is more natural (and fair to the Student) to adopt the belief-interpretation $surprise_B$. Thus, we are lead to consider now its self-referential version. Interestingly enough, in this case *there is only one possible limit model!*

Proposition 4. *(Existence and Uniqueness of Fixed-Point) As before, let* $\mathbf{S} = (S, \leq, \|.\|)$ *be any natural model for the Surprise Story, let* T *be any soft positive upgrade, and put* $\mathbf{M} := Texam(\mathbf{S})$. *If the fixed-point limit model* $T^{\infty} surprise_B(\mathbf{M})$ *exists, then it is given by:*

In this unique fixed-point model, the Student believes that the exam will be on Monday (and if not on Monday, then on Tuesday, etc.), and he knows that the exam will not be a surprise: $Be_1 \wedge K \neg surprise_B$ holds in $T^{\infty} surprise_B(\mathbf{M})$.

Moreover, this fixed point is realized as a limit-model when iterating some soft positive upgrades: e.g. if we take $T = \Uparrow$ *(strong trust) then, for any initial natural model* $\mathbf{S}$ *(and putting as before* $\mathbf{M} = \Uparrow surprise_B(\mathbf{S})$*), the fixed-point limit model* $T^{\infty} surprise_B(\mathbf{M})$ *does exist, and it is reached in at most 5 steps:* $T^5 surprise_B(\mathbf{M}) = T^{\infty} surprise_B(\mathbf{M})$ *is the above model.*

Proof. For the first part: note that soft upgrades preserve the set of states and the valuation, so all models $T^{\infty} surprise_B(\mathbf{M})$ are 'natural' models. Let $T^{\infty} surprise_B(\mathbf{M}) = (S, \leq^{\infty}, \| \bullet \|)$ (with the same S and valuation as $\mathbf{S}$).

Suppose now that $\|surprise_B\|_{T^{\infty}(\mathbf{M})} \neq \emptyset$. Since T is positive, this gives us that $Max_{Tsurprise_B(\leq^{\infty})}(S) \subseteq \|surprise_B\|_{T^{\infty}(\mathbf{M})}$. By definition, $T^{\infty} surprise_B(\mathbf{M})$ is a fixed point of $T^{\infty} surprise_B$, so $Tsurprise_B(\leq^{\infty}) = \leq^{\infty}$, and we get that $Max_{\leq^{\infty}}(S) \subseteq \|surprise_B\|_{T^{\infty}(\mathbf{M})}$, i.e. $Bsurprise_B$ holds in $T^{\infty} surprise_B(\mathbf{M})$, contradicting the validity of $\neg Bsurprise_B$ (cf. Proposition 1).

Thus, our assumption is false: we have $\|surprise_B\|_{T^{\infty}(\mathbf{M})} = \emptyset$, i.e. $K \neg surprise_B$ holds in $T^{\infty} surprise_B(\mathbf{M})$. By part 2 of Proposition 2, the plausibility order in $T^{\infty} surprise_B(\mathbf{M})$ must be $s_1 >^{\infty} s_2 >^{\infty} s_3 >^{\infty} s_4 >^{\infty} s >^{\infty} s_6$.

The second part is an easy verification: after $\Uparrow exam$, we have $s_6 < s_i$ in $\mathbf{M} = Texam(\mathbf{S})$, for all $i = 1,5$. The first $\Uparrow surprise_B$ makes $s_6 < s_5 < s_i$ in $Tsurprise_B(\mathbf{M})$, for all $i = 1, 4$; etc.

The 'True' Solution? As mentioned above, the belief-interpretation $surprise_B$ seems to be the natural one whenever the Teacher's fallibility is accepted. Hence, we favor Proposition 4 as the "true solution" to the paradox. After the (iterated) revision, we reach the fixed-point model $T^\infty surprise_B(\mathbf{M})$, according to which the *Teacher did indeed lie* (and the *Student knows it*): the sentence $surprise_B$ is *now false in all possible worlds*. But this is only because of the change of plausibility induced by the iterated upgrades (given the Student's willingness to learn from the Teacher). After that, the Student cannot be surprised: in any day when no exam was given, he believes the exam is tomorrow; while on Friday evening (if there was no exam by then), he knows there is no exam. No matter what happens, *there is no surprise: the Teacher has been 'proven' wrong!*

At the Border of Paradox. The iterated revisions involved in our solution closely resemble the informal backwards-induction argument of the Student in the standard version of the Paradox. For example, if we start with the model $\mathbf{S}$ in Example 1 and take $T \in \{\Uparrow, \uparrow\}$, then after $Texam$, we reach the model

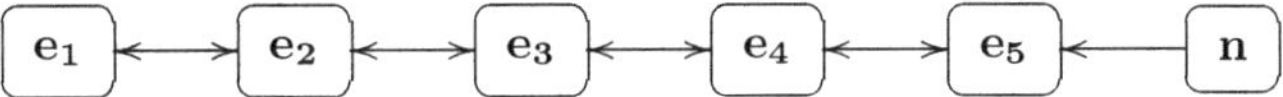

After that, the successive upgrades $Tsurprise_B$ yield the models:

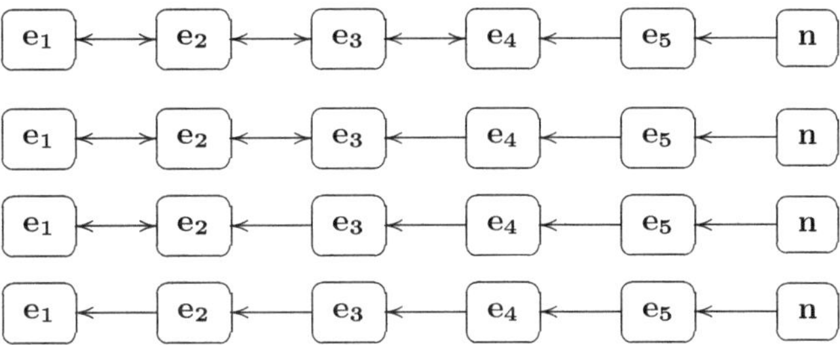

We reached the fixed-point: there is no contradiction, but the path towards the final resolution is a 'soft' dynamic simulation of the 'paradoxical' argument. In this sense, *our solution lies just "at the border of paradox"*: it provides a *partial vindication of the Student's informal 'proof' that the Teacher lied!*

4   Conclusion

Our analysis of the Surprise Exam 'paradox' shows that in fact no real paradox is involved. The non-self-referential versions are easily solvable, while the self-referential versions can be interpreted using a sequence of upgrades, at the end of which the Teacher's announcement comes up as *false* (regardless of which reading we adopt: $surprise_K$ or $surprise_B$). Since in all cases the announcement has a definite truth value, *the Surprise Exam is not a Liar-like paradox*. But something odd does happen here, something more interesting and deeper than a 'paradox'.

An 'Informative Lie'. *The falsehood of the Teacher's announcement is only guaranteed as long as the Student adopts a (soft) positive attitude towards the* Teacher: i.e., if he considers her a reliable (but not infallible) source of information. In that case, we saw that *he can indeed 'prove' that the Teacher 'lied'.* But *her lie is 'informative'*: after hearing it, the Student comes to believe that there will be an exam and also comes to *know* that the exam will *not* be a surprise.

Willingness to Believe Leads to Falsification. Note though that *the correct version of the Student's 'proof' is not a simple deductive argument, but a process of belief revision* that invalidates the Teacher's claim: the Student changes his beliefs, in such a way that no surprise exam can take place after this!

Negative Loop Between Truth Value and Attitude. What happens if the Student adopts instead a negative or neutral attitude? In that case, the above conclusion is not guaranteed: the *surprise* announcement *may still be true*! E.g., if in Example 1 the Student adopted the *neutral* attitude *id*, then the announcement (in both its readings, $surprise_K$ and $surprise_B$) would be *true* if and only if there will be an exam. The persistent "feeling of paradox" is due to the *negative feedback loop between the announcement's truth value and the Student's attitude*: the only way for the Student to 'falsify' the announcement is to be willing to believe it! *To know that the Teacher lied, he has to adopt a soft positive attitude towards her announcements.* This sounds "almost paradoxical", but there is no contradiction, due to the *softness* of the Student's attitude: being "willing to believe" the Teacher does *not* imply knowing (or even believing) that she is telling the truth! On the contrary, the negative loop yields the opposite conclusion.

No Going Back. So, after 'proving' that the announcement was a lie, why can't the Student just dismiss it, and revert to his initial plausibility? This was the last step in the original 'paradoxical' Surprise Exam argument, and this was *the only mistake*, according to our diagnosis. The answer is that *there is no logical justification for the Student to abandon his willingness to believe*: the Teacher's "informative lie" *does not prove that she is not to be trusted at all, or that nothing can be learnt by being willing to believe her.* Concluding that the announcement was false is not a warranty for dismissing it altogether. On the contrary, it confirms the appropriateness of the Student's positive attitude and reinforces it, in a "virtuous circle": only by adopting it, the Student has learnt enough to be able to 'falsify' this particular announcement. As we saw, the conclusion that the Teacher lied is entirely dependent on the Student's positive attitude, and thus it cannot constitute a reason for dropping this attitude: that would invalidate the conclusion, reopening the possibility that the Teacher told the truth!

Robustness of Our Solution. As noted, *the fixed-point model $T^\infty surprise_B(\mathbf{M})$ is unique (if existing at all), regardless of which soft positive attitude we choose, and regardless of which initial plausibility we start with* (on the set of 6 worlds). This means that our solution is in some sense 'canonical', modulo the adoption of a soft positive attitude by the Student: *any Student who is willing to learn from his Teacher (while realizing her fallibility) will come to the same solution.*

The Lying Teacher has the Last Word. As a final irony, note that the only way for the Student to 'prove' the Teacher 'wrong' is to keep expecting that the exam is tomorrow! We find it very pleasant that this solution agrees with every Teacher's true intentions: have her Students always prepared for an exam.[13]

Acknowledgments. We thank our audiences (at many venues where precursors of this work have been informally presented since 2010) for their helpful feedback, and thank the anonymous referees for their useful comments.

Disclosure of Interests. The authors have no competing interests to declare.

References

1. Baltag, A., Bezhanishvili, N., Fernández-Duque, D.: The Topology of Surprise. In: Proceedings of 19th International Conference on Principles of Knowledge Representation and Reasoning. Series KR, pp. 33–42, KR Ray Reiter Best Paper Award (2022)
2. Baltag, A., Moss, L.S.: Logics for Epistemic Programs. In: Symons, J., Hintikka, J. (eds.) Synthese, vol. 139, no. 2, pp. 165–224 (2004)
3. Baltag, A., Moss, L.S., Solecki, S.: The logic of public announcements, common knowledge, and private suspicions. In: Proceedings TARK, vol. 98, pp. 43–56 (1998)
4. Baltag, A., Moss, L.S., Solecki, S.: Logics for epistemic actions: completeness, decidability, expressivity. Logics **1**(2), 97–147 (2023)
5. Baltag, A., Renne, B.: Dynamic Epistemic Logic. In: Stanford Encyclopedia of Philosophy (2016)
6. Baltag, A., Smets, S.: A qualitative theory of dynamic interactive belief revision. Texts Logic Games **3**, 9–58 (2008)
7. Baltag, A., Smets, S.: Group belief dynamics under iterated revision: fixed points and cycles of joint upgrades. In: Proceedings of TARK, vol. 12, pp. 41–50 (2009)
8. Baltag, A., Smets, S.: Learning by questions and answers: from belief-revision cycles to doxastic fixed points. In: Proceedings of WOLLIC 2009, LNAI, vol. 5514, pp. 124–139 (2009b)
9. Binmore, K.: Playing for Real: A Text on Game Theory. Oxford University Press (2007)
10. Benthem, J.: Dynamic logic of belief revision. J. Appl. Non Class. Logics **17**(2), 129–155 (2007)
11. van Benthem, J.: Logical Dynamics of Information and Interaction. Cambridge University Press (2011)
12. Binkley, R.: The surprise examination in modal logic. J. Phil. **65**, 127–136 (1968)

---

[13] We find it equally pleasant that our epistemic solution agrees with Binmore's game-theoretic analysis [9]. Formalizing the Surprise Exam scenario as a sequential game, he notes that the Student has an obvious (and unique) winning strategy: always expect the exam tomorrow! No surprise: the Teacher's announcement is false!

13. Bosch, J.: The examination paradox and formal prediction. Logique et Anal. (N.S.) **15**, 505–525 (1972)
14. Chow, T.Y.: The surprise examination or unexpected hanging paradox. Am. Math. Monthly **105**, 41–51 (1998). arXiv:math/9903160 (2011)
15. van Ditmarsch, H., van der Hoek, W., Kooi, B.: Dynamic Epistemic Logic. Springer (2007)
16. Fitch, F.: A goedelized formulation of the prediction paradox. Amer. Phil. Quart. **1**, 161–164 (1964)
17. Halpern, J., Moses, Y.: Taken by surprise: the paradox of the surprise test revisited. J. Philos. Log. **15**, 281–304 (1986)
18. Gerbrandy, J.: The surprise examination in dynamic epistemic logic. Synthese **155**, 21–33 (2007)
19. Kvart, I.: The paradox of surprise examination. Logique et Anal. (N.S.) **21**(82–83), 337–344 (1978)
20. Moore, G.E.: A reply to my critics. In: The Philosophy of G.E. Moore, The Library of Living Philosophers. Northwestern University, pp. 535–677 (1942)
21. Plaza, J.: Logics of Public Communication. In: Proceedings 4th International Symposium on Methodologies for Intelligent Systems, pp. 201–216 (1989)
22. Quine, W.V.O.: On a so-called paradox. Mind **62**, 65–67 (1953)
23. Rott, H.: Shifting priorities: simple representations for twenty-seven iterated theory change operators. In: Makinson, D., Malinowski, J., Wansing, H. (eds.) Towards Mathematical Philosophy (Trends in Logic), vol. IV, pp. 269–296. Springer (2006)
24. Sorensen, R.A.: Blindspotting and choice variations of the prediction paradox. Am. Philos. Q. **23**(4), 337–352 (1986)
25. Sorensen, R.A.: Conditional blindspots and the knowledge squeeze: a solution to the prediction paradox. Australasian J. Phil **62**, 126–135 (1984)
26. Sorensen, R.: Epistemic Paradoxes. In: Zalta E., Nodelman, U. (eds.) The Stanford Encyclopedia of Philosophy. Fall edition (2024)

Probabilistic Causal Kripke Models

Yiwen Ding[1](✉) ⒾⒹ, Krishna Manoorkar[1](✉) ⒾⒹ, Apostolos Tzimoulis[2](✉) ⒾⒹ,
and Ruoding Wang[1,3](✉) ⒾⒹ

[1] Department of Ethics, Governance and Society, Vrije Universiteit Amsterdam,
Amsterdam, The Netherlands
`dyiwen666@gmail.com`, `krishna.manoorkar@gmail.com`, `rodinew0915@gmail.com`
[2] Department of Computer Science, University of Luxembourg, Esch-sur-Alzette,
Luxembourg
`apostolos@tzimoulis.eu`
[3] Department of Philosophy, Xiamen University, Xiamen, China

Abstract. We extend the framework of causal Kripke models in [8] to a
probabilistic setting, by allowing a quantitative representation of a causal
agent's uncertainty. This framework incorporates probabilities into the
Halpern-Pearl model of causality, enabling the evaluation of how likely
an event is to be the actual cause of another. It also provides a structured
approach to reason about causality in scenarios involving multiple possibilities, uncertainty, and knowledge. Furthermore, we illustrate that
this framework is suitable for causal analysis in different probabilistic
scenarios by providing several examples.

Keywords: Causal model · Kripke models · Probabilistic causality ·
Counterfactual reasoning

1 Introduction

Understanding and analyzing causality in a formal framework has been a significant area of research in philosophy and science [9,11,19,21]. With the advent of
AI and the growing role of automated decision-making systems, understanding
the causal mechanisms of these algorithms has become a problem of significant
research interest. One of the key areas of AI research is designing algorithms
capable of comprehending causal information and performing causal reasoning
[5,26,27]. Causal reasoning is fundamental to understanding notions such as
responsibility, blame, harm, and explanation, which are important aspects in
designing ethical and responsible AI systems [2].

Actual causality refers to the causality of a specific event which has actually
happened (e.g. "John died because Alice shot him") rather than general causes
(e.g. "smoking causes cancer"). Several formal approaches have been used for
modelling actual causality [13,14,21,22]. One of the most prominent formalizations of actual causation was developed by Halpern and Pearl [17,18,25]. This
model describes dependencies between *endogenous variables* and *exogenous variables* using *structural equations*. Based on causal models Halpern and Pearl have

V. Goranko et al. (Eds.): LORI 2025, LNCS 16010, pp. 51–64, 2026.
https://doi.org/10.1007/978-981-95-2481-5_4

given three different definitions of actual causality known as *original, updated and modified* definitions [15,17,18] of actual causality using counterfactual reasoning. The formal language developed to describe actual causality in this model is used to define several notions like normality, blame, accountability and responsibility. This model has been used in several applications in law [24], database theory [23], model checking [4,6,7], and AI [2,9,20].

In [8], this framework was extended to the possible world setting to formalize the role of notions, such as knowledge, temporality, possibility, normality (or typicality) and uncertainty, in causal reasoning and related applications. In causal Kripke models, even though the agents involved in events may have some uncertainty, the causal agent (i.e. the agent doing the causal analysis) is assumed to have the full information. In this work, we extend this work to probabilistic causal Kripke models to formalize causal analysis when the causal agents operate under uncertainty (i.e. incomplete information). This uncertainty may arise either from lack of certainty about the sequence of events involved or from the uncertainty regarding the structural equations that model dependence between variables.

While the probabilistic extensions of Halpern-Pearl models have been studied [1,3,12], our work integrates probabilistic reasoning and possible world semantics to formalize reasoning about causality when both the modal notions, such as knowledge, temporality, possibility, normality (or typicality) and uncertainty of causal agents play a role. Our approach is similar to the one proposed by Halpern in [16, Chapter 2.5], where we "pull out" the uncertainty and represent it via probability functions over deterministic causal Kripke models.

Probabilistic causal Kripke models are obtained by defining a probability measure on a set of causal Kripke models. Each causal Kripke model corresponds to a possible scenario considered by the causal agent, and the probability measure on it describes the likelihood of these scenarios according to the causal agent. This framework is then used to generalize the modified HP definition[1] of causality to the probabilistic setting to define 'the probability that an event A is an actual cause of event B' and apply it to analyze example scenarios for causal analysis under uncertainty.

Structure of the Paper. In Sect. 2, we recall some preliminaries on causal Kripke models and the modified HP definition of causality within this framework. Section 3 introduces probabilistic causal Kripke models, a modal language for reasoning about them, and an HP definition of probabilistic causality. In Sect. 4, we present several examples and analyze them using the probabilistic causal Kripke models introduced in the previous section. Finally, Sect. 5 concludes the paper and discusses potential directions for future research.

---

[1] We choose the modified definition as it is usually more intuitive in different scenarios.

2   Preliminaries

In this section, we revisit the key concepts of causal Kripke models, the modal language for causality, and the modified Halpern-Pearl definition of causality within causal Kripke models, presented in [8].

2.1   Causal Kripke Models

A *causal Kripke model* is a tuple $\mathcal{K} = (\mathcal{S}, W, R, \mathcal{F})$, where W is a finite set of possible worlds, $R \subseteq W \times W$ is an accessibility relation, and $\mathcal{S} = (\mathcal{U}, \mathcal{V}, \mathcal{R})$ is the signature s.t. $\mathcal{U}$ and $\mathcal{V}$ are the disjoint sets of exogenous variables (i.e., variables whose value is independent of other variable in the model) and endogenous variables (i.e., variables whose value is determined by other variables in the model), and $\mathcal{R}$ is a function assigning each $\Gamma \in \mathcal{U} \cup \mathcal{V}$ and a world $w \in W$ a set of possible values that Γ can take at w, and $\mathcal{F} = (f_{(X_i, w_j)} \mid X_i \in \mathcal{V}, w_j \in W)$ assigns to each endogenous variable X_i and each world w_j a map such that

$$f_{(X_i, w_j)} : \prod_{(Z,w)\in((\mathcal{U}\cup\mathcal{V})\times W)\setminus\{(X_i,w_j)\}} \mathcal{R}(Z, w) \to \mathcal{R}(X_i, w_j).$$

For any variable Γ and world w we use (Γ, w) to denote the restriction of the variable Γ to the world w. That is, (Γ, w) is a variable that takes a value c iff the variable Γ takes the value c in the world w. For any $\Gamma \in \mathcal{U}$ (resp. $\Gamma \in \mathcal{V}$) and any world $w \in W$, we say (Γ, w) is an exogenous (resp. endogenous) variable. Note that we allow the same endogenous variable to have different structural equations associated with it in different worlds. A *context* over a causal Kripke model $\mathcal{K} = (\mathcal{S}, W, R, \mathcal{F})$ is a function $\bar{t}$ s.t. for any $w \in W$, and $U \in \mathcal{U}$, it assigns a value in $\mathcal{R}(U, w)$. A *causal Kripke setting* is a pair $(\mathcal{K}, \bar{t})$, where $\mathcal{K}$ is a causal Kripke model and $\bar{t}$ is a context for it.

For any variables $X \in \mathcal{V}$, and $\Gamma \in \mathcal{U} \cup \mathcal{V}$, and any $w, w' \in W$, we say (X, w) *directly depends on* (Γ, w') (or (Γ, w') is a parent of (X, w)) if there exist $\gamma, \gamma' \in \mathcal{R}(\Gamma, w')$, and $\bar{z} \in \mathcal{R}((\mathcal{U} \cup \mathcal{V}) \times W \setminus \{(X, w), (\Gamma, w')\})$, s.t. $f_{(X,w)}(\bar{z}, \gamma) \neq f_{(X,w)}(\bar{z}, \gamma')$. A causal Kripke model is said to be *recursive* if it contains no cyclic dependencies. In recursive models, as there are no cyclic dependencies the values of all endogenous variables at all the worlds are completely determined by the context. If $\mathcal{V}$ only contains binary variables (i.e. the variables which take values either 0 or 1), then for any context $\bar{t}$, and any world w, we use $\bar{t}(w)$ to denote the set of endogenous variables set to value 1 at w by $\bar{t}$. In this paper, we only consider recursive causal Kripke models.

Given a causal Kripke model $\mathcal{K} = (\mathcal{S}, W, R, \mathcal{F})$, an *assignment* over $\mathcal{K}$ is a function on some subset $\mathcal{Y} \subseteq \mathcal{V} \times W$ s.t., for every $(X, w) \in \mathcal{Y}$, it assigns some value in $\mathcal{R}(X, w)$. Let $\mathcal{Y}$ be a finite subset of $\mathcal{V} \times W$. Let $\overline{Y}$ be an injective (possibly empty) listing of all variables in $\mathcal{Y}$. Let $\overline{Y} = \bar{y}$ be an assignment s.t. for any $Y_i \in \mathcal{Y}$, $y_i \in \mathcal{R}(Y_i)$. The causal Kripke model obtained from *intervention*, which sets the values of the variables in $\overline{Y}$ to $\bar{y}$, is given by $\mathcal{K}_{\overline{Y} \leftarrow \bar{y}} = (\mathcal{S}, W, R, \mathcal{F}_{\overline{Y} \leftarrow \bar{y}})$, where $\mathcal{F}_{\overline{Y} \leftarrow \bar{y}}$ is obtained by replacing, for each $Y_i \in \mathcal{Y}$, the structural equation f_{Y_i} with $Y_i = y_i$.

2.2 Modal Language for Describing Causality

In this section, we recall the formal logical framework for describing causality in causal Kripke models. The language is hybrid, containing not only modal operators but also a countable set of names, denoted by W. In principle, each model $\mathcal{M}$ comes with an assignment from W to worlds of $\mathcal{M}$, however, in practice, we often conflate names with elements of Kripke models.

The language associated with model $\mathcal{M}$ is denoted by $\mathsf{L}_{\mathcal{M}}(W)$. We often omit W and just write $\mathsf{L}_{\mathcal{M}}$ when W is clear from the context. $\mathcal{S}$ is a given signature, and all X, Y, x, y come from $\mathcal{S}$. In what follows, we will consistently use Y to denote a variable parametrized with a name for a world, such as $Y = (X, w)$. It is important to note that in our language, interventions only involve such variables. Any event α and formula ϕ of the language $\mathsf{L}_{\mathcal{M}}$ are defined by the following recursion.

$$\alpha ::= X = x \mid (X, w) = x \mid \neg\alpha \mid \alpha \wedge \alpha \mid \Box\alpha \quad \text{where, } X \in \mathcal{V}, w \in W$$

$$\phi ::= X = x \mid (X, w) = x \mid \neg\phi \mid \phi \wedge \phi \mid \Box\phi \mid [\overline{Y} \leftarrow \overline{y}]\alpha,$$

where $\overline{Y} \leftarrow \overline{y}$ is an intervention. In particular, the language $\mathsf{L}_{\mathcal{M}}$ has two types of atomic propositions, denoted by the variables of the form X and of the form (X, w). The satisfaction relation $\Vdash$ on the causal Kripke models is defined as follows: Given a causal Kripke setting $(\mathcal{K}, \overline{t})$ with $\mathcal{K} = (\mathcal{S}, W, R, \mathcal{F})$, for any worlds $w, w' \in W$, and any primitive event $X = x$ (resp. $(X, w') = x$), $(\mathcal{K}, \overline{t}, w) \Vdash X = x$ (resp. $(\mathcal{K}, \overline{t}, w) \Vdash (X, w') = x$) iff the value of X is set to x at w (resp. at w') by the context $\overline{t}$. Note that the satisfaction of $(X, w') = x$ is independent of the world in which it is evaluated. The satisfaction relation for Boolean connectives is defined by the standard recursion. For the $\Box$ operator, $(\mathcal{K}, \overline{t}, w) \Vdash \Box\alpha$ iff for all w', wRw' implies $(\mathcal{K}, \overline{t}, w') \Vdash \alpha$. For the formulas including intervention, for any event α, $(\mathcal{K}, \overline{t}, w) \Vdash [\overline{Y} \leftarrow \overline{y}]\alpha$ iff $(\mathcal{K}_{[\overline{Y} \leftarrow \overline{y}]}, \overline{t}, w) \Vdash \alpha$.

2.3 HP Definition of Actual Causality

In this section, we provide a definition of actual causality based on the modified HP definition of actual causality in Halpern-Pearl models.

Definition 1. *Let α be any event. For $\overline{Y} \subseteq \mathcal{V} \times W$, $\overline{Y} = \overline{y}$ is an actual cause of α in a causal Kripke setting $(\mathcal{K}, \overline{t})$ in a world w if the following conditions hold.*

> *AC1. $(\mathcal{K}, \overline{t}, w) \Vdash \alpha$, and for every $w_j \in W$, $(\mathcal{K}, \overline{t}, w_j) \Vdash (X_i, w_j) = y_{ij}$, for every $(X_i, w_j) = y_{ij} \in \overline{Y} = \overline{y}$.*
>
> *AC2a^m. There exists a set of variables $\overline{N} \subseteq \mathcal{V} \times W$, and a setting $\overline{y}'$ of the variables in $\overline{Y}$ s.t. if $\overline{n}^*$ is s.t. $(\mathcal{K}, \overline{t}, w') \Vdash X = n^*$, for every $(X, w') = n^* \in \overline{N} = \overline{n}^*$, then $(\mathcal{K}, \overline{t}, w) \Vdash [\overline{Y} \leftarrow \overline{y}', \overline{N} \leftarrow \overline{n}^*]\neg\alpha$.*
>
> *AC3. $\overline{Y}$ is the minimal set of variables that satisfy AC1 and AC2a^m.*

For any set of variables $\mathcal{Y} \subseteq \mathcal{V} \times W$, we use $cause(\overline{Y} = \overline{y}, \alpha)$ to say that $\overline{Y} = \overline{y}$ is an actual cause of α. Moreover, we write $(\mathcal{K}, \overline{t}, w) \Vdash \Box cause(\overline{Y} = \overline{y}, \alpha)$ if for all w' s.t. wRw', $(\mathcal{K}, \overline{t}, w') \Vdash cause(\overline{Y} = \overline{y}, \alpha)$, and $(\mathcal{K}, \overline{t}, w) \Vdash \Diamond cause(\overline{Y} = \overline{y}, \alpha)$ if there exists w' s.t. wRw' as well as $(\mathcal{K}, \overline{t}, w') \Vdash cause(\overline{Y} = \overline{y}, \alpha)$.

3   Probabilistic Causal Kripke Models

The probabilistic causal Kripke models are intended to model scenarios where the agent performing the causal analysis (we refer to this agent as the *causal agent* to distinguish from the other agents that may be involved in a scenario being analyzed) is uncertain about the events and/or structural equations. The different possible scenarios considered by the causal agents can be represented by different causal Kripke models. Different possible worlds in each causal Kripke model may describe different states of affairs relevant to that scenario, and the relation R describes some relationship, such as temporal order, indiscernibility, accessibility, etc. To discuss the probability, we can define a probability measure μ on these causal Kripke models to describe the likelihood assigned by the causal agent to each possible scenario. We assume that the causal agent has full information regarding each possible scenario (i.e. causal Kripke model), however, there is uncertainty in determining which scenario actually occurred. That is, there may be possible worlds across different scenarios which are indistinguishable to the causal agent. We can introduce *names* to refer to those possible worlds that are indistinguishable to the causal agent across different possible scenarios. As such case, we posit that the analysis of the causal agent would often be treated as a single unit and involve the interventions which occur over all such worlds simultaneously. The inclusion of names in the formal framework allows us to naturally capture such set of worlds and interventions on them (named interventions).

Note that the relation R in causal Kripke model may denote the indiscernibility between different worlds as per the agents who are involved in actions in that scenario (i.e. Alice and Bob in next section's examples). However, the agent performing causal analysis (i.e. the causal agent) has complete information regarding all the worlds in a given scenario. Thus, in our framework, the uncertainty of agents involved in actions is captured by the possible world framework inside the causal Kripke model, while the uncertainty of the agent performing causal analysis is captured by the probability measure μ. In this way, the probabilistic causal Kripke model separates the uncertainty (or lack of information) of different types of agents involved.

In the following, we give the formal definitions of probabilistic causal Kripke models, contexts and interventions over them, and also define probabilistic actual causality in these models.

A *probabilistic causal Kripke model* is a tuple $\mathcal{M} = (\mathcal{C}, \mu, N, G)$, where $\mathcal{C}$ is a finite set of causal Kripke models $\mathcal{K} = (\mathcal{S}, \mathcal{K}(W), \mathcal{K}(R), \mathcal{K}(\mathcal{F}))$ with the signature $\mathcal{S} = (\mathcal{U}, \mathcal{V}, R_{\mathcal{K}})$, $\mu : \mathcal{P}(\mathcal{C}) \to [0,1]$ is a probability measure on $\mathcal{C}$, N is a set of *names*, and G is a set of maps $\{g_{\mathcal{K}} : N \to \mathcal{K}(W) \mid \mathcal{K} \in \mathcal{C}\}$. We assume that the sets of possible worlds in different causal Kripke models are disjoint, that is, for any $\mathcal{K}$, $\mathcal{K}'$ in $\mathcal{C}$, $\mathcal{K}(W) \cap \mathcal{K}'(W) = \emptyset$.

Each causal Kripke model $\mathcal{K}$ in $\mathcal{C}$ is the same as defined in Sect. 2.1, which represents a possible scenario considered by a causal agent. μ assigns probabilities to different scenarios according to the causal agent. Each name $n \in N$ is assigned to a unique world in each possible scenario. Intuitively, each name n represents the same state of affairs across different possible scenarios which are indistinguishable to the causal agent, allowing us to intervene on the values of variables across these scenarios. For any name n, $g_{\mathcal{K}}(n)$ represents the possible world named by n (or representing n) in the possible scenario described by causal Kripke model $\mathcal{K}$.

A *context* over $\mathcal{M}$ is a map $\overline{T}$ which maps every causal Kripke model $\mathcal{K} \in \mathcal{C}$ to a context $\bar{t}$ over $\mathcal{K}$. A *probabilistic causal Kripke setting* is a pair $(\mathcal{M}, \overline{T})$, where $\mathcal{M}$ is a probabilistic causal Kripke model and $\overline{T}$ is a context over it. A probabilistic causal Kripke model $\mathcal{M} = (\mathcal{C}, \mu, N, G)$ is said to be *recursive* if every causal Kripke model $\mathcal{K}$ in $\mathcal{C}$ is recursive. We let $\mathcal{M}(W) = \bigcup_{\mathcal{K} \in \mathcal{C}} \mathcal{K}(W)$. An *assignment* over $\mathcal{M}$ is a function on some subset $\mathcal{Y} \subseteq \mathcal{V} \times \mathcal{M}(W)$ s.t., for every $(X, w) \in \mathcal{Y}$ with $w \in \mathcal{K}(W)$, it assigns some value in $\mathcal{R}_{\mathcal{K}}(X, w)$. A *name assignment* over $\mathcal{M}$ is a function on some subset $\mathcal{Z} \subseteq \mathcal{V} \times N$, s.t., for every $(X, n) \in \mathcal{Z}$, it assigns a value in $\mathcal{R}_{\mathcal{K}}(X, g_{\mathcal{K}}(n))$ for all $\mathcal{K} \in \mathcal{C}$.

Let $\mathcal{M} = (\mathcal{C}, \mu, N, G)$ be a probabilistic causal Kripke model and $\mathcal{Y}$ be a finite subset of $\mathcal{V} \times \mathcal{M}(W)$ and $\overline{Y}$ be an injective (possibly empty) listing of all variables in $\mathcal{Y}$. Let $\overline{Y} = \overline{y}$ be an assignment s.t., for each $Y_i = (X, w) \in \mathcal{Y}$ with $w \in \mathcal{K}(W)$, it assigns $y_i \in \mathcal{R}_{\mathcal{K}}(X, w)$. The probabilistic causal Kripke model obtained from *intervention*, which sets values of variables in $\overline{Y}$ to $\overline{y}$, is given by $\mathcal{M}_{\overline{Y} \leftarrow \overline{y}} = (\mathcal{C}', \mu, N, G)$, where $\mathcal{C}'$ is obtained from $\mathcal{C}$ as follows: For any $\mathcal{K} \in \mathcal{C}$, we replace $\mathcal{K}$ with $\mathcal{K}_{\overline{Y_{\mathcal{K}}} \leftarrow \overline{y_{\mathcal{K}}}}$, where $Y_{\mathcal{K}} = \mathcal{Y} \cap (\mathcal{V} \times \mathcal{K}(W))$.

Let $\mathcal{Z}$ be a finite subset of $\mathcal{V} \times N$ and $\overline{Z}$ be an injective (possibly empty) listing of all variables in $\mathcal{Z}$. For any name assignment $\overline{Z} = \overline{z}$, and $\mathcal{K} \in \mathcal{C}$, we use $\mathcal{K}(\overline{Z} = \overline{z})$ to denote the assignment that maps variable $(X, g_{\mathcal{K}}(n))$ to z_i, for each variable $Z_i = (X, n) \in \overline{Z}$. Let $\mathcal{M}(\overline{Z} = \overline{z})$ be the assignment obtained by taking the concatenation of all the assignments $\mathcal{K}(\overline{Z} = \overline{z})$ over all $\mathcal{K}$ in $\mathcal{C}$. Then, the probabilistic causal Kripke model obtained from *named intervention*, which sets values of variables in $\overline{Z}$ to $\overline{z}$, is given by $\mathcal{M}_{\overline{Z} \leftarrow \overline{z}} = \mathcal{M}_{\mathcal{M}(\overline{Z} = \overline{z})}$. Intuitively, this says that the intervention on an endogenous variable associated with some name is understood as the simultaneous application of interventions on the value of that endogenous variable at all the worlds represented by the same name across different possible scenarios (i.e. causal Kripke models).

The interventions on variables of the form (X, w) allow us to do counterfactual reasoning on variable X at some possible world w even when the causal agent finds it indistinguishable from other possible worlds represented by the same name as w. This can be seen as the causal agent performing intervention in the hypothetical (or counterfactual) scenario in which they know that $\mathcal{K}$ containing the possible world w is the real scenario.

3.1 Probabilistic Modal Language for Describing Causality

In this section, we define the formal logical framework for describing causality in a probabilistic setting. Since our aim is to discuss the values of variables across different possible worlds, as well as the values of variables at the worlds represented by the same name in different possible scenarios, our hybrid language consists of two disjoint countable sets W and N. In principle, each model $\mathcal{M}$ is associated with a map that assigns elements of W to worlds within $\mathcal{M}(W)$, and a collection of maps G that map elements in N to worlds in different causal Kripke models. We will denote the language associated with model $\mathcal{M}$ as $L_{\mathcal{M}}(W, N)$. For simplicity, we often write $L_{\mathcal{M}}$ when W and N are clear from the context.

For a given probabilistic causal Kripke model $\mathcal{M} = (\mathcal{C}, \mu, N, G)$, the language $L_{\mathcal{M}}$ associated with it is given by the following recursion.

$$\alpha ::= X = x \mid (X, w) = x \mid (X, n) = x \mid \neg\alpha \mid \alpha \wedge \alpha \mid \Box\alpha,$$
$$\phi ::= X = x \mid (X, w) = x \mid (X, n) = x \mid \neg\phi \mid \phi \wedge \phi \mid \Box\phi \mid [\overline{Y} \leftarrow \overline{y}]\alpha,$$
$$\Delta ::= P_a\phi \mid \Delta \wedge \Delta \mid \neg\Delta, \text{ for } a \in [0, 1],$$

where $X \in \mathcal{V}$, $w \in \mathcal{M}(W)$, $n \in N$, and $[\overline{Y} \leftarrow \overline{y}]$ is an intervention. Here, we have added atomic propositions of the form $(X, n) = x$ to the modal language of causality described in Sect. 2.2, which intuitively stands for '$X = x$ holds in all the possible worlds named by n across all the different scenarios'. Let $\overline{T}$ be a context over $\mathcal{M}$. We define the satisfaction for any formula $\psi \in L_{\mathcal{M}}$ at a world $w \in \mathcal{M}(W)$ (denoted by $\mathcal{M}, w \Vdash \psi$), and a name $n \in N$ (denoted by $\mathcal{M}, n \Vdash \psi$)[2] by the following recursion:

(1) For any atomic proposition of the form $(X, n) = x$, $\mathcal{M}, w \Vdash (X, n) = x$, iff for any $\mathcal{K} \in \mathcal{C}$, $(\mathcal{K}, \overline{T}(\mathcal{K}), g_{\mathcal{K}}(n)) \Vdash X = x$, where satisfaction on the causal Kripke setting is defined as in Sect. 2.2. The satisfaction at the world w for atomic propositions of the form $X = x$ or $(X, w) = x$, and that for the connectives $\neg$, $\wedge$, $\Box$, and the intervention $[\overline{Y} \leftarrow \overline{y}]$ are defined as in Sect. 2.2. Note that we only allow interventions on the variables of the form (X, w).

(2) For any name n, and any formula $\psi \in L_{\mathcal{M}}$, $\mathcal{M}, n \Vdash \psi$ iff for any $\mathcal{K} \in \mathcal{C}$, $(\mathcal{M}, g_{\mathcal{K}}(n)) \Vdash \psi_{\mathcal{K}}$, where $\psi_{\mathcal{K}}$ is obtained from ψ by replacing every occurrence of n with $g_{\mathcal{K}}(n)$. For any formula ψ, $\psi_{\mathcal{K}}$ represents the instantiation of ψ in the possible scenario $\mathcal{K}$.

(3) For any formula ϕ, and any $a \in [0, 1]$, $\mathcal{M}, w \Vdash P_a\phi$ iff $\mu(\{\mathcal{K} \mid \mathcal{K} \in \mathcal{C} \ \& \ (\mathcal{K}, w) \Vdash \phi_{\mathcal{K}}\}) \geq a$, and $\mathcal{M}, n \Vdash P_a\phi$ iff $\mu(\{\mathcal{K} \mid \mathcal{K} \in \mathcal{C} \ \& \ (\mathcal{K}, g_{\mathcal{K}}(n)) \Vdash \phi_{\mathcal{K}}\}) \geq a$. Intuitively, $P_a\phi$ stands for 'ϕ is satisfied with the probability at least a'. The satisfaction relation for propositional connectives on formulas of the type Δ is defined in a straightforward manner.

---

[2] Here, we use w (resp. n) to represent both the world (resp. name) and the variable which refers to that world (resp. name).

3.2   HP Definition of Probabilistic Causality

We now generalize the modified HP definition of actual causality presented in Sect. 2.3 to the probabilistic setting. That is, we describe how to answer the questions of the form 'What is the probability that A is the actual cause of event B?' for a given probabilistic causal Kripke model $\mathcal{M} = (\mathcal{C}, \mu, N, G)$. Here, A is of the form $\overline{Y} = \overline{y} \wedge \overline{Z} = \overline{z}$, for some assignment $\overline{Y} = \overline{y}$ and name assignment $\overline{Z} = \overline{z}$, and B is of the form α discussed in Sect. 3.1. For any assignment $\overline{Y} = \overline{y}$, and any $\mathcal{K} \in \mathcal{C}$, we use $\overline{Y}_{\mathcal{K}}$ to denote the listing of variables in $\overline{Y}$ of the form (X, w) for some $w \in \mathcal{K}(W)$.

Let $(\mathcal{M}, \overline{T})$ be a probabilistic causal Kripke setting. Given an assignment $\overline{Y} = \overline{y}$, a name assignment $\overline{Z} = \overline{z}$, an event α, and a world $w \in \mathcal{K}(W)$ for some $\mathcal{K} \in \mathcal{C}$, $(\mathcal{M}, \overline{T}, w) \Vdash cause(\overline{Y} = \overline{y} \wedge \overline{Z} = \overline{z}, \alpha)$, iff for all $\mathcal{K}' \in \mathcal{C}$, $(\mathcal{K}, \overline{T}(\mathcal{K}), w) \Vdash cause(\overline{Y}_{\mathcal{K}'} = \overline{y}_{\mathcal{K}'} \wedge \mathcal{K}'(\overline{Z} = \overline{z}), \alpha_{\mathcal{K}'})$, where $\alpha_{\mathcal{K}'}$ is obtained from α by replacing every occurrence of n with $g_{\mathcal{K}'}(n)$. That is, the actual cause at a world $w \in \mathcal{K}(W)$ is determined by restricting to the possible scenario $\mathcal{K}$, and then applying the definition of actual causality on causal Kripke models.

Finally, for any name n, any assignment $\overline{Y} = \overline{y}$ and any name assignment $\overline{Z} = \overline{z}$ over $\mathcal{M}$, we define $P(cause(\overline{Y} = \overline{y} \wedge \overline{Z} = \overline{z}, \alpha))$ at a name n to be equal to $\mu(\{\mathcal{K} \in \mathcal{C} \mid \mathcal{K}, \overline{T}(\mathcal{K}), g_{\mathcal{K}}(n) \Vdash cause(\overline{Y}_{\mathcal{K}} = \overline{y}_{\mathcal{K}} \wedge \mathcal{K}(\overline{Z} = \overline{z}), \alpha_{\mathcal{K}})\})$. That is, the probability of $\overline{Y} = \overline{y} \wedge \overline{Z} = \overline{z}$ being the actual cause of α is obtained by calculating the probability of the set of possible scenarios in which the instantiation of $\overline{Y} = \overline{y} \wedge \overline{Z} = \overline{z}$ to that scenario is the cause of $\alpha_{\mathcal{K}}$.

Note that we allow causes to be conjunctions of the form $\overline{Y} = \overline{y} \wedge \overline{Z} = \overline{z}$. This allows us to consider the values of variables at a name (i.e. at the set of worlds indistinguishable to the causal agent) and/or at a specific world (i.e. in some specific scenario) to be actual cause.

4   Modelling Probabilistic Causality and Examples

We now give examples to demonstrate the application of probabilistic causal Kripke models in modeling causal analysis in scenarios with uncertainty.

Example 1. Alice plans to visit her friend Betty tomorrow. Betty decides between taking Alice to Oxford or Cambridge. After checking the weather, Alice finds a high chance of rain in Oxford but not in Cambridge. Considering both options, she carries an umbrella. The next day, they visit Cambridge, but unexpectedly, it rains. Fortunately, Alice does not catch a cold. Assume that Alice's likelihood of catching a cold depends only on whether it rains and whether she has an umbrella, and that without rain, Alice does not catch a cold. There are three possible dependence scenarios: (1) Alice never catches a cold, (2) she avoids a cold in the rain only if she has an umbrella, and (3) she always catches a cold in the rain. Suppose that initially (i.e. before the above sequence of events) the causal agent believes that the probabilities of the dependencies given by (1), (2), and (3) are 0.1, 0.8 and 0.1, respectively. What is the probability that Alice's consideration of the possibility of rain in Oxford is the cause of her not catching a cold? We model this scenario as follows.

Note that the causal agent is fully aware of all the events occurring and the only source of uncertainty for them comes from the uncertainty about the dependence between variables. As Alice does not catch cold even when it rains in Cambridge in this example, the causal agent knows this and disregards the possibility of dependence being given by (3). Hence, the agent only considers two possible scenarios in which the dependence is given either by (1) or (2) represented by two causal Kripke models $\mathcal{K}_1$, and $\mathcal{K}_2$, corresponding to dependence given by (1), and (2), respectively. The probability measure μ for each of these scenarios according to the causal agent is given by conditioning on the original probability distribution on dependence relationships with the fact that Alice did not catch cold despite the rain, i.e. $\mu(\{\mathcal{K}_1\}) = 0.1/(0.8 + 0.1) = 0.111$, and $\mu(\{\mathcal{K}_2\}) = 0.8/(0.8 + 0.1) = 0.889$.

Let S be the signature with binary endogenous variables 'Alice carries an umbrella' (o), 'Alice is in Oxford' (p), 'Alice is in Cambridge' (q), 'It rains' (r), and 'Alice catches cold' (s). In each possible scenario $\mathcal{K}_i$, we have four possible worlds, the world w_{i0} representing the world yesterday when Alice makes her decision, the possible future worlds considered by Alice w_{i1} and w_{i2} in which she visits Oxford and Cambridge w_{i1} and w_{i2}, and the world w_{i3} today in which she visits Cambridge and it rains there. Dependence among the variables o, r, and s in the worlds w_{13}, and w_{23} is given by (1), and (2), respectively. That is, we have structural equations $Eq(s, w_{13}) = 0$, and $Eq(s, w_{23}) = (r, w_{23}) \wedge \neg(o, w_{23})$, respectively. We have four names n_0, n_1, n_2, and n_3, s.t. for any possible scenario $\mathcal{K}_i$, $g_{\mathcal{K}_i}(n_j) = w_{ij}$. This reflects the idea that each name refers to the worlds across the different scenarios which are indistinguishable to the causal agent.

For any possible scenario $\mathcal{K}_i$, let R_i be the relation s.t. $w_{ij} R_i w_{ik}$ if at the world w_{ij} Alice believes world w_{ik} is the possible future world. Here, we have $R_i = \{(w_{i0}, w_{i1}), (w_{i0}, w_{i2})\}$ for all i. Other structural equations for the scenarios are as follows: $Eq(o, w_{i0}) = \Diamond r$ for all i, i.e., in all possible scenarios, Alice takes an umbrella if she thinks it is possible that tomorrow she visits a place where it is highly likely to rain, and $Eq(o, w_{ij}) = (o, w_{i0})$ for $j \neq 0$ and for all i, i.e. in all possible scenarios, if she decides to take an umbrella, then she has it with her wherever she visits on the next day, $Eq(s, w_{i0}) = Eq(s, w_{i1}) = Eq(s, w_{i2}) = 0$ for all i, i.e. in all possible scenarios, Alice does not have a cold on the day before or in the possible future worlds she considers.

Let $U = (U_1, U_2) \in \{0, 1\}^2$ be an exogenous variable s.t. for $i = 1, 2$, the structural equations for p, q, and are $Eq(p, w_{ij}) = (U_1, w_{ij})$, $Eq(q, w_{ij}) = \neg(U_1, w_{ij})$ for $j \neq 0$, and $Eq(p, w_{i0}) = Eq(q, w_{i0}) = 0$, and $Eq(r, w_{ij}) = (U_2, w_{ij})$ for all j. Let $\overline{T}$ be a context s.t. U is set to $(1, 1)$ in w_{i0} and w_{i1}, to $(0, 0)$ in w_{i2}, and to $(0, 1)$ in w_{i3} for all i. Thus, we get $\overline{T}(w_{i0}) = \{o, r\}$, $\overline{T}(w_{i1}) = \{o, p, r\}$, $\overline{T}(w_{i2}) = \{o, q\}$, and $\overline{T}(w_{i3}) = \{o, q, r\}$ for all i.

The causal agent aims to determine whether Alice's consideration of possible rain in Oxford caused her not to catch a cold. Since both $\mathcal{K}_1$ and $\mathcal{K}_2$ are possible, this cannot be modeled purely in terms of indistinguishable possible worlds.

Instead, we use names to represent worlds that are indistinguishable to the agent across scenarios. Specifically, the probability that Alice's consideration of the possibility of rain in Oxford is the actual cause of her not catching a cold is expressed by $P(cause((r, n_1) = 1, (s, n_3) = 1)^3$. Here, n_1 represents the hypothetical world Alice considers in which she visits Oxford, and n_3 represents the real world where she visits Cambridge and remains healthy despite the rain.

Here, it is easy to check that $(r, w_{11}) = 1$ is not the actual cause of $(s, w_{13}) = 0$ in $\mathcal{K}_1$ since $Eq(s, w_{13}) = 0$ regardless of the value of (r, w_{11}), but $(r, w_{21}) = 1$ is the actual cause of $(s, w_{23}) = 0$ in $\mathcal{K}_2$. Therefore, $P(cause((r, n_1) = 1, (s, n_3) = 1)) = 0.889$. Thus, according to the causal agent, in this scenario, the probability that Alice's consideration of the possibility of rain in Oxford is the actual cause of her not catching a cold is 0.889.

Example 2 (Navigation). Bob is heading to village A and arrives at a location that is actually B. Bob is sure that A is to the east of B, but he is uncertain whether he is at B or C. He must choose to move East or West at his current location. In reality, A is to the east of both B and C. If he is unsure about the direction to A from his current location, he chooses either direction with probability 0.5. The question is: what is the probability that A being east of C is the cause of Bob moving East? We model this scenario as follows.

Let $\mathcal{S}$ be the signature with binary endogenous variables 'A is to the east of the current location' (p), and 'Bob goes East from the current location' (q). Note that the causal agent is fully aware of all the events occurring, and the only uncertainty for them is whether Bob goes East depends on his knowledge of A being east of B and C, or his choice to go East without knowing the direction of A. The dependence between Bob's decision to go East and his current state of knowledge is given by the structural equation (1) $Eq(q) = \Box p$ (Bob moves East if he is sure that A is to the east) or (2) $Eq(q) = \Diamond p$ (Bob moves East if he thinks it possible that A is to the east) with probability 0.5. Thus, the probabilistic causal Kripke model for this scenario would consist of two causal Kripke models $\mathcal{K}_1$, and $\mathcal{K}_2$, each corresponding to dependencies given by (1) and (2), respectively, each assigned probability 0.5.

Each possible scenario $\mathcal{K}_i$ consists of two possible worlds w_{i0} and w_{i1} representing the locations B and C, respectively. Dependence between the variables p and q in the worlds w_{10}, w_{11}, w_{20}, and w_{21} is $Eq((q, w_{1j})) = \Box p$, and $Eq((q, w_{2j})) = \Diamond p$, for $j = 0, 1$. We have two names n_0 and n_1, s.t. for any possible scenario $\mathcal{K}_i$, $g_{\mathcal{K}_i}(n_j) = w_{ij}$. That is, names n_0 and n_1 refer to the locations B and C, respectively, across two possible scenarios.

For any possible scenario $\mathcal{K}_i$, let R_i be the relation s.t. $w_{ij} R_i w_{ik}$ if the location w_{ij} is indiscernible from w_{ik} for Bob. As Bob is not sure whether he is at B or C, we have $R_i = \{(w_{i0}, w_{i0}), (w_{i1}, w_{i1}), (w_{i0}, w_{i1}), (w_{i1}, w_{i0})\}$ for all i.

---

[3] Since all the variables in this formula are hybrid, i.e. they explicitly refer to the values of variables at specific worlds, the meaning of this formula is independent of the name or the world where it is evaluated.

Let $U \in \{0, 1\}$ be an exogenous variable s.t. the structural equation for p is $Eq(p, w_{ij}) = (U, w_{ij})$ for all i and j. Let $\overline{T}$ be a context s.t. U is set to 1 in all the possible worlds. Thus, we get $\overline{T}(w_{ij}) = \{p, q\}$ for all i and j.

The causal agent seeks to determine whether A being East of C is the cause of Bob's decision to move East from B. As the agent considers both $\mathcal{K}_1$ and $\mathcal{K}_2$ to be possible scenarios, we can not formally model this in terms of possible worlds which may be indistinguishable to the causal agent, but in terms of names which represent the possible worlds that represent the same location in two possible scenarios, i.e. the probability that A being to the east of C being the cause of Bob moving to the east from his current location B is given by $P(cause((p, n_1) = 1, (q, n_0) = 1)$. This is because in each scenario considered possible by the causal agents, n_0 represents the possible world in which Bob is at B (remember that we assume that the causal agent has full information of events, we assume that the causal agent has full information of events so he knows Bob is at B).

Here, it is easy to check that $(p, w_{11}) = 1$ is the actual cause of $(q, w_{10}) = 1$ in scenario $\mathcal{K}_1$, but $(p, w_{21}) = 1$ is not the actual cause of $(q, w_{20}) = 1$ in scenario $\mathcal{K}_2$. Therefore, by definition, $P(cause((p, n_1) = 1, (q, n_0) = 1)) = 0.5$. Thus, in this scenario A being to the East of C is the cause of Bob moving to the East in his current location B with probability 0.5.

In the previous two examples, the only uncertainty for the causal agent comes from the uncertainty in structural equations. However, the next example shows that our framework is applicable also in the case where causal agent is uncertain about the facts and sequence of events occurring.

Example 3. Consider a slight variation of the above scenario in which Bob receives a message from John through an unreliable channel indicating that he is at location B. The message has a 50% chance of reaching Bob. If the message is received, Bob knows he is at B; otherwise, he remains uncertain whether he is at B or C. Every other aspect of the scenario, including Bob's decision-making process, remains the same as before. We model this scenario as follows.

Here, the causal agent has two sources of uncertainty, one whether Bob receives the message, and the second about the dependence among the variables. We assume these uncertainties are independent, leading to four possible causal Kripke models $\mathcal{K}_1$, $\mathcal{K}_2$, $\mathcal{K}_3$, and $\mathcal{K}_4$ corresponding to (a) dependence being given by (1) and Bob receiving the message, (b) dependence being given by (1) and Bob not receiving the message, (c) dependence being given by (2) and Bob receiving the message, and (d) dependence being given by (2) and Bob not receiving the message, respectively. The probability μ for each of these scenarios according to the causal agent is 0.25.

Let $\mathcal{S}$ be the same signature as before. In each possible scenario $\mathcal{K}_i$, there are two possible worlds w_{i0} and w_{i1} that represent the locations B and C, respectively. Structural equations for the variables p and q in the worlds w_{i0} and w_{i1} are $Eq((q, w_{i0})) = \Box p$, for $i = 1, 2$ and by $Eq((q, w_{i0})) = \Diamond p$ for $i = 3, 4$. We also have two names n_0 and n_1, s.t. for any possible scenario $\mathcal{K}_i$, $g_{\mathcal{K}_i}(n_j) = w_{ij}$. That is, names n_0 and n_1 refer to the locations (across different possible scenarios) B and C, respectively. Here, Bob's ability to distinguish locations depends on whether he receives the message. Thus, $R_i = \{(w_{i0}, w_{i0}), (w_{i1}, w_{i1}), (w_{i0}, w_{i1}), (w_{i1}, w_{i0})\}$ for $i = 2, 4$; and $R_i = \{(w_{i0}, w_{i0}), (w_{i1}, w_{i1})\}$ for $i = 1, 3$.

Let $U \in \{0, 1\}$ be an exogenous variable s.t. the structural equation for p is $Eq((p, w_{ij})) = (U, w_{ij})$ for all i and j. Let $\overline{T}$ be a context s.t. U is set to 1 in all the possible worlds. Thus, we get $\overline{T}(w_{ij}) = \{p, q\}$ for all i and j. In this case, the causal agent is interested in whether A being to the east of C is the cause of Bob moving to the east in his current location B. Since multiple scenarios are possible, we model this using names, with the probability given by $P(cause((p, n_1) = 1, (q, n_0) = 1))$.

Here, we can show that $(p, w_{21}) = 1$ is the actual cause of $(q, w_{20}) = 1$ in scenario $\mathcal{K}_2$, but for $i = 1, 3, 4$, $(p, w_{i1}) = 1$ is not the actual cause of $(q, w_{i0}) = 1$ in scenario $\mathcal{K}_i$. Therefore, by definition, $P(cause((p, n_1) = 1, (q, n_0) = 1)) = 0.25$. Thus, in this scenario A being to the East of C being the cause of Bob moving to the East in his current location B with probability 0.25. The contributions to this probability by each scenario $\mathcal{K}_i$ is given by $P(cause((p, w_{i1}) = 1, (q, w_{i0}) = 1))$. In this case, we get $P(cause((p, w_{i1}) = 1, (q, w_{i0}) = 1)) = 0$ for $i \neq 2$, and $P(cause((p, w_{i1}) = 1, (q, w_{i0}) = 1)) = 0.25$ for $i = 2$. That is, if the causal agent determines $\mathcal{K}_2$ is not a possible scenario, then the probability that A being to the East of C is the cause of Bob moving to the East in his current location B would be 0.

In all the examples discussed in this paper, all the endogenous variables have the same values in all the worlds represented by the same name. However, this need not be the case in general, i.e. an agent may find two worlds in which some endogenous variable has different values indistinguishable. We leave this to future work due to space constraints.

5   Conclusion and Future Directions

In this paper, we have extended the causal Kripke model framework developed in [8] to a probabilistic setting. This framework allows for the quantitative modeling of causality in the presence of uncertainty while incorporating modal notions such as knowledge, temporality, and possibility. This work opens several directions for future research.

A first step will be to provide an axiomatization for the logic we describe. In [8] a sound and (weakly) complete axiomatization for causal Kripke frames is provided. We conjecture that combining that axiomatization with systems for probabilistic logics, such as the ones developed in [10], will be enough to axiomatize the logic introduced in this article.

A natural extension of this work is to generalize concepts such as explanation, harm, and blame within the probabilistic modal setting introduced in this paper. This framework provides a foundation for formalizing these notions in a broad range of real-world causal scenarios.

In our approach, we account for the agent's uncertainty by assigning probabilities to deterministic causal Kripke models. Previous research has explored generalized Halpern-Pearl models that incorporate non-deterministic structural equations directly [1,3]. Extending such models to the possible-world setting would be an intriguing direction for future work.

Acknowledgments. This project has received funding from the European Union's Horizon 2020 research and innovation programme under the Marie Skłodowska-Curie grant agreement No. 101007627. Yiwen Ding is supported by the China Scholarship Council No. 202206220057. Krishna Manoorkar is supported by the NWO grant KIVI.2019.001 awarded to Alessandra Palmigiano. Apostolos Tzimoulis is supported by the Luxembourg National Research Fund (FNR) (INTER/D-FG/23/17415164/LODEX) and by the Key Project of the Chinese Ministry of Education (22JJD720021). Ruoding Wang is supported by the China Scholarship Council No. 202206310072.

Disclosure of Interests. The authors have no competing interests to declare that are relevant to the content of this article.

References

1. Barbero, F.: On the logic of interventionist counterfactuals under indeterministic causal laws. In: International Symposium on Foundations of Information and Knowledge Systems, pp. 203–221. Springer (2024). https://doi.org/10.1007/978-3-031-56940-1_11
2. Beckers, S.: Causal explanations and XAI. In: Conference on Causal Learning and Reasoning, pp. 90–109. PMLR (2022)
3. Beckers, S.: Nondeterministic causal models. arXiv:2405.14001 (2024)
4. Beer, I., Ben-David, S., Chockler, H., Orni, A., Trefler, R.: Explaining counterexamples using causality. Formal Methods Syst. Des. **40**, 20–40 (2012)
5. Bergstein, B.: What AI still can't do. MIT Technol. Rev. **123**(2), 1–7 (2020). https://www.technologyreview.com/2020/02/19/868178/what-ai-still-cant-do/
6. Chockler, H., Halpern, J.Y., Kupferman, O.: What causes a system to satisfy a specification? ACM Trans. Comput. Logic (TOCL) **9**(3), 1–26 (2008)
7. Datta, A., Garg, D., Kaynar, D., Sharma, D., Sinha, A.: Program actions as actual causes: a building block for accountability. In: 2015 IEEE 28th Computer Security Foundations Symposium, pp. 261–275 (2015)
8. Ding, Y., Manoorkar, K., Tzimoulis, A., Wang, R., Wang, X.: Causal kripke models. In: Verbrugge, R. (ed.) Proceedings 19th Conference on Theoretical Aspects of Rationality and Knowledge, Oxford, United Kingdom, 28–30th June 2023, volume 379 of Electronic Proceedings in Theoretical Computer Science, pp. 185–200. Open Publishing Association (2023)
9. Dubois, D., Prade, H.: A Glance at Causality Theories for Artificial Intelligence, pp. 275–305. Springer International Publishing, Cham (2020). https://doi.org/10.1007/978-3-030-06164-7_9
10. Fagin, R., Halpern, J.Y., Megiddo, N.: A logic for reasoning about probabilities. Inf. Comput. **87**(1), 78–128 (1990). Special Issue: Selections from 1988 IEEE Symposium on Logic in Computer Science

11. Falcon, A.: Aristotle on causality. In: Stanford Encyclopedia of Philosophy (2008). https://plato.stanford.edu/entries/aristotle-causality/
12. Fenton-Glynn, L.: A proposed probabilistic extension of the Halpern and Pearl definition of 'actual cause'. Br. J. Philos. Sci. **0**, 1–64 (2017)
13. Glymour, C., Wimberly, F.: Actual causes and thought experiments. In: Causation and Explanation. The MIT Press (2007)
14. Hall, N.: Structural equations and causation. Philos. Stud. **132**, 109–136 (2007)
15. Halpern, J.Y.: A modification of the Halpern-Pearl definition of causality. In: Proceedings of the 24th International Conference on Artificial Intelligence, IJCAI 2015, pp. 3022–3033. AAAI Press (2015)
16. Halpern, J.Y.: Actual Causality. The MIT Press, USA (2016)
17. Halpern, J.Y., Pearl, J.: Causes and explanations: a structural-model approach. part I: Causes. Br. J. Philos. Sci. **56**, 843–847 (2005)
18. Halpern, J.Y., Pearl, J.: Causes and explanations: a structural-model approach. Part II: Explanations. Br. J. Philos. Sci. **56**, 889-991 (2005)
19. Hume, D.: An Enquiry Concerning Human Understanding (Edited by Peter Millican). Oxford University Press, New York (2020)
20. Ibrahim, A., Klesel, T., Zibaei, E., Kacianka, S., Pretschner, A.: Actual causality canvas: a general framework for explanation-based socio-technical constructs. In: ECAI 2020, pp. 2978–2985. IOS Press (2020)
21. Lewis, D.: Causation. J. Philos. **70**(17), 556–567 (1973)
22. Lewis, D.: Causation as influence. J. Philos. **97**(4), 182–197 (2004)
23. Meliou, A., Gatterbauer, W., Halpern, J.Y., Koch, C., Moore, K.F., Suciu, D.: Causality in databases. IEEE Data Eng. Bull. **33**(ARTICLE), 59–67 (2010)
24. Moore, M.S.: Causation and Responsibility: An Essay in Law, Morals, and Metaphysics. Oxford University Press (2009)
25. Pearl, J.: Causality, 2nd edn. Cambridge University Press, Cambridge (2009)
26. Pearl, J., Mackenzie, D.: The Book of Why: The New Science of Cause and Effect, 1st edn. Basic Books Inc, USA (2018)
27. Schölkopf, B.: Causality for Machine Learning, 1st edn., pp. 765–804. Association for Computing Machinery, New York (2022)

Generalized Causal Models
with Ontological Dependencies

Jingzhi Fang[1] and Jiji Zhang[2]($\boxtimes$)

[1] Institute of Logic and Cognition, Department of Philosophy,
Sun Yat-sen University, Guangzhou, China
fangjzh5@mail.sysu.edu.cn
[2] Department of Philosophy, The Chinese University of Hong Kong,
Shatin, Hong Kong, NT, China
jijizhang@cuhk.edu.hk

Abstract. Causal models consisting of structural equations have proved a powerful formalism for approaching theoretical and applied questions on causal reasoning. The standard framework assumes that all variables that appear in a causal model must represent properties or events that are metaphysically independent, with no dependencies more intimate than causal relations, such as conceptual, constitutive, or other ontological dependencies. However, there are important contexts that call for lifting this restriction, such as the debate over high-level causation in philosophy and the task of causal representation learning in artificial intelligence. This paper proposes a natural generalization of structural causal models to accommodate (asymmetric) ontological dependence and provides an interpretation of causal counterfactuals based on such models. We discuss several implications of this generalization and present an axiomatization of the resulting counterfactual logic.

Keywords: Causal models · Ontological dependence ·
Counterfactuals · Intervention

1 Introduction

Causal modelling ([20,24]) has found promising applications in both philosophy and artificial intelligence. In philosophy, it is used as a powerful formal tool to address questions ranging from conceptual analysis of causation to epistemological inquiries about the possibility and extent of causal discovery ([19]). In artificial intelligence, it provides important machinery for knowledge representation and causal reasoning ([8]), and has recently found interesting integrations with representation learning ([23]).

The standard framework assumes that all variables that appear in a causal model must represent properties or events that are metaphysically independent, with no dependencies more intimate than causal relations, such as conceptual,

© The Author(s), under exclusive license to Springer Nature Switzerland AG 2026
V. Goranko et al. (Eds.): LORI 2025, LNCS 16010, pp. 65–78, 2026.
https://doi.org/10.1007/978-981-95-2481-5_5

constitutive, or other ontological dependencies ([16,17]). This restriction simplifies many modelling and inferential tasks by focusing purely on causal dependencies. However, there are important contexts that call for lifting this restriction. In the philosophy of science, for example, an important problem is how causation in the special sciences such as biology and psychology is related to causation in more fundamental sciences such as physics, or how causation between higher-level variables is related to that between lower-level variables ([5]). The problem of mental causation in the philosophy of mind is a famous example of this sort ([13]). To address this problem using causal models probably requires including variables at different levels in the same model, though they may bear non-causal dependencies ([27]).

Similarly, in artificial intelligence, representation learning aims to discover meaningful features or patterns from raw data and often involves non-causal relationships between different levels of abstraction. In particular, in the growing field of causal feature or representation learning, a crucial aim is to construct high-level causal variables from low-level measurements ([4,23]). For this and related tasks, it would be highly beneficial to have a more expressive framework to accommodate both causal and non-causal dependencies.

More generally, it is commonplace for big datasets to contain variables that have dependencies that are more robust than causal; even a standard blood test report, for example, routinely includes such 'redundant' variables as total-bilirubin, direct-bilirubin, and indirect-bilirubin, where total-bilirubin, by definition, records the sum of the latter two. This issue becomes especially prominent with multi-modal datasets; it is plausible to think that when measurements from different modalities are correlated, they are sometimes due to causal connections (e.g., between text and image, or audio and video), but other times due to non-causal dependencies of some sort (e.g., the same target being measured at different levels of granularity). It is therefore potentially productive to develop modelling tools that are suitable for such datasets.

There have been some attempts to generalize causal modelling to represent non-causal relations along with causal ones. One line of attack is to reinterpret or generalize the machinery of causal Bayesian network ([2,6,25]), where a graphical representation is used to represent causal dependencies and ontological dependencies simultaneously, and an appropriate Markov condition is formulated between the graphical structure and the joint probability distribution. This approach has the potential to adapt the techniques for discovery of graphical causal structure ([21]) to the task of inferring dependency structures that feature both causal and non-causal dependencies. However, it is not yet clear if logical and statistical theories of counterfactual reasoning can be developed in this approach; at least in Judea Pearl's influential work ([20]), such theories are naturally developed using structural equation models rather than causal Bayesian networks.

On the other hand, Beckers et al. ([1]) generalized structural causal models by adding non-causal relationships as "constraints" in models. In particular, ontological dependencies can be formalized as constraints, which are required to remain invariant even when the involved variables are manipulated by interventions. They then formulated logical theories of counterfactual conditionals in the generalized framework. Despite the admirable generality of this work, however, it is not tailored to represent *asymmetric* non-causal dependencies, because their constraints do not differentiate between dependent variables and the variables they depend on. But some important kinds of ontological dependence are widely regarded as asymmetric. A dominant view on the relation of metaphysical grounding, for example, is that it is asymmetric and irreflexive ([3]). In fact, some philosophers (e.g. [11,22,26]) have advocated using recursive structural equation models to model grounding relations, in the exact same formal manner in which asymmetric causal relations are represented by such models, so that the asymmetry and irreflexivity of these relations are made explicit. Similarly, even though the relation of supervenience is not in general required to be asymmetric, the more significant instances of supervenience are usually taken to hold asymmetrically and are particularly interesting because of such asymmetry, such as the alleged supervenience of mental properties or states on physical properties or states, which is asymmetric due to the supposed multiple realizability of the former by the latter. As we will show in this paper, once we restrict ourselves to asymmetric ontological dependencies (as opposed to constraints in general), there is a more natural way to generalize structural causal models, which keeps a simple graphical representation of the overall structure as a directed *acyclic* graph and hence also entails a generalization of the causal Bayesian network machinery mentioned above.

The rest of the paper will proceed as follows. We describe in the next section structural causal models and our generalization to accommodate ontological dependencies. Section 4 presents an axiomatization of the resulting logic for counterfactual reasoning, building on Joseph Halpern's seminal work ([7]). Section 3 discusses a few implications of our generalization and compares it with Beckers et al.'s generalization. We conclude with some remarks about future work in Sect. 5.

2   Generalizing Causal Models

In the standard framework ([20]), a structural causal model consists of a pair $M = \langle S, F \rangle$, where $S = \langle \mathbf{U}, \mathbf{V}, \mathcal{R} \rangle$ includes *exogenous* variables $\mathbf{U}$ and *endogenous* variables $\mathbf{V}$, together with their ranges ($\mathcal{R}(X)$ denotes the set of possible values for every $X \in \mathbf{U} \cup \mathbf{V}$), and F is a set of functions or equations representing dependencies among variables. Each endogenous variable $Y \in \mathbf{V}$ is associated with a function or structural equation f_Y that maps $\times_{Z \in (\mathbf{U} \cup \mathbf{V} \setminus \{Y\})} \mathcal{R}(Z)$ to $\mathcal{R}(Y)$. Exogenous variables do not have such functions in the model; their values are supposed to be determined by factors not represented by the model. A value setting $\mathbf{u}$ of all the exogenous variables $\mathbf{U}$ is called a *context*.[1] A requirement often imposed on such models is that given a context, there is a *unique solution*

---

[1] We usually use boldface upper case letters to denote sets of variables and the corresponding lower case letters to denote value settings.

of all the equations in the model, that is, a context determines the values of all endogenous variables. A pair $\langle M, \mathbf{u} \rangle$ consisting of a model and a context is called a *causal setting*.

An intervention on a set of endogenous variables $\mathbf{X}$ is understood to be forcing $\mathbf{X}$ to take on a certain value setting $\mathbf{x}$, without directly affecting other variables. In a causal model M, this is represented by replacing the original equation of each element X of $\mathbf{X}$ with $X = x$, where x is the value of X in $\mathbf{x}$, and keeping other equations intact. We call the model after the intervention $\mathbf{X} = \mathbf{x}$ a *submodel* of M, denoted as $M_{\mathbf{X}=\mathbf{x}}$.

In this paper, we explore a simple idea of generalizing this setup to accommodate ontological dependencies between levels of reality, and as a first step in this direction, we focus on what we call *bipartite* models, in which endogenous variables are partitioned into two parts, a set of lower-level or bottom-level variables $\mathbf{V}_b$ and a set of higher-level variables $\mathbf{V}_h$. For each $X \in \mathbf{V}_b$, there is a "nomological" function, whereas for each $Y \in \mathbf{V}_h$, there is a "metaphysical" function. Intuitively, "nomological" structural equations are just the good old equations in standard causal models, which can be modified by (causal) interventions, but "metaphysical" structural equations represent ontological dependencies that cannot be modified by (causal) interventions.[2]

Formally, a bipartite generalized structural equation model is a pair $M = \langle S, F \rangle$, in which $S = \langle \mathbf{U}, \mathbf{V}_b \dot{\cup} \mathbf{V}_h, \mathcal{R} \rangle$, and $F = F_n \dot{\cup} F_m$, where F_n contains a function $f_X : \times_{Z \in \mathbf{U} \cup \mathbf{V}_b \setminus \{X\}} \mathcal{R}(Z) \to \mathcal{R}(X)$ for each $X \in \mathbf{V}_b$, and F_m contains a function $f_Y : \times_{Z \in \mathbf{V}_b} \mathcal{R}(Z) \to \mathcal{R}(Y)$ for each $Y \in \mathbf{V}_h$. Note that the nomological functions F_n only involve bottom-level variables together with exogenous variables, reflecting an assumption of "nomological sufficiency" at the bottom-level. In addition, the metaphysical functions F_m incorporate the assumption that each higher-level variable is ontologically determined by a set of bottom-level variables. Both assumptions can be relaxed, which will be explored in future work.

Interventions on bottom-level variables, i.e., any $\mathbf{X} \subseteq \mathbf{V}_b$, are represented in the same way as before. They amount to replacing the original equations for $\mathbf{X}$ and keep other equations the same. For simplicity, we impose Pearl's "unique solution" requirement on interventions of bottom-level variables. That is, after any intervention (including the null intervention) on variables in $\mathbf{V}_b$, there is a unique solution to all the (new and old) equations relative to any context.

How about interventions on higher-level variables? Since metaphysical equations are supposed to be unmodifiable by interventions, manipulating a higher-level variable necessarily involves changing some lower-level variables in order to maintain the original metaphysical equation for that higher-level variable. The idea that differences at the higher level must be accompanied by those at the bottom level is, of course, very similar to the standard formulation of supervenience ([18]). There are multiple ways to understand interventions on higher-level variables while respecting this feature, and we explore what we think a most natural

---

[2] Some philosophers have proposed a stronger notion of intervention that can break metaphysical dependencies (e.g., [22]), but we will not consider that in this paper.

way in this paper. Define an *("ontological") parent* of $Y \in \mathbf{V}_h$ as a variable (in $\mathbf{V}_b$) that directly contributes to ontologically determining Y. That is, we say X is a parent of Y if there is a value configuration $\mathbf{z}$ of $\mathbf{V}_b \setminus \{X\}$ such that some change in the value of X is associated with a change in the value of Y relative to $\mathbf{z}$, or specifically, $f_Y(\mathbf{z}, x) \neq f_Y(\mathbf{z}, x')$ where $x \neq x' \in \mathcal{R}(X)$. For each $Y \in \mathbf{V}_h$, we use $\mathbf{Pa}(Y)$ to denote all the parents of Y.

We propose to understand an intervention on $Y \in \mathbf{V}_h$ as amounting to interventions on its parents. Specifically, an intervention to fix $Y = y$ for $Y \in \mathbf{V}_h$ and $y \in \mathcal{R}(Y)$ is represented by a set of submodels, each of them, $M_{\mathbf{Pa}(Y)=\mathbf{pa_i}}$, satisfies that $y = f_Y(\mathbf{pa_i}, \mathbf{z})$ for any $\mathbf{z} \in \mathcal{R}(\mathbf{V}_b \setminus \mathbf{Pa}(Y))$. More generally, for any set $\mathbf{Y} \subseteq \mathbf{V}_b \cup \mathbf{V}_h$, an intervention to set $\mathbf{Y} = \mathbf{y}$ is represented as a set of submodels, each of them, $M_{\mathbf{X}=\mathbf{x}}$ in which $\mathbf{X} = (\mathbf{Y} \cap \mathbf{V}_b) \cup \mathbf{Pa}(\mathbf{Y} \cap \mathbf{V}_h)$, satisfies that $\mathbf{X} = \mathbf{x}$ is consistent with $\mathbf{Y} = \mathbf{y}$ and the metaphysical equations of $\mathbf{Y} \cap \mathbf{V}_h$. Note that this set of submodels may be empty, which means that the intervention in question is "metaphysically impossible". The simplest example is of course when the intervention requires a simultaneous change in a higher-level variable and in its (bottom-level) parents in a way that is inconsistent with a metaphysical equation.

Recall that in standard causal models, the modifiability of structural equations is an important device for encoding asymmetry. A structural equation like $Y = X$ may look symmetric between X and Y, but the equation's holding under an intervention of X and breaking under that of Y reflects the asymmetry. In the generalization proposed above, the asymmetry between higher-level variables and their bottom-level determiners is not encoded as such, as metaphysical equations are not modifiable. The asymmetry is instead built in through the distinction of levels and by taking interventions on higher-level variables as (usually multiply) realized by interventions on bottom-level variables.

Consider now statements of the form: if $\mathbf{Y}$ were (had been) intervened to take value setting $\mathbf{y}$, then it would be (have been) the case that φ. Call such sentences *causal counterfactuals*, which will be symbolized as $\mathbf{Y} = \mathbf{y} \; \square\!\!\rightarrow \varphi$ (as a shorthand for $(Y_1 = y_1 \wedge ... \wedge Y_k = y_k) \; \square\!\!\rightarrow \varphi$ where $Y_1, ..., Y_k$ are distinct members of $\mathbf{Y}$). Given the proposed understanding of interventions relative to a bipartite model, a natural semantics for such causal counterfactuals is: $\mathbf{Y} = \mathbf{y} \; \square\!\!\rightarrow \varphi$ is true in a causal setting $\langle M, \mathbf{u} \rangle$ if and only if for each submodel $M_{\mathbf{X}=\mathbf{x}}$ in which $\mathbf{X} = (\mathbf{Y} \cap \mathbf{V}_b) \cup \mathbf{Pa}(\mathbf{Y} \cap \mathbf{V}_h)$ and $\mathbf{X} = \mathbf{x}$ is consistent with $\mathbf{Y} = \mathbf{y}$, φ is satisfied by the solution of $M_{\mathbf{X}=\mathbf{x}}$. A causal counterfactual is vacuously true if the antecedent describes a metaphysically impossible intervention.

Given the proposed tracing of interventions on higher-level variables to relevant bottom-level variables, this understanding of causal counterfactuals that involve higher-level variables is formally similar to what is known as backtracking counterfactuals (for treatments of backtracking counterfactuals using causal models, see, e.g., [9,10,14]), as our semantics for counterfactuals whose antecedents involve higher-level variables is effectively "downtracking".[3] We will leave it to future work to compare "downtracking" counterfactuals to "backtrack-

---

[3] We thank a reviewer for noting the formal similarity to backtracking counterfactuals.

ing" ones, but it is worth pointing out a possible difference that is potentially significant. Intuitively, and as Lewis ([15]) famously emphasized, backtracking counterfactuals are not suitable for defining causation, due to obvious worries about confounding. However, the jury is still out on the suitability of downtracking counterfactuals for defining causation. Influential interventionist treatments of mental causation (e.g., [27]) suggest that they are, though this remains a matter of controversy in the philosophical literature. We are inclined to agree that high-level causation, if it makes sense, must embrace appropriate kinds of downtracking counterfactual as reflecting causal dependence, but the details will need to be supplied in future work.

Here is a toy example to illustrate the generalization. Consider the (largely stipulated) scenario that appears frequently in the discussion of Jaegwon Kim's causal exclusion argument about mental causation ([12]): a physical state is causally sufficient for another physical state, and each physical state ontologically realizes a mental state that is multiply realizable in that sense that it may be realized by some other physical states. The simplest bipartite model to incorporate these features is the following. Let $\mathbf{U} = \{U\}$, $\mathbf{V}_b = \{P_1, P_2\}$, $\mathbf{V}_h = \{M_1, M_2\}$, $\mathcal{R}(U) = \mathcal{R}(P_1) = \mathcal{R}(P_2) = \{0, 1, 2, 3\}$ and $\mathcal{R}(M_1) = \mathcal{R}(M_2) = \{0, 1\}$; The nomological equations at the bottom-level are: $P_1 =_n U$ and $P_2 =_n P_1$, and the metaphysical equations are

$$M_1 =_m \begin{cases} 0 & \text{if } P_1 = 0 \text{ or } 1 \\ 1 & \text{if } P_1 = 2 \text{ or } 3 \end{cases}$$

and

$$M_2 =_m \begin{cases} 0 & \text{if } P_2 = 0 \text{ or } 1 \\ 1 & \text{if } P_2 = 2 \text{ or } 3 \end{cases}.$$

The overall parental structure among the endogenous variables may be represented by the following directed acyclic graph, where the horizontal arrow corresponds to a nomological equation and the vertical arrows correspond to metaphysical equations.

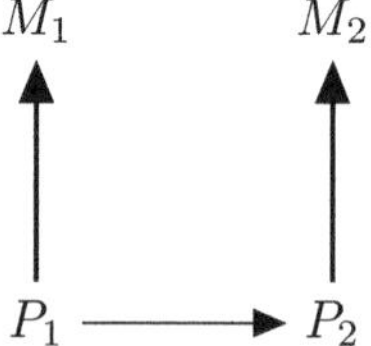

Consider now an intervention to force $M_1 = 0$. In our proposal, this is represented by the set of submodels $\{M_{P_1=0}, M_{P_1=1}\}$. Since both satisfy that $M_2 = 0$ (relative to any context), the causal counterfactual $M_1 = 0 \,\square\!\!\rightarrow M_2 = 0$ is true in the model (relative to any context). Similarly for $M_1 = 1 \,\square\!\!\rightarrow M_2 = 1$. There is thus counterfactual dependence of different values of M_2 on different values of M_1 in David Lewis [15]'s sense, which implies that a counterfactual definition of (actual) causation can vindicate mental causation with this model. As noted earlier, whether such "downtracking" counterfactuals can be relied on for defining or revealing causal dependence is a matter of debate, but it is at least a promising approach to make sense of mental or other kinds of high-level causation. We can also consider more complex interventions, such as an intervention to force $M_1 = 1 \wedge P_2 = 0$. This would be represented by the set of submodels $\{M_{P_1=2,P_2=0}, M_{P_1=3,P_2=0}\}$. As a result, the model satisfies the counterfactual $(M_1 = 1 \wedge P_2 = 0) \,\square\!\!\rightarrow M_2 = 0$.

3   Axiomatization of Reasoning About Counterfactuals

In this section, we present an axiomatization of the counterfactual logic resulting from the proposed interpretation for causal counterfactuals in bipartite models. For simplicity, we assume the signature S satisfies that $\mathbf{V}_b$ is finite and $\mathcal{R}(X)$ is finite for every $X \in \mathbf{V}_b \dot{\cup} \mathbf{V}_h$. Relative to S, a causal counterfactual formula is of the form $\mathbf{Y} = \mathbf{y} \,\square\!\!\rightarrow \varphi$, where $\mathbf{Y} = \mathbf{y}$ stands for a formula of the form $(Y_1 = y_1 \wedge ... \wedge Y_k = y_k) \,\square\!\!\rightarrow \varphi$ where $Y_1, ..., Y_k$ are distinct, $Y_i \in \mathbf{V}_b \dot{\cup} \mathbf{V}_h$, $y_i \in \mathcal{R}(Y_i)$, and φ is a Boolean combination of formulas of the form $X = x$ $(X \in \mathbf{V}_b \dot{\cup} \mathbf{V}_h, x \in \mathcal{R}(X))$. The special case when $k = 0$ is simply φ. Following [7], we consider a language relative to S, denoted as $\mathcal{L}(S)$, which consists of Boolean combinations of causal counterfactual formulas. As described previously, the key semantic rules are:

$\langle M, \mathbf{u} \rangle \vDash X = x$ iff X takes value x in the solution of M relative to $\mathbf{u}$;

$\langle M, \mathbf{u} \rangle \vDash \mathbf{Y} = \mathbf{y} \,\square\!\!\rightarrow \varphi$ iff for each submodel $M_{\mathbf{X}=\mathbf{x}}$ such that $\mathbf{X} = (\mathbf{Y} \cap \mathbf{V}_b) \cup \mathbf{Pa}(\mathbf{Y} \cap \mathbf{V}_h)$ and $\mathbf{X} = \mathbf{x}$ is consistent with $\mathbf{Y} = \mathbf{y}$, $\langle M_{\mathbf{X}=\mathbf{x}}, \mathbf{u} \rangle \vDash \varphi$.

The rules for Boolean combinations of formulas are as usual. In the following axiomatic system AX_S, the semantics of counterfactuals is captured by D9. To represent a parent set of an endogenous variable in $\mathbf{V}_h$, we write $Y \mapsto X$ $(Y \in \mathbf{V}_b, X \in \mathbf{V}_h)$ to denote that Y is a parent of X, which is an abbreviation of $\bigvee_{\mathbf{z} \in \mathcal{R}(\mathbf{Z}), y \neq y' \in \mathcal{R}(Y), x \neq x' \in \mathcal{R}(X)} ((\mathbf{Z} = \mathbf{z} \wedge Y = y) \,\square\!\!\rightarrow X = x \wedge (\mathbf{Z} = \mathbf{z} \wedge Y = y') \,\square\!\!\rightarrow X = x')$ where $\mathbf{Z} = \mathbf{V}_b \setminus \{X, Y\}$.[4] We then present the following axiom schemata and the rule MP as follows.

D0 All instances of propositional tautologies
D1 $\bigvee_{x \in \mathcal{R}(X)} \mathbf{Y} = \mathbf{y} \,\square\!\!\rightarrow X = x$, where $\mathbf{Y} \subseteq \mathbf{V}_b$
D2 $\mathbf{Y} = \mathbf{y} \,\square\!\!\rightarrow X = x \Rightarrow \neg\mathbf{Y} = \mathbf{y} \,\square\!\!\rightarrow X = x'$, where $x \neq x'$
D3 $\mathbf{Y} = \mathbf{y} \,\square\!\!\rightarrow (X = x \wedge \alpha) \Rightarrow (\mathbf{Y} = \mathbf{y} \wedge X = x) \,\square\!\!\rightarrow \alpha$, where $X \in \mathbf{V}_b$
D4 $(\mathbf{Y} = \mathbf{y} \wedge X = x) \,\square\!\!\rightarrow X = x$
D5 $(\mathbf{Y} = \mathbf{y} \,\square\!\!\rightarrow Z = z) \wedge ((\mathbf{Y} = \mathbf{y} \wedge Z = z) \,\square\!\!\rightarrow X = x) \Rightarrow (\mathbf{Y} = \mathbf{y} \,\square\!\!\rightarrow X = x)$
D6 $((\mathbf{Y} = \mathbf{y} \wedge X = x) \,\square\!\!\rightarrow Z = z) \wedge ((\mathbf{Y} = \mathbf{y} \wedge Z = z) \,\square\!\!\rightarrow X = x) \Rightarrow \mathbf{Y} = \mathbf{y} \,\square\!\!\rightarrow X = x$, where $\mathbf{Y} \cup \{X, Z\} \subseteq \mathbf{V}_b$, and X and Z are distinct
D7 $\mathbf{Y} = \mathbf{y} \,\square\!\!\rightarrow \neg\alpha \Leftrightarrow \neg(\mathbf{Y} = \mathbf{y} \,\square\!\!\rightarrow \alpha)$, where $\mathbf{Y} \subseteq \mathbf{V}_b$
D8 $\mathbf{Y} = \mathbf{y} \,\square\!\!\rightarrow (\alpha \wedge \beta) \Leftrightarrow (\mathbf{Y} = \mathbf{y} \,\square\!\!\rightarrow \alpha) \wedge (\mathbf{Y} = \mathbf{y} \,\square\!\!\rightarrow \beta)$

---

[4] This formula is inspired by the formula used to represent "Y affects X" in [8]. The difference is that in our formula $\mathbf{Z}$ doesn't include $\mathbf{U}$. This is because exogenous variables cannot directly affect $\mathbf{V}_h$ variables in our setup.

D9 $(\mathbf{Y}_1 = \mathbf{y}_1 \wedge \mathbf{Y}_2 = \mathbf{y}_2) \;\Box\!\!\to\; \alpha \Leftrightarrow \bigvee_{\mathbf{P} \subseteq \mathbf{V}_b}((\bigwedge_{P \in \mathbf{P}} \bigvee_{Y \in \mathbf{Y}_2} P \mapsto Y) \wedge (\neg \bigvee_{P' \in \mathbf{V}_b \backslash \mathbf{P}} \bigvee_{Y \in \mathbf{Y}_2} P' \mapsto Y) \wedge \bigwedge_{\mathbf{p} \in \mathcal{R}(\mathbf{P})}(\bar{\mathbf{P}} = \mathbf{p} \;\Box\!\!\to\; \mathbf{Y}_2 = \mathbf{y}_2 \Rightarrow (\mathbf{Y}_1 = \mathbf{y}_1 \wedge \mathbf{P} = \mathbf{p}) \;\Box\!\!\to\; \alpha))$, where $\mathbf{Y}_1 \subseteq \mathbf{V}_b$, $\mathbf{Y}_2 \subseteq \mathbf{V}_h$

MP From $\vdash \varphi \Rightarrow \psi$ and $\vdash \varphi$, infer $\vdash \psi$

AX_S is analogous to Halpern [7, pp. 325-6]'s axiomatic system for causal models with the "unique solution" property. D1-D4 are similar to Halpern's axioms "definiteness", "functionality", "composition" and "effectiveness", respectively, and D7 and D8 are also needed for proving completeness in Halpern's system.

D3 has the restriction that "X" in the axiom is a bottom-level variable. The unrestricted formulation of D3 is not sound. D5 is the "cut" axiom and is needed here due to a special feature of the bipartite model.[5] D6 is known as the "reversibility" axiom in Halpern's system. D9 expresses the semantics for causal counterfactuals relative to bipartite models. This axiom translates the truth condition of $(\mathbf{Y}_1 = \mathbf{y}_1 \wedge \mathbf{Y}_2 = \mathbf{y}_2) \;\Box\!\!\to\; \alpha$ $(\mathbf{Y}_1 \subseteq \mathbf{V}_b, \mathbf{Y}_2 \subseteq \mathbf{V}_h)$ in bipartite models as follows: there is a parent set $\mathbf{P}$ for $\mathbf{Y}_2$ such that each value setting $\mathbf{p}$ of $\mathbf{P}$ that can realize $\mathbf{Y}_2 = \mathbf{y}_2$ satisfies that α can be derived under intervention $\mathbf{P} = \mathbf{p}$ and $\mathbf{Y}_1 = \mathbf{y}_1$.

As an axiom, D9 is cumbersome. There are probably more elegant axiomatizations to be discovered. However, although D9 greatly facilitates the completeness proof below, and the basic proof strategy follows Halpern's seminal work, the completeness result is not simply a corollary of Halpern's results or other completeness results in the literature. The extra need to use D5 is an illustration.[6]

Theorem 1. *AX_S is sound and complete with respect to $\mathcal{M}_S$.*

Proof. The soundness of the axioms is easy to verify. For completeness, we adopt the canonical model approach. To show that a valid formula is provable, we show that an AX_S-consistent formula φ can be satisfied in a canonical model. φ can be extended into a maximally AX_S-consistent set C. We use this set to construct a canonical model M^C as follows. For any $X \in \mathbf{V}_b \dot\cup \mathbf{V}_h$ and $\mathbf{u} \in R(\mathbf{U})$, $f_X^C(\mathbf{y}, \mathbf{u}) = x$ if $\mathbf{Y} = \mathbf{y} \;\Box\!\!\to\; X = x \in C$ where $\mathbf{Y} = \mathbf{V}_b \backslash \{X\}$, which is guaranteed by the fact that D1 and D2 are in C.

We need to show that M^C has the unique-solution property, that is, $M^C_{\mathbf{Y}=\mathbf{y}}$ has a unique solution under any context $\mathbf{u}$ for any intervention $\mathbf{Y} = \mathbf{y}$ where $\mathbf{Y} \subseteq \mathbf{V}_b$. Then we can prove that $\psi \in C$ iff $\langle M^C, \mathbf{u} \rangle \vDash \psi$ for any $\psi \in \mathcal{L}(S)$ and any context $\mathbf{u}$. We first show the following claims step by step: 1) for any $\mathbf{Y} \subseteq \mathbf{V}_b$, $\mathbf{Y} = \mathbf{y} \;\Box\!\!\to\; X = x \in C$ iff $X = x$ in all solutions for $\langle M^C_{\mathbf{Y}=\mathbf{y}}, \mathbf{u} \rangle$, and $\langle M^C_{\mathbf{Y}=\mathbf{y}}, \mathbf{u} \rangle$ has a unique solution by induction on $|\mathbf{V}_b \backslash \mathbf{Y}|$; 2) for any $\mathbf{Y} \subseteq \mathbf{V}_b$, $\mathbf{Y} = \mathbf{y} \;\Box\!\!\to\; \alpha \in C$ iff α is true in $\langle M^C_{\mathbf{Y}=\mathbf{y}}, \mathbf{u} \rangle$; 3) for any Boolean combination ψ'

---

[5] In the completeness proof, D5 is used in the truth lemma for the case of $\mathbf{Y} = \mathbf{y} \;\Box\!\!\to\; X = x$ in which $X \in \mathbf{V}_h$ and $|\mathbf{V}_b \backslash \mathbf{Y}| = 1$.

[6] We thank a reviewer for raising this question.

of $\mathbf{Y} = \mathbf{y} \,\Box\!\!\rightarrow \alpha$ ($\mathbf{Y} \subseteq \mathbf{V}_b$), $\psi' \in C$ iff ψ' is true in $\langle M^C_{\mathbf{Y}=\mathbf{y}}, \mathbf{u} \rangle$ by induction on the structure of ψ'.

$|\mathbf{V}_b \setminus \mathbf{Y}| = 0$, as the value of $\mathbf{V}_h$ is determined by $\mathbf{V}_b$, assume that $\mathbf{v}_h$ is the value of $\mathbf{V}_h$ derived by $\mathbf{Y} = \mathbf{y}$. The model $M^C_{\mathbf{Y}=\mathbf{y}}$ has a unique solution $(\mathbf{y}, \mathbf{v}_h)$ under $\mathbf{u}$. Suppose $\mathbf{Y} = \mathbf{y} \,\Box\!\!\rightarrow X = x \in C$. If $X \in \mathbf{V}_b$, then $X = x$ is consistent with $\mathbf{Y} = \mathbf{y}$ as any instances of D2 and D4 are in C. Thus $X = x$ is in the solution $(\mathbf{y}, \mathbf{v}_h)$. If $X \in \mathbf{V}_h$, according to the definition of f^C, X has value x in the solution $(\mathbf{y}, \mathbf{v}_h)$. Conversely, if $(\mathbf{y}, \mathbf{v}_h)$ is consistent with $X = x$, we have $\mathbf{Y} = \mathbf{y} \,\Box\!\!\rightarrow X = x \in C$ based on the fact that D4 is in C and the definition of f^C.

$|\mathbf{V}_b \setminus \mathbf{Y}| = 1$, assume that Z is the variable in $\mathbf{V}_b \setminus \mathbf{Y}$, z is the value of Z determined by $\mathbf{Y} = \mathbf{y}$ and $\mathbf{u}$, and $\mathbf{v}_h$ is the value of $\mathbf{V}_h$ determined by $\mathbf{Y} = \mathbf{y}$ and $Z = z$ in M^C. Then the unique solution of $\langle M^C_{\mathbf{Y}=\mathbf{y}}, \mathbf{u} \rangle$ is $(\mathbf{y}, z, \mathbf{v}_h)$. Suppose $\mathbf{Y} = \mathbf{y} \,\Box\!\!\rightarrow X = x \in C$. If $X \in \mathbf{Y}$, $X = x$ is consistent with $(\mathbf{y}, z, \mathbf{v}_h)$ as D2 and D4 are in C. If X is Z, according to the definition of f^C, $\mathbf{Y} = \mathbf{y} \,\Box\!\!\rightarrow Z = z \in C$, then $X = x$ is the same as $Z = z$ as D2 is in C. If $X \in \mathbf{V}_h$, according to the definition of f^C, $(\mathbf{Y} = \mathbf{y} \wedge Z = z) \,\Box\!\!\rightarrow X = x' \in C$ where x' is the value of X in $\mathbf{v}_h$. As $\mathbf{Y} = \mathbf{y} \,\Box\!\!\rightarrow Z = z \in C$ and instances of D8 are in C, we have $\mathbf{Y} = \mathbf{y} \,\Box\!\!\rightarrow (X = x \wedge Z = z) \in C$. Due to D3, $(\mathbf{Y} = \mathbf{y} \wedge Z = z) \,\Box\!\!\rightarrow X = x \in C$. Then because of D2, x and x' must be the same. So in the unique solution $(\mathbf{y}, z, \mathbf{v}_h)$, X takes value x.

Conversely, we suppose that X takes x in $(\mathbf{y}, z, \mathbf{v}_h)$. If $X \in \mathbf{Y}$, then $\mathbf{Y} = \mathbf{y} \,\Box\!\!\rightarrow X = x \in C$ as any instance of D4 is in C. When X is Z, as $Z = z$ is derived by the fact that $\mathbf{Y} = \mathbf{y} \,\Box\!\!\rightarrow Z = z \in C$, so $\mathbf{Y} = \mathbf{y} \,\Box\!\!\rightarrow X = x \in C$. When $X \in \mathbf{V}_h$, as $(\mathbf{Y} = \mathbf{y} \wedge Z = z) \,\Box\!\!\rightarrow X = x \in C$ and $\mathbf{Y} = \mathbf{y} \,\Box\!\!\rightarrow Z = z \in C$, due to D5, we have $\mathbf{Y} = \mathbf{y} \,\Box\!\!\rightarrow X = x \in C$.

$|\mathbf{V}_b \setminus \mathbf{Y}| = k$ ($k \geq 2$), as D1 and D2 are in C, we can determine the value of every endogenous variable after the intervention $\mathbf{Y} = \mathbf{y}$. Assume $\mathbf{v}$ is the determined value setting, we first show that $\mathbf{v}$ is a solution of $M^C_{\mathbf{Y}=\mathbf{y}}$ under $\mathbf{u}$. According to D8, we have $\mathbf{Y} = \mathbf{y} \,\Box\!\!\rightarrow \mathbf{V}_{b\mathbf{Y}} = \mathbf{v} \in C$ where $\mathbf{V}_{b\mathbf{Y}}$ is the set $\mathbf{V}_b$ excluding $\mathbf{Y}$. Because of D3, for any $Z \in \mathbf{V}_{b\mathbf{Y}}$, we have $(\mathbf{Y} = \mathbf{y} \wedge \mathbf{V}_{b\mathbf{Y} \cup \{Z\}} = \mathbf{v}) \,\Box\!\!\rightarrow Z = z$ where $\mathbf{V}_{b\mathbf{Y} \cup \{Z\}}$ is the set $\mathbf{V}_b$ excluding $\mathbf{Y}$ and Z, and z is the value of Z in $\mathbf{v}$. For any $W \in \mathbf{V}_h$, similarly we have $(\mathbf{Y} = \mathbf{y} \wedge \mathbf{V}_{b\mathbf{Y}} = \mathbf{v}) \,\Box\!\!\rightarrow W = w$ where w is the value of W in $\mathbf{v}$. According to the definition of f^C, $\mathbf{v}$ solves all the equations in $M^C_{\mathbf{Y}=\mathbf{y}}$ under $\mathbf{u}$.

Assume that there is another solution $\mathbf{v}'$, then there must be some variable $T \in \mathbf{V}_b \setminus \mathbf{Y}$ whose value t in $\mathbf{v}$ is different from its value t' in $\mathbf{v}'$. We can find that T in $\mathbf{V}_b \setminus \mathbf{Y}$ because there must be some variables taking different values in $\mathbf{v}$ and $\mathbf{v}'$ once a variable in $\mathbf{V}_h$ taking different values in the two solutions. As $\mathbf{Y} = \mathbf{y} \,\Box\!\!\rightarrow T = t$ and any instance of D2 are in C, $\mathbf{Y} = \mathbf{y} \,\Box\!\!\rightarrow T = t' \notin C$. Let Q be another variable in $\mathbf{V}_b \setminus \mathbf{Y}$ and q be its value in $\mathbf{v}'$. $\mathbf{v}'$ is a solution of $M^C_{\mathbf{Y}=\mathbf{y} \wedge Q=q}$ under $\mathbf{u}$, and $M^C_{\mathbf{Y}=\mathbf{y} \wedge Q=q}$ has a unique solution under $\mathbf{u}$ according to inductive hypothesis. So $\mathbf{v}'$ is the unique solution of $M^C_{\mathbf{Y}=\mathbf{y} \wedge Q=q}$ under $\mathbf{u}$, it follows that $T = t'$ in the solution of $M^C_{\mathbf{Y}=\mathbf{y} \wedge Q=q}$ under $\mathbf{u}$. Thus, we have

$(\mathbf{Y} = \mathbf{y} \wedge Q = q) \;\Box\!\!\rightarrow\; T = t' \in C$. Similarly, $(\mathbf{Y} = \mathbf{y} \wedge T = t') \;\Box\!\!\rightarrow\; Q = q \in C$. Due to D6, we have $\mathbf{Y} = \mathbf{y} \;\Box\!\!\rightarrow\; T = t' \in C$, contradiction.

Suppose that $\mathbf{Y} = \mathbf{y} \;\Box\!\!\rightarrow\; X = x \in C$, x is the value of X in $\mathbf{v}$, so $X = x$ in the solution for $\langle M^{C}_{\mathbf{Y}=\mathbf{y}}, \mathbf{u}\rangle$. Conversely, if $X = x$ in $\mathbf{v}$, then $\mathbf{Y} = \mathbf{y} \;\Box\!\!\rightarrow\; X = x \in C$ according to the definition of $\mathbf{v}$.

The cases of $\mathbf{Y} = \mathbf{y} \;\Box\!\!\rightarrow\; \neg\alpha'$ and $\mathbf{Y} = \mathbf{y} \;\Box\!\!\rightarrow\; (\alpha_1 \wedge \alpha_2)$ are easy to prove by using axioms D7 and D8 respectively, together with the inductive hypothesis and the property of maximally AX_S-consistent sets. Therefore, we have $\mathbf{Y} = \mathbf{y} \;\Box\!\!\rightarrow\; \alpha$ iff $\langle M^{C}, \mathbf{u}\rangle \vDash \mathbf{Y} = \mathbf{y} \;\Box\!\!\rightarrow\; \alpha$ for $\mathbf{Y} \subseteq \mathbf{V}_b$.

The cases of Boolean combinations of $\mathbf{Y} = \mathbf{y} \;\Box\!\!\rightarrow\; \alpha$ ($\mathbf{Y} \subseteq \mathbf{V}_b$) are easy to verify according to the property of maximally AX_S-consistent sets and inductive hypothesis.

Now we prove that $\psi \in C$ iff $\langle M^{C}, \mathbf{u}\rangle \vDash \psi$ for any $\psi \in \mathcal{L}(S)$ and context $\mathbf{u}$ by induction on the structure of ψ.

We first show the case of $(\mathbf{Y}_1 = \mathbf{y}_1 \wedge \mathbf{Y}_2 = \mathbf{y}_2) \;\Box\!\!\rightarrow\; \alpha$ in which $\mathbf{Y}_1 \subseteq \mathbf{V}_b$, $\mathbf{Y}_2 \subseteq \mathbf{V}_h$. Suppose it is in C, according to D9, we have $\bigvee_{\mathbf{P} \subseteq \mathbf{V}_b} ((\bigwedge_{P \in \mathbf{P}} \bigvee_{Y \in \mathbf{Y}_2} P \mapsto Y) \wedge (\neg \bigvee_{P' \in \mathbf{V}_b \backslash \mathbf{P}} \bigvee_{Y \in \mathbf{Y}_2} P' \mapsto Y) \wedge \bigwedge_{\mathbf{p} \in \mathcal{R}(\mathbf{P})} (\mathbf{P} = \mathbf{p} \;\Box\!\!\rightarrow\; \mathbf{Y}_2 = \mathbf{y}_2 \Rightarrow (\mathbf{Y}_1 = \mathbf{y}_1 \wedge \mathbf{P} = \mathbf{p}) \;\Box\!\!\rightarrow\; \alpha)) \in C$. According to the above results, $\langle M^{C}_{\mathbf{Y}=\mathbf{y}}, \mathbf{u}\rangle \vDash \bigvee_{\mathbf{P} \subseteq \mathbf{V}_b} ((\bigwedge_{P \in \mathbf{P}} \bigvee_{Y \in \mathbf{Y}_2} P \mapsto Y) \wedge (\neg \bigvee_{P' \in \mathbf{V}_b \backslash \mathbf{P}} \bigvee_{Y \in \mathbf{Y}_2} P' \mapsto Y) \wedge \bigwedge_{\mathbf{p} \in \mathcal{R}(\mathbf{P})} (\mathbf{P} = \mathbf{p} \;\Box\!\!\rightarrow\; \mathbf{Y}_2 = \mathbf{y}_2 \Rightarrow (\mathbf{Y}_1 = \mathbf{y}_1 \wedge \mathbf{P} = \mathbf{p}) \;\Box\!\!\rightarrow\; \alpha))$. That means there is a parent set $\mathbf{P}$ of $\mathbf{Y}_2$ such that for any value setting $\mathbf{p}$ of $\mathbf{P}$, if $\mathbf{P} = \mathbf{p}$ derives $\mathbf{Y}_2 = \mathbf{y}_2$, $\langle M^{C}_{\mathbf{Y}_1=\mathbf{y}_1 \wedge \mathbf{P}=\mathbf{p}}, \mathbf{u}\rangle \vDash \alpha$. Therefore, $\langle M^{C}, \mathbf{u}\rangle \vDash (\mathbf{Y}_1 = \mathbf{y}_1 \wedge \mathbf{Y}_2 = \mathbf{y}_2) \;\Box\!\!\rightarrow\; \alpha$. Conversely, suppose $\langle M^{C}, \mathbf{u}\rangle \vDash (\mathbf{Y}_1 = \mathbf{y}_1 \wedge \mathbf{Y}_2 = \mathbf{y}_2) \;\Box\!\!\rightarrow\; \alpha$, according to the soundness of AX_S, $\langle M^{C}, \mathbf{u}\rangle \vDash \bigvee_{\mathbf{P} \subseteq \mathbf{V}_b} ((\bigwedge_{P \in \mathbf{P}} \bigvee_{Y \in \mathbf{Y}_2} P \mapsto Y) \wedge (\neg \bigvee_{P' \in \mathbf{V}_b \backslash \mathbf{P}} \bigvee_{Y \in \mathbf{Y}_2} P' \mapsto Y) \wedge \bigwedge_{\mathbf{p} \in \mathcal{R}(\mathbf{P})} (\mathbf{P} = \mathbf{p} \;\Box\!\!\rightarrow\; \mathbf{Y}_2 = \mathbf{y}_2 \Rightarrow (\mathbf{Y}_1 = \mathbf{y}_1 \wedge \mathbf{P} = \mathbf{p}) \;\Box\!\!\rightarrow\; \alpha))$. It follows that $\bigvee_{\mathbf{P} \subseteq \mathbf{V}_b} ((\bigwedge_{P \in \mathbf{P}} \bigvee_{Y \in \mathbf{Y}_2} P \mapsto Y) \wedge (\neg \bigvee_{P' \in \mathbf{V}_b \backslash \mathbf{P}} \bigvee_{Y \in \mathbf{Y}_2} P' \mapsto Y) \wedge \bigwedge_{\mathbf{p} \in \mathcal{R}(\mathbf{P})} (\mathbf{P} = \mathbf{p} \;\Box\!\!\rightarrow\; \mathbf{Y}_2 = \mathbf{y}_2 \Rightarrow (\mathbf{Y}_1 = \mathbf{y}_1 \wedge \mathbf{P} = \mathbf{p}) \;\Box\!\!\rightarrow\; \alpha)) \in C$. As any instance of D9 is in C, thus $(\mathbf{Y}_1 = \mathbf{y}_1 \wedge \mathbf{Y}_2 = \mathbf{y}_2) \;\Box\!\!\rightarrow\; \alpha \in C$.

According to the property of maximally AX_S-consistent sets and inductive hypothesis, the Boolean combinations of $(\mathbf{Y}_1 = \mathbf{y}_1 \wedge \mathbf{Y}_2 = \mathbf{y}_2) \;\Box\!\!\rightarrow\; \alpha$ are easy to verify.

4　Potential Advantages of Bipartite Models

As mentioned at the beginning, our proposal is closely related to Beckers et al. [1]'s causal models with non-causal constraints. In a sense, the proposed bipartite models can be seen as a special case of causal models with constraints, where the constraints are encoded by what we call metaphysical equations. Specifically, given a bipartite model $M = \langle S, F_n \dot{\cup} F_m\rangle$ where $S = \langle \mathbf{U}, \mathbf{V}_b \dot{\cup} \mathbf{V}_h, \mathcal{R}\rangle$, the corresponding causal model with constraints T is a triple $\langle S, F_n, \mathcal{C}\rangle$ in which $\mathcal{C}$ is the set of all value settings of all variables that are consistent with the metaphysical equations. Beckers et al. call a value setting of all variables an "extended state", denoted as $(\mathbf{u}, \mathbf{v})$. Interventions on endogenous variables in T are defined as usual, but a new manipulation operation *disconnects* is introduced to disassociate an endogenous variable from its equations. In our language of leveled variables, this new operation makes it possible that the parents of a higher-level variable that is intervened on could take whatever values to maintain the constraints between them without following their equations. The corresponding counterfactuals have the form $(disc(\mathbf{X}) \wedge \mathbf{Y} = \mathbf{y}) \,\square\!\!\rightarrow \varphi$, in which $\mathbf{X}$ and $\mathbf{Y}$ are disjoint. The relevant semantic rules relative to a causal model with constraints T are the following:

$\langle T, \mathbf{u} \rangle \Vdash \mathbf{Y} = \mathbf{y} \,\square\!\!\rightarrow \varphi$ iff $\langle T_{\mathbf{Y}=\mathbf{y}}, \mathbf{u}, \mathbf{v} \rangle \Vdash \varphi$ for all $\mathbf{v}$ such that $(\mathbf{u}, \mathbf{v}) \in \mathcal{C}$ and $(\mathbf{u}, \mathbf{v})$ satisfies the causal equations in $T_{\mathbf{Y}=\mathbf{y}}$;

$\langle T, \mathbf{u} \rangle \Vdash (disc(\mathbf{X}) \wedge \mathbf{Y} = \mathbf{y}) \,\square\!\!\rightarrow \varphi$ iff $\langle T_{-\mathbf{X}}, \mathbf{u} \rangle \Vdash \mathbf{Y} = \mathbf{y} \,\square\!\!\rightarrow \varphi$ in which $T_{-\mathbf{X}}$ is the model after T removes the equations of the variables in $\mathbf{X}$.

If we model the example in Sect. 2 using causal models with constraints, the metaphysical equations are treated as constraints. That means M_1 and M_2 do not have equations and the constraint only contains the possible extended states that satisfy the constraints. Then, in context $U = 0$, for example, $M_1 = 1 \,\square\!\!\rightarrow M_2 = 1$ is *vacuously* true because there is no state that is consistent with the constraint about M_1 and at the same time satisfies the causal equation of P_1, whereas $M_1 = 0 \,\square\!\!\rightarrow M_2 = 0$ is non-vacuously true as the unique state $(U = 0, P_1 = 0, P_2 = 0, M_1 = 0, M_2 = 0)$ satisfies $M_2 = 0$. By contrast, in our bipartite model, both counterfactuals are non-vacuously true. This discrepancy indicates that what corresponds to "$M_1 = 1 \,\square\!\!\rightarrow M_2 = 1$" in our framework is actually "$(disc(P_1), M_1 = 1) \,\square\!\!\rightarrow M_2 = 1$" in Beckers et al.'s. In context $U = 0$, $(disc(P_1), M_1 = 1) \,\square\!\!\rightarrow M_2 = 1$ is non-vacuously true because after we remove the equation of P_1, there are two feasible states, $(U = 0, P_1 = 2, P_2 = 2, M_1 = 1, M_2 = 1)$ and $(U = 0, P_1 = 3, P_2 = 3, M_1 = 1, M_2 = 1)$, that satisfy the remaining equation and the constraints, and are both consistent with $M_2 = 1$.

In general, for a causal counterfactual $(\mathbf{Y}_1 = \mathbf{y}_1 \wedge \mathbf{Y}_2 = \mathbf{y}_2) \,\square\!\!\rightarrow \alpha$ where $\mathbf{Y}_1 \subseteq \mathbf{V}_b, \mathbf{Y}_2 \subseteq \mathbf{V}_h$, the corresponding counterfactual in the framework of causal models with constraints is of the form $(\mathbf{Y}_1 = \mathbf{y}_1 \wedge \mathbf{Y}_2 = \mathbf{y}_2 \wedge disc(\mathbf{Pa}(\mathbf{Y}_2))) \,\square\!\!\rightarrow \alpha$. This is one way in which our framework, though less general than Beckers et al.'s, simplifies the treatment of causal counterfactuals in the presence of ontological dependencies.

More important, by explicitly representing the asymmetric nature of onto-logical dependencies in the bipartite setup, our framework still allows a *directed* graphical representation of the overall dependency structure and so opens the door to adapting various powerful learning algorithms in the causal discovery literature for inferring directed graphical structures from data ([21]) to struc-tural learning in the context of bipartite causal models. In particular, given the unique solution property of a bipartite model, any joint probability distribution over the exogenous variables induces a joint probability distribution over the endogenous variables. Like for standard causal models, if we assume that no two variables in $\mathbf{V}_b$ shares an exogenous parent, that the exogenous variables are jointly probabilistically independent, and that the graphical representation of the overall dependency structure is acyclic, then the induced joint distribution over the endogenous variables will satisfy the Markov condition with the graphical representation. That is, our generalization of structural causal models also points to a generalization of causal Bayesian networks.

On the other hand, with causal counterfactuals in place, we can also explore various ways to generalize the existing definitions of actual causation based on standard causal models to definitions based on bipartite models, which are likely to be helpful in the philosophical debates over high-level causation. As the simple example in Sect. 2 illustrates, such definitions are likely to vindicate high-level causation in many cases.

5 Concluding Remarks

This paper proposes bipartite causal models to accommodate asymmetric onto-logical dependencies in causal modelling, and provides an axiomatization of reasoning about causal counterfactuals in the new framework. Compared to Beckers et al.'s elegant work on causal models with constraints ([1]), our proposal is less general but is simpler in its treatment of interventions on high-level variables and makes the asymmetric nature of ontological dependencies explicit, which is potentially advantageous for developing structural learning algorithms.

To generalize bipartite causal models further, we could consider three or more levels of variables so that the relations across multiple levels like physics-biology-psychology can be investigated in an integrated causal model. If we adopt a similar definition of interventions on higher-level variables, an intervention on any higher level may be transformed into interventions on the bottom level, and so a model with three or more levels will have a reduced model that takes the bipartite form. Therefore, an interesting question is under what conditions can features of the underlying, multi-level structure be recovered from a reduced bipartite form. Alternatively, we can investigate definitions of interventions on higher-level variables that cannot be reduced to interventions on the bottom level.

Another direction of further generalization is to allow nomological equations at higher levels or across levels. A major challenge in this direction is to find ways to maintain consistency of nomological relations at different levels. In this paper, we quarantined this challenge by placing all nomological equations at the bottom level, but it will need to be addressed if a more non-reductive framework proves to be useful.

As described in the previous section, it is easy to make bipartite models probabilistic by adding a joint probability distribution over exogenous variables. In such probabilistic models, the probabilistic effect of an intervention is in general set-valued. The calculation of *interventional probabilities* (of the form $P(\varphi_{\mathbf{Y}=\mathbf{y}})$ or $P(\mathbf{Y} = \mathbf{y} \; \Box\!\!\rightarrow \varphi))$ involves a set of probability quantities, namely $\{P(\varphi_{\mathbf{X}=\mathbf{x}}) \mid \mathbf{X} = (\mathbf{Y} \cap \mathbf{V}_b) \cup \mathbf{Pa}(\mathbf{Y} \cap \mathbf{V}_h) \text{ and } \mathbf{X} = \mathbf{x} \text{ is consistent with } \mathbf{Y} = \mathbf{y}\}$, in which each $P(\varphi_{\mathbf{X}=\mathbf{x}})$ is the sum of contexts that make $\mathbf{X} = \mathbf{x} \; \Box\!\!\rightarrow \varphi$ true. Likewise, the calculation of *counterfactual probabilities* (of the form $P(\varphi_{\mathbf{Y}=\mathbf{y}} \mid e)$ or $P(\mathbf{Y} = \mathbf{y} \; \Box\!\!\rightarrow \varphi \mid e))$ refers to quantities $P(\varphi_{\mathbf{X}=\mathbf{x}} \mid e)$ where $\mathbf{X} = (\mathbf{Y} \cap \mathbf{V}_b) \cup \mathbf{Pa}(\mathbf{Y} \cap \mathbf{V}_h)$ and $\mathbf{X} = \mathbf{x}$ is consistent with $\mathbf{Y} = \mathbf{y}$. This feature of probabilistic bipartite models will provide opportunities to explore how *imprecise probabilities* interact with counterfactuals and causation. It is also worthwhile to explore generalized causal Bayesian networks with ontological dependence in accordance with the idea of bipartite models. In the existing attempts to generalize causal Bayesian networks (e.g., [6,25]), probability quantities involving interventions are not explicitly defined, but they are necessary to make full use of *causal* Bayesian networks. Once our idea in this paper is extended to a probabilistic framework, it is likely to shed light on causal reasoning in Bayesian networks that include both causal and non-causal arrows.

Finally, an obvious line of further research is to develop and compare accounts of actual causation based on bipartite or multi-level causal models. Such an investigation promises to further clarify and contribute to resolving the philosophical debates over the adequacy of interventionism in accommodating high-level causation.

Acknowledgements. JF's research was supported in part by the National Social Science Foundation Key Program in China (No. 23&ZD240). JZ's research was supported in part by the Research Grants Council of Hong Kong (HSSPFS 34000324).

Disclosure of Interests. Neither author has any conflicts of interest to declare.

References

1. Beckers, S., Halpern, J., Hitchcock, C.: Causal models with constraints. In: Conference on Causal Learning and Reasoning, pp. 866–879. PMLR (2023)
2. Blanchard, T., Hüttemann, A.: Causal modeling in multilevel settings: a new proposal. Philos. Phenomenol. Res. **109**(2), 433–457 (2024)
3. Bliss, R., Trogdon, K.: Metaphysical Grounding. In: Zalta, E.N., Nodelman, U. (eds.) The Stanford Encyclopedia of Philosophy. Metaphysics Research Lab, Stanford University, Summer 2024 edn. (2024)
4. Chalupka, K., Eberhardt, F., Perona, P.: Causal feature learning: an overview. Behaviormetrika **44**, 137–164 (2017)
5. Fang, W., Zhang, J.: Proportionality, determinate intervention effects, and high-level causation. Erkenntnis (2024). https://doi.org/10.1007/s10670-024-00859-8
6. Gebharter, A.: Causal exclusion and causal Bayes nets. Philos. Phenomenol. Res. **95**(2), 353–375 (2017)
7. Halpern, J.Y.: Axiomatizing causal reasoning. J. Artif. Intell. Res. **12**, 317–337 (2000)
8. Halpern, J.Y.: Actual Causality. MIT Press, Cambridge, MA (2016)
9. Hao, G.Y., Zhang, J., Huang, B., Wang, H., Zhang, K.: Natural counterfactuals with necessary backtracking. In: Globerson, A., Mackey, L., Belgrave, D., Fan, A., Paquet, U., Tomczak, J., Zhang, C. (eds.) Advances in Neural Information Processing Systems. vol. 37, pp. 14962–14995. Curran Associates, Inc. (2024)
10. Hiddleston, E.: A causal theory of counterfactuals. Noûs **39**(4), 632–657 (2005)
11. Jansson, L.: When are structural equation models apt? Causation versus grounding. In: Reutlinger, A., Saatsi, J. (eds.) Explanation Beyond Causation: Philosophical Perspectives on Non-Causal Explanations, pp. 250–266. Oxford University Press (2018)

12. Kim, J.: Current Issues in Philosophy of Mind. Cambridge University Press, New York (1998)
13. Kim, J.: Mind in a Physical World: An Essay on the Mind-Body Problem and Mental Causation. MIT Press (1998)
14. Kügelgen, J., Mohamed, A., Beckers, S.: Backtracking counterfactuals. Proc. Mach. Learn. Res. **213**, 1–20 (2023)
15. Lewis, D.: Causation. J. Philos. **70**, 556–567 (1973)
16. McDonald, J.: Causal models and metaphysics - part 1: Using causal models. Philosophy Compass **19**(4) (2024). https://doi.org/10.1111/phc3.12975
17. McDonald, J.: Causal models and metaphysics–part 2: interpreting causal models. Philos. Compass **19**(7), e13007 (2024). https://doi.org/10.1111/phc3.13007
18. McLaughlin, B., Bennett, K.: Supervenience. In: Zalta, E.N., Nodelman, U. (eds.) The Stanford Encyclopedia of Philosophy. Metaphysics Research Lab, Stanford University, Winter 2023 edn. (2023)
19. Paul, L.A., Hall, N.: Causation: A User's Guide. Oxford University Press (2013)
20. Pearl, J.: Causality: Models, 2nd edn. Reasoning and Inference. Cambridge University Press, USA (2009)
21. Peters, J., Janzing, D., Schölkopf, B.: Elements of causal inference: foundations and learning algorithms. The MIT Press (2017)
22. Schaffer, J.: Grounding in the image of causation. Philos. Stud. **173**(1), 49–100 (2016)
23. Schölkopf, B., et al.: Toward causal representation learning. Proc. IEEE **109**(5), 612–634 (2021)
24. Spirtes, P., Glymour, C., Scheines, R.: Causation, Prediction, and Search, 2nd Edition (2000)
25. Stern, R., Eva, B.: Anti-reductionist interventionism. Br. J. Philos. Sci. **74**(1), 241–267 (2023)
26. Wilson, A.: Metaphysical causation. Noûs **52**(4), 723–751 (2018)
27. Woodward, J.: Interventionism and causal exclusion. Philos. Phenomenol. Res. **91**(2), 303–347 (2015)

Interval Temporal Logic HS with Path Quantifiers

Yanjun Li[(✉)] and Shuyuan Li

Nankai University, Tianjin, China
lyjlogic@nankai.edu.cn

Abstract. To enhance the expressiveness of the interval temporal logic HS over branching interval models, this paper extends the language of HS by introducing quantifiers over paths. We also modify the original semantics of HS by evaluating formulas on pairs of intervals and paths, rather than on intervals alone. Our extension of HS aligns with the extension of CTL to CTL*. The resulting logic is called HS* in this paper. We show that HS* is strictly more expressive than HS, and we also demonstrate that HS* is a fragment of Monadic Path Logic, a monadic second-order logic where set quantification is restricted to paths.

Keywords: Interval temporal logic · Branching interval models · Expressivity

1 Introduction

Interval temporal logics are temporal logics that focus on interval entities. An interval represents a period of time. Unlike point-based temporal logics, such as LTL (Linear Temporal Logic) and CTL (Computation Tree Logic), interval-based temporal logics can naturally express statements involving durative events or continuous processes. For example: "I woke up several times during the night." Since a time point is a special case of a time interval, and there are more possible relations between time intervals than between time points, interval-based temporal logics are generally more expressive than point-based temporal logics. Interval-based temporal logic is an active area of research in fields such as philosophy, linguistics, and artificial intelligence (see [15]).

In AI community, an influential early work on interval-based temporal logic is Allen's paper [2], within which Allen considered 13 binary relations between time intervals in a linearly ordered time flow. For example, one of the Allen's 13 relations is that "the interval j is begun by the interval i", which can be depicted as follows:

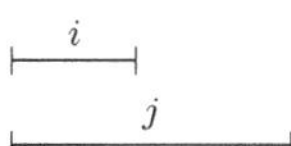

In [11], Halpern and Shoham proposed a propositional modal logic for reasoning about Allen's interval relations, which is known as the interval temporal logic HS.

© The Author(s), under exclusive license to Springer Nature Switzerland AG 2026
V. Goranko et al. (Eds.): LORI 2025, LNCS 16010, pp. 79–94, 2026.
https://doi.org/10.1007/978-981-95-2481-5_6

In the formal language of HS, there are 12 unary modalities, each corresponding to one of Allen's interval relations, except the equivalence relation. It is shown that over non-strict interval models, the expressivity of the whole HS language is equivalent to its fragment that includes only the following four modalities: $B, \overline{B}, E$ and $\overline{E}$, which respectively correspond to the interval relation *begin, begun by, end* and *ended by* (see [18]). The expressivity of other fragments has been extensively studied in the literature (see [1,6,9]).

In addition to HS and its fragments, several other interval temporal logics are discussed in the literature. The sub-interval logic which focuses on reasoning about the sub-interval relation was studied from the perspective of both philosophy and computer science [12,16]. The interval logics based on the interval relation *meet* and its inverse were introduced and studied in the framework of first-order logic and its propositional variants [5,8]. Among all these unary modal interval temporal logics, HS is the most expressive one. The first interval temporal logic with binary operators was introduced by [14], which contains the binary *chop* operator C. It is proved in [13] that such an operator is not definable in HS. The framework of [14] was later extended in [18] to the logic CDT, which adds two more binary operators D and T. A generalization of CDT to branching interval models was investigated in [7].

Eventhough HS is originally interpreted over branching interval models, however, it cannot freely quantify over paths. As in CTL, only a restricted combination of interval-relation operators and path quantifiers can be expressed in HS. For example, the HS formula $\langle \overline{B} \rangle \varphi$ over branching interval models expresses that "there *exists* a path and *exists* an interval in this path such that it is begun by the current interval and it satisfies φ", which can be depicted in Fig. 1. The combination of quantifiers in $\langle \overline{B} \rangle \varphi$ is similar to that in the CTL formula $\mathbf{E}F\varphi$ (which means "there *exists* a path on which there *exists* a future state satisfying φ"). As the CTL* formula $\mathbf{A}F\varphi$ cannot be expressed in CTL, the sentence "for *each* path there *exists* an interval on this path such that it is begun by the current interval and it satisfies φ", depicted in Fig. 2, cannot be naturally expressed in HS.

Fig. 1. ∃∃.   **Fig. 2.** ∀∃.

In this paper, we extend the interval temporal logic HS by introducing path quantifiers. Our extension follows a similar approach to the one used in extending CTL to CTL*, and we refer to the resulting logic as HS*. To enable quantification over paths, we modify the semantics of HS by evaluating formulas on pairs of intervals and paths, rather than on intervals alone. As a result, in this framework, interval relations can be expressed without relying on the linear interval property (see Definition 5). In the literature, the linear interval property is considered an important assumption for interval temporal logics, as it ensures that each interval corresponds to a unique linear time period. In this paper, we show that HS is a fragment of HS*. We also show that the logic HS* can be translated into Monadic Path Logic (MPL) which is a monadic second-order logic where set quantification is restricted to paths (see [10]).

The rest of the paper is organized as follows. In Sect. 2, we introduce the basic framework of the logic HS. In Sect. 3, we extend HS with path quantifiers, which leads to the logic HS*. Section 4 shows that HS* is strictly more expressive than HS. In Sect. 5, we show that the logic HS* can be translated into Monadic Path Logic (MPL). Section 6 concludes with some remarks.

2 Preliminary

In this section, we briefly introduce the language, model, and semantics of the interval temporal logic HS.

Let **P** be a set of propositional letters. The language of the logic HS is defined as follows.

Definition 1 (Language of HS). *The language $\mathcal{L}^{HS}$ is defined by the following BNF rules:*

$$\varphi ::= p \mid \neg\varphi \mid (\varphi \wedge \varphi) \mid \langle B\rangle\varphi \mid \langle \overline{B}\rangle\varphi \mid \langle E\rangle\varphi \mid \langle \overline{E}\rangle\varphi$$

where $p \in \mathbf{P}$.

The language $\mathcal{L}^{HS}$ is interpreted on interval models.

Definition 2 (Interval). *A temporal structure S is a strict partial order $S = \langle D, < \rangle$, where D is a set of time points and $<$ is a strict partial order on D. An interval in S is a pair $[d, d']$ such that $d, d' \in D$ and $d \leq d'$. If d is strictly less than d', i.e. $d < d'$, the interval $[d, d']$ is a strict interval. If $d = d'$, the interval $[d, d'](= [d, d])$ is also called a point-interval.*

In the literature, strict intervals and non-strict intervals are carefully distinguished, as the logic can behave differently over each type. In this paper, we focus on non-strict intervals. Moreover, in some literature, the temporal structure is not merely a strict partial order but also a tree-like structure, where branching occurs only in the future and not in the past. In this paper, we explore a generalized extension of the logic HS, so we allow time to branch both in the past and in the future.

Definition 3 (Interval model). *An interval frame is a pair $\langle S, \mathbb{I}(S)\rangle$ where S is a temporal structure and $\mathbb{I}(S)$ is the set of all intervals on S. An interval model is a triple $\mathcal{M} = \langle S, \mathbb{I}(S), V\rangle$ where $\langle S, \mathbb{I}(S)\rangle$ is an interval frame and V is an assignment function from $\mathbf{P}$ into $2^{\mathbb{I}(S)}$, that is, $V : \mathbf{P} \rightarrow 2^{\mathbb{I}(S)}$.*

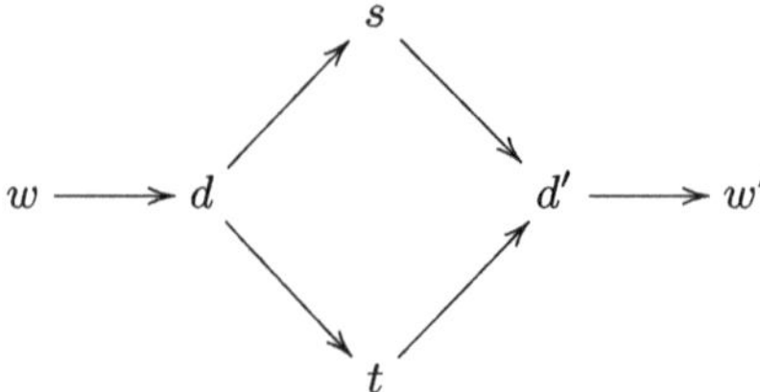

Fig. 3. A temporal structure without linear interval property.

Definition 4 (Semantics of HS). *Given an interval model $\mathcal{M} = \langle S, \mathbb{I}(S), V \rangle$ and an interval $[d, d']$ in $\mathcal{M}$, the satisfaction relation of HS is defined as follows:*

$$
\begin{aligned}
\mathcal{M}, [d, d'] &\vDash p &&\Longleftrightarrow [d, d'] \in V(p) \\
\mathcal{M}, [d, d'] &\vDash \neg\varphi &&\Longleftrightarrow \mathcal{M}, [d, d'] \nvDash \varphi \\
\mathcal{M}, [d, d'] &\vDash \varphi \wedge \psi &&\Longleftrightarrow \mathcal{M}, [d, d'] \vDash \varphi \text{ and } \mathcal{M}, [d, d'] \vDash \psi \\
\mathcal{M}, [d, d'] &\vDash \langle B \rangle \varphi &&\Longleftrightarrow \text{there exists } t \in S \text{ s.t. } d \leq t < d' \text{ and } \mathcal{M}, [d, t] \vDash \varphi \\
\mathcal{M}, [d, d'] &\vDash \langle \overline{B} \rangle \varphi &&\Longleftrightarrow \text{there exists } t \in S \text{ s.t. } d' < t \text{ and } \mathcal{M}, [d, t] \vDash \varphi \\
\mathcal{M}, [d, d'] &\vDash \langle E \rangle \varphi &&\Longleftrightarrow \text{there exists } t \in S \text{ s.t. } d < t \leq d' \text{ and } \mathcal{M}, [t, d'] \vDash \varphi \\
\mathcal{M}, [d, d'] &\vDash \langle \overline{E} \rangle \varphi &&\Longleftrightarrow \text{there exists } t \in S \text{ s.t. } t < d \text{ and } \mathcal{M}, [t, d'] \vDash \varphi
\end{aligned}
$$

In [17], an interval is defined as $[d, d'] = \{s \in D \mid d \leq s \leq d'\}$. Note that an interval $[d, d']$ in S does not determine a linearly ordered subset of S, such as the interval $[d, d']$ in Fig. 3. However, on temporal structures with linear interval property, it does so.

Definition 5 (Linear interval property). *A temporal structure $S = \langle D, < \rangle$ is said to have the* linear interval property *if for each interval $[d, d']$ in S, the set $\{s \in D \mid d \leq s \leq d'\}$ forms a linearly ordered set with respect to the order $<$.*

Note that a temporal structure with the linear interval property does not necessarily have to be a linear structure. The linear interval property is not a crucial assumption for the logic HS, as the logic remains well-defined without it. However, it is significant in that it ensures each interval corresponds to a unique linear time period, making it possible to express Allen's interval relations in HS over branching interval models.

3   Language and Semantics of HS*

In this section, we will extend the logic HS with path quantifiers, and the resulting logic is called the logic HS* in this paper.

Definition 6 (Language of HS*). *The language $\mathcal{L}^{HS^*}$ is defined by the following BNF rules:*

$$\varphi ::= p \mid \neg\varphi \mid (\varphi \wedge \varphi) \mid \langle \mathcal{B} \rangle \varphi \mid \langle \overline{\mathcal{B}} \rangle \varphi \mid \langle \mathcal{E} \rangle \varphi \mid \langle \overline{\mathcal{E}} \rangle \varphi \mid \mathbf{A}\varphi$$

where $p \in \mathbf{P}$.

The connectives $\lor, \rightarrow, \leftrightarrow$ are defined using the standard abbreviations. The formula $\mathbf{E}\varphi$ is defined as $\neg\mathbf{A}\neg\varphi$, and $[x]\varphi$ is defined as $\neg\langle x\rangle\neg\varphi$ where $x \in \{\mathcal{B},\overline{\mathcal{B}},\mathcal{E},\overline{\mathcal{E}}\}$.

The intuitive readings of the modal formulas in $\mathcal{L}^{\mathsf{HS}^*}$ are as follows:

- $\langle\mathcal{B}\rangle\varphi$ means there exists a subinterval that begins the current interval of the current path and the subinterval satisfies φ;
- $\langle\overline{\mathcal{B}}\rangle\varphi$ means there exists an interval begun by the current interval of the current path and it satisfies φ;
- $\langle\mathcal{E}\rangle\varphi$ means there exists a subinterval that ends the current interval of the current path and the subinterval satisfies φ;
- $\langle\overline{\mathcal{E}}\rangle\varphi$ means there exists an interval ended by the current interval of the current path and it satifies φ;
- $\mathbf{A}\varphi$ means for each path through the current interval, φ holds on the current interval of this path.

There is a slight difference between the intuitive meaning of the HS^*-formula $\langle\mathcal{B}\rangle\varphi$ (respectively, $\langle\overline{\mathcal{B}}\rangle\varphi, \langle\mathcal{E}\rangle\varphi, \langle\overline{\mathcal{E}}\rangle\varphi$) and the HS-formula $\langle B\rangle\varphi$ (respectively, $\langle\overline{B}\rangle\varphi, \langle E\rangle\varphi, \langle\overline{E}\rangle\varphi$). This difference arises because the truth of a HS^*-formula is evaluated on pairs of paths and intervals, rather than on intervals alone. This distinction is crucial, as HS^* is designed to reason about both paths and intervals.

Definition 7 (Path). *Let $S = \langle D, <\rangle$ be a temporal structure. A (full) path in S is a maximal linear subset of S. We say that a path ρ passes through an interval $[d, d']$ if $d, d' \in \rho$.*

Let $[d, d']$ be an interval and ρ be a path that passes through it. The pair $(\rho, [d, d'])$ refers to the unique linear interval $\{s \in \rho \mid d \leq s \leq d'\}$. In this way, we can discuss linear intervals without requiring the linear interval property. For example, in Fig. 3, the interval $[d, d']$ refers to the set $\{d, s, t, d'\}$, and $(\rho, [d, d'])$ refers to the linear set $\{d, s, d'\}$ where ρ is the path passing through s.

A *pointed interval model* is a triple $(\mathcal{M}, \rho, [d, d'])$ where $\mathcal{M}$ is an interval model, ρ is a path of $\mathcal{M}$, and $[d, d']$ is an interval that ρ passes through.

Definition 8 (Semantics of HS^*). *The satisfaction relation $\models$ between pointed interval models and $\mathcal{L}^{\mathsf{HS}^*}$-formulas is defined as follows:*

$$\mathcal{M}, \rho, [d, d'] \models p \iff [d, d'] \in V(p)$$
$$\mathcal{M}, \rho, [d, d'] \models \neg\varphi \iff \mathcal{M}, \rho, [d, d'] \not\models \varphi$$
$$\mathcal{M}, \rho, [d, d'] \models \varphi \land \psi \iff \mathcal{M}, \rho, [d, d'] \models \varphi \text{ and } \mathcal{M}, \rho, [d, d'] \models \psi$$
$$\mathcal{M}, \rho, [d, d'] \models \langle\mathcal{B}\rangle\varphi \iff \text{there exists } t \in \rho \text{ s.t. } d \leq t < d' \text{ and } \mathcal{M}, \rho, [d, t] \models \varphi$$
$$\mathcal{M}, \rho, [d, d'] \models \langle\overline{\mathcal{B}}\rangle\varphi \iff \text{there exists } t \in \rho \text{ s.t. } d' < t \text{ and } \mathcal{M}, \rho, [d, t] \models \varphi$$
$$\mathcal{M}, \rho, [d, d'] \models \langle\mathcal{E}\rangle\varphi \iff \text{there exists } t \in \rho \text{ s.t. } d < t \leq d' \text{ and } \mathcal{M}, \rho, [t, d'] \models \varphi$$
$$\mathcal{M}, \rho, [d, d'] \models \langle\overline{\mathcal{E}}\rangle\varphi \iff \text{there exists } t \in \rho \text{ s.t. } t < d \text{ and } \mathcal{M}, \rho, [t, d'] \models \varphi$$
$$\mathcal{M}, \rho, [d, d'] \models \mathbf{A}\varphi \iff \text{for each path } \rho' \text{ through } [d, d'] : \mathcal{M}, \rho', [d, d'] \models \varphi.$$

Note that over branching interval models, the semantics of the HS-formula $\langle \overline{B} \rangle \varphi$ is equivalent to the following:

$$\mathcal{M}, [d, d'] \vDash \langle \overline{B} \rangle \varphi \iff \text{there exists a path } \rho \text{ and exists a point } t \in \rho$$
$$\text{s.t. } d' < t \text{ and } \mathcal{M}, [d, t] \vDash \varphi.$$

Hence, it can seen that in HS, the quantification over path an quantification over interval are bounded together.

The logic HS* allows arbitrary nesting and combinations of path quantifiers and interval-relation operators, making it more expressive than HS. For example, the situation discussed in Fig. 2 can be expressed by the formula $\mathbf{A}\langle \overline{B} \rangle \varphi$:

$$\mathcal{M}, \rho, [d, d'] \vDash \mathbf{A}\langle \overline{B} \rangle \varphi \iff \text{for each path } \rho' \text{ through } [d, d'] \text{ there exists } t \in \rho' :$$
$$d' < t \text{ and } \mathcal{M}, \rho', [d, t] \vDash \varphi.$$

The formula $\mathbf{A}\langle \overline{\mathcal{E}} \rangle \varphi$ expresses a similar property, but with branching in the past rather than branching in the future, which can be depicted as follows:

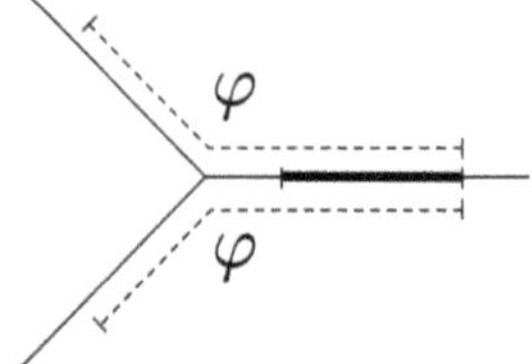

In [4], they investigated a combined operator $E[A]\varphi$ which expresses that "there exists a path such that on each interval that the current interval meets, φ holds". This can be expressed in HS* by the formula $\mathbf{E}[\mathcal{A}]\varphi$, where $\mathcal{A}$ is the operator for the interval relation of *meet*, defined as follows:

$$[\mathcal{A}]\varphi := ([\mathcal{B}]\bot \wedge \langle \overline{\mathcal{B}} \rangle \varphi) \vee \langle \mathcal{E} \rangle ([\mathcal{B}]\bot \wedge \langle \overline{\mathcal{B}} \rangle \varphi).$$

Moreover, it is known that the HS-formula

$$(\langle B \rangle \varphi \wedge \langle B \rangle \psi) \to (\langle B \rangle (\varphi \wedge \langle B \rangle \psi) \vee \langle B \rangle (\varphi \wedge \psi) \vee \langle B \rangle (\langle B \rangle \varphi \wedge \psi))$$

defines the class of interval frames with the linear interval property (see [17]). The following proposition shows that it can also be defined by simpler HS*-formulas.

Proposition 1. $\langle \mathcal{B} \rangle p \to \mathbf{A}\langle \mathcal{B} \rangle p$ *defines the class of interval frames with the linear interval property, and so does* $\langle \mathcal{E} \rangle p \to \mathbf{A}\langle \mathcal{E} \rangle p$.

Proof. We will only show the case of $\langle \mathcal{B} \rangle p \to \mathbf{A}\langle \mathcal{B} \rangle p$; for the case of $\langle \mathcal{E} \rangle p \to \mathbf{A}\langle \mathcal{E} \rangle p$, it can be proved similarly.

Firstly, we show that if a temporal frame $F = \langle S, \mathbb{I}(S) \rangle$ has the linear interval property, then $F \vDash \langle \mathcal{B} \rangle p \to \mathbf{A}\langle \mathcal{B} \rangle p$.

Given an assignment $V : \mathbf{P} \to 2^{\mathbb{I}(S)}$, Let $\mathcal{M} = \langle F, V \rangle$. If $\mathcal{M}, \rho, [d, d'] \vDash \langle \mathcal{B} \rangle p$, by Definition 8, we know that there exists $d \leq t < d'$ such that $\mathcal{M}, \rho, [d, t] \vDash p$. Since $\mathcal{M}$ has the linear interval property, by Definition 5, we know that $\{s \in D \mid d \leq s \leq d'\}$ is linearly ordered. Thus, for each path ρ' through $[d, d']$, it follows that $t \in \rho'$. Thus, $\mathcal{M}, \rho', [d, d'] \models \langle \mathcal{B} \rangle p$. Then, we have $\mathcal{M}, \rho, [d, d'] \models \mathbf{A} \langle \mathcal{B} \rangle p$. Thus, $F \models \langle \mathcal{B} \rangle p \to \mathbf{A} \langle \mathcal{B} \rangle p$.

Secondly, we will show that if a temporal frame $F = \langle S, \mathbb{I}(S) \rangle$ does not have the linear interval property, then $F \nvDash \langle \mathcal{B} \rangle p \to \mathbf{A} \langle \mathcal{B} \rangle p$.

Since $F = \langle S, \mathbb{I}(S) \rangle$ does not have the linear interval property, it follows that there is some interval $[d, d']$ in F such that $X = \{s \in D \mid d \leq s \leq d'\}$ is not linearly ordered, which implies that there are $s', s'' \in X$ such that $s' \nleq s''$ and $s'' \nleq s'$. It follows that $d < s' < d'$ and $d < s'' < d'$. Let ρ' and ρ'' be two paths that respectively pass through $\{d, s', d'\}$ and $\{d, s'', d'\}$. Now, we define the assignment V as

$$V(p) = \{[d, s']\}.$$

Let $\mathcal{M} = \langle F, V \rangle$. Due to $\mathcal{M}, \rho', [d, s'] \models p$, we have $\mathcal{M}, \rho', [d, d'] \models \langle \mathcal{B} \rangle p$. However, since ρ'' does not pass through $[d, s']$, it follows $\mathcal{M}, \rho'', [d, d'] \nvDash \langle \mathcal{B} \rangle p$. Hence, $\mathcal{M}, \rho', [d, d'] \nvDash \mathbf{A} \langle \mathcal{B} \rangle p$, and then $\mathcal{M}, \rho', [d, d'] \nvDash \langle \mathcal{B} \rangle p \to \mathbf{A} \langle \mathcal{B} \rangle p$. Thus, $F \nvDash \langle \mathcal{B} \rangle p \to \mathbf{A} \langle \mathcal{B} \rangle p$.

In the remainder of this section, we explore some valid formulas in HS*.

Proposition 2. $\models p \to \mathbf{A}p$.

Given that $\models \mathbf{A}\varphi \to \varphi$ holds, it follows that $\models p \leftrightarrow \mathbf{A}p$. This indicates that the truth value of a propositional atom over an interval is independent of the path containing that interval. Such invariance arises from the definition of interval models, where assignments are functions from $\mathbf{P}$ into $2^{\mathbb{I}(S)}$. However, if the assignment is redefined differently—for example, as a function from $\mathbf{P}$ to the power set of $\{(\rho, [d, d']) \mid [d, d'] \in \mathbb{I}(S) \text{ and } \rho \text{ contains } [d, d']\}$—then Proposition 2 no longer holds.

Intuitively, the following proposition states that if every path has an interval ρ begins with the current interval, then for any such interval ρ' on any path, the relationship between ρ and ρ' must satisfy: $\rho = \rho'$, ρ begins ρ', or ρ' begins ρ.

Proposition 3. $\models \mathbf{A} \langle \overline{\mathcal{B}} \rangle p \to \mathbf{A} [\overline{\mathcal{B}}] (p \vee \langle \mathcal{B} \rangle p \vee \langle \overline{\mathcal{B}} \rangle p)$.

Proof. Let $\mathcal{M}, \rho, [d, d'] \models \mathbf{A} \langle \overline{\mathcal{B}} \rangle p$. By Definition 8, we know that for all paths ρ' through $[d, d']$, there exists $t \in \rho'$ such that $d' < t$ and $\mathcal{M}, \rho', [d, t] \models p$. It follows that for all paths ρ' and every point $d' < v$ in ρ', there are three distinct cases: $t < v$ along with $\mathcal{M}, \rho', [d, v] \models \langle \mathcal{B} \rangle p$ or $t = v$ along with $\mathcal{M}, \rho', [d, v] \models p$ or $v < t$ along with $\mathcal{M}, \rho', [d, v] \models \langle \overline{\mathcal{B}} \rangle p$. Therefore, for all paths ρ' and every point $d' < v$ in ρ', $\mathcal{M}, \rho', [d, v] \models p \vee \langle \mathcal{B} \rangle p \vee \langle \overline{\mathcal{B}} \rangle p$. By Definition 8, it follows that $\mathcal{M}, \rho, \langle d, d' \rangle \models \mathbf{A} [\overline{\mathcal{B}}] (p \vee \langle \mathcal{B} \rangle p \vee \langle \overline{\mathcal{B}} \rangle p)$.

The following proposition is similar to Proposition 3 that concerns the 'begun-by' relation, but this proposition concerns the 'ended-by' relation. It states that if every path has an interval ρ is ended by the current interval, then for any such interval ρ' on any path, the relationship between ρ and ρ' must satisfy: $\rho = \rho'$, ρ ends ρ', or ρ' ends ρ.

Proposition 4. $\vDash \mathbf{A}\langle\overline{\mathcal{E}}\rangle p \to \mathbf{A}[\overline{\mathcal{E}}](p \vee \langle\mathcal{E}\rangle p \vee \langle\overline{\mathcal{E}}\rangle p)$.

Proof. The proof of this is similar to Proposition 3.

The key distinction between Propositions 3 and 4 lies in their treatment of path branching: Proposition 3 concerns path branching in the future while Proposition 4 concerns path branching in the past.

The following proposition concerns the 'begin' relation. It states that if every path has an interval ρ beginning the current interval, then for any such interval ρ' on any path, the relationship between ρ and ρ' must satisfy: $\rho = \rho'$, ρ begins ρ', or ρ' begins ρ.

Proposition 5. $\vDash \mathbf{A}\langle\mathcal{B}\rangle p \to \mathbf{A}[\mathcal{B}](p \vee \langle\mathcal{B}\rangle p \vee \langle\overline{\mathcal{B}}\rangle p)$.

Proof. Let $\mathcal{M}, \rho, [d, d'] \vDash \mathbf{A}\langle\mathcal{B}\rangle p$. By Definition 8, we know that for all paths ρ' through $[d, d']$, there exists $t \in \rho'$ such that $d \leq t < d'$ and $\mathcal{M}, \rho', [d, t] \vDash p$. It follows that for all paths ρ' and every point $d' \leq v < d'$ in ρ', there are three distinct cases: $t < v$ along with $\mathcal{M}, \rho', [d, v] \vDash \langle\mathcal{B}\rangle p$ or $t = v$ along with $\mathcal{M}, \rho', [d, v] \vDash p$ or $v < t$ along with $\mathcal{M}, \rho', [d, v] \vDash \langle\overline{\mathcal{B}}\rangle p$. Therefore, for all paths ρ' and every point $d \leq v < d'$ in ρ', $\mathcal{M}, \rho', [d, v] \vDash p \vee \langle\mathcal{B}\rangle p \vee \langle\overline{\mathcal{B}}\rangle p$. By Definition 8, it follows that $\mathcal{M}, \rho, [d, d'] \vDash \mathbf{A}[\mathcal{B}](p \vee \langle\mathcal{B}\rangle p \vee \langle\overline{\mathcal{B}}\rangle p)$.

The following proposition concerns the 'end' relation. It states that if every path has an interval ρ ending the current interval, then for any such interval ρ' on any path, the relationship between ρ and ρ' must satisfy: $\rho = \rho'$, ρ ends ρ', or ρ' ends ρ.

Proposition 6. $\vDash \mathbf{A}\langle\mathcal{E}\rangle p \to \mathbf{A}[\mathcal{E}](p \vee \langle\mathcal{E}\rangle p \vee \langle\overline{\mathcal{E}}\rangle p)$.

Proof. The proof of this is similar to Proposition 5.

4   Comparision with HS

The following proposition demonstrates that the logic HS is equivalent to a fragment of HS* in which only restricted combinations of interval-relation operators and path quantifiers are permitted.

$$
\begin{aligned}
f(p) &= p \\
f(\neg\varphi) &= \neg f(\varphi) \\
f(\varphi \wedge \psi) &= f(\varphi) \wedge f(\psi) \\
f(\langle B\rangle\varphi) &= \mathbf{E}\langle\mathcal{B}\rangle f(\varphi) \\
f(\langle E\rangle\varphi) &= \mathbf{E}\langle\mathcal{E}\rangle f(\varphi) \\
f(\langle\overline{B}\rangle\varphi) &= \mathbf{E}\langle\overline{\mathcal{B}}\rangle f(\varphi) \\
f(\langle\overline{E}\rangle\varphi) &= \mathbf{E}\langle\overline{\mathcal{E}}\rangle f(\varphi)
\end{aligned}
$$

Fig. 4. $f : \mathcal{L}^{\mathrm{HS}} \to \mathcal{L}^{\mathrm{HS}^*}_{\mathbf{E}\langle\rangle}$

$$
\begin{aligned}
g(p) &= p \\
g(\neg\varphi) &= \neg g(\varphi) \\
g(\varphi \wedge \psi) &= g(\varphi) \wedge g(\psi) \\
g(\mathbf{E}\langle\mathcal{B}\rangle\varphi) &= \langle B\rangle g(\varphi) \\
g(\mathbf{E}\langle\mathcal{E}\rangle\varphi) &= \langle E\rangle g(\varphi) \\
g(\mathbf{E}\langle\overline{\mathcal{B}}\rangle\varphi) &= \langle\overline{B}\rangle g(\varphi) \\
g(\mathbf{E}\langle\overline{\mathcal{E}}\rangle\varphi) &= \langle\overline{E}\rangle g(\varphi)
\end{aligned}
$$

Fig. 5. $g : \mathcal{L}^{\mathrm{HS}^*}_{\mathbf{E}\langle\rangle} \to \mathcal{L}^{\mathrm{HS}}$

Proposition 7. *Let the fragment of* $\mathcal{L}^{HS^*}$*, denoted by* $\mathcal{L}^{HS^*}_{\mathbf{E}\langle\rangle}$*, be defined as follows:*

$$\varphi ::= p \mid \neg\varphi \mid (\varphi \wedge \varphi) \mid \mathbf{E}\langle\mathcal{B}\rangle\varphi \mid \mathbf{E}\langle\overline{\mathcal{B}}\rangle\varphi \mid \mathbf{E}\langle\mathcal{E}\rangle\varphi \mid \mathbf{E}\langle\overline{\mathcal{E}}\rangle\varphi.$$

Let the translation function $f : \mathcal{L}^{HS} \rightarrow \mathcal{L}^{HS^*}_{\mathbf{E}\langle\rangle}$ *be defined in Fig. 4 and the translation function* $g : \mathcal{L}^{HS^*}_{\mathbf{E}\langle\rangle} \rightarrow \mathcal{L}^{HS}$ *be defined in Fig. 5. We then have the following results:*

(1.) For each $\varphi \in \mathcal{L}^{HS}$*,* $\mathcal{M}, [d, d'] \vDash \varphi$ *if and only if* $\mathcal{M}, \rho, [d, d'] \vDash f(\varphi)$*.*
(2.) For each $\varphi \in \mathcal{L}^{HS^*}_{\mathbf{E}\langle\rangle}$*,* $\mathcal{M}, \rho, [d, d'] \vDash \varphi$ *if and only if* $\mathcal{M}, [d, d'] \vDash g(\varphi)$*.*

Proof. We prove them by induction on the structure of φ. The cases for atomic propositions and boolean connectives are straightforward. We focus on the other cases. Firstly, we prove *(1.)*:

- $\varphi = \langle B\rangle\psi$. Let $\mathcal{M}, [d, d'] \vDash \langle B\rangle\psi$, by Definition 4, it follows that there exists a t such that $d \leq t < d'$ and $\mathcal{M}, [d, t] \vDash \psi$. By inductive hypothesis, we know that $\mathcal{M}, \rho, \langle d, t\rangle \vDash f(\psi)$. By Definition 8, it follows that $\mathcal{M}, \rho, [d, d'] \vDash \langle\mathcal{B}\rangle f(\psi)$. Since there exists a path $\rho' = \rho$ through $[d, d']$ such that $\mathcal{M}, \rho', [d, d'] \vDash \langle\mathcal{B}\rangle f(\psi)$, by Definition 8, we have that $\mathcal{M}, \rho, [d, d'] \vDash \mathbf{E}\langle\mathcal{B}\rangle f(\psi)$. By Fig. 4, we have $\mathcal{M}, \rho, [d, d'] \vDash f(\langle B\rangle\psi)$. Therefore, we have shown that if $\mathcal{M}, [d, d'] \vDash \langle B\rangle\psi$, then $\mathcal{M}, \rho, [d, d'] \vDash f(\langle B\rangle\psi)$. The converse holds in a similar way.
- $\varphi = \langle E\rangle\psi$. The proof of this is similar to the case of $\langle B\rangle\psi$.
- $\varphi = \langle\overline{B}\rangle\psi$. Let $\mathcal{M}, [d, d'] \vDash \langle\overline{B}\rangle\psi$, by Definition 4, it follows that there exists a t such that $d' < t$ and $\mathcal{M}, [d, t] \vDash \psi$. By inductive hypothesis, we know that $\mathcal{M}, \rho, \langle d, t\rangle \vDash f(\psi)$. By Definition 8, we know that $\mathcal{M}, \rho, [d, d'] \vDash \langle\overline{\mathcal{B}}\rangle f(\psi)$. Since there exists a path $\rho' = \rho$ through $[d, d']$ such that $\mathcal{M}, \rho', [d, d'] \vDash \langle\overline{\mathcal{B}}\rangle f(\psi)$, by Definition 8, we know that $\mathcal{M}, \rho, [d, d'] \vDash \mathbf{E}\langle\overline{\mathcal{B}}\rangle f(\psi)$. By Fig. 4, we have $\mathcal{M}, \rho, [d, d'] \vDash f(\langle\overline{B}\rangle\psi)$. Therefore, we have shown that if $\mathcal{M}, [d, d'] \vDash \langle\overline{B}\rangle\psi$, then $\mathcal{M}, \rho, [d, d'] \vDash f(\langle\overline{B}\rangle\psi)$. The converse holds in a similar way.
- $\varphi = \langle\overline{E}\rangle\psi$. The proof of this is similar to the case of $\langle\overline{B}\rangle\psi$.

Then, we prove *(2.)*:

- $\varphi = \mathbf{E}\langle\mathcal{B}\rangle\psi$. Let $\mathcal{M}, \rho, [d, d'] \vDash \mathbf{E}\langle\mathcal{B}\rangle\psi$. By Definition 8, there exists a path ρ' through $[d, d']$ and a point t such that $d \leq t < d'$ and $\mathcal{M}, \rho', [d, t] \vDash \psi$. By inductive hypothesis, we know that $\mathcal{M}, [d, t] \vDash g(\psi)$. By Definition 4, we have $\mathcal{M}, [d, d'] \vDash \langle B\rangle g(\psi)$ By Fig. 5, we have $\mathcal{M}, [d, d'] \vDash g(\mathbf{E}\langle\mathcal{B}\rangle\psi)$. Therefore, we have shown that if $\mathcal{M}, \rho, [d, d'] \vDash \mathbf{E}\langle\mathcal{B}\rangle\psi$, then $\mathcal{M}, [d, d'] \vDash g(\mathbf{E}\langle\mathcal{B}\rangle\psi)$. The converse holds in a similar way.
- $\varphi = \mathbf{E}\langle\mathcal{E}\rangle\psi$. The proof of this is similar to the case of $\mathbf{E}\langle\mathcal{B}\rangle\psi$.
- $\varphi = \mathbf{E}\langle\overline{\mathcal{B}}\rangle\psi$. Let $\mathcal{M}, \rho, [d, d'] \vDash \mathbf{E}\langle\overline{\mathcal{B}}\rangle\psi$. By Definition 8, we know that there exists a path ρ' through $[d, d']$ and a point t such that $d' < t$ and $\mathcal{M}, \rho', [d, t] \vDash \psi$. By inductive hypothesis, we know that $\mathcal{M}, [d, t] \vDash g(\psi)$. By Definition 4, we have $\mathcal{M}, [d, t] \vDash \langle\overline{B}\rangle g(\psi)$ By Fig. 5, we have $\mathcal{M}, [d, t] \vDash g(\mathbf{E}\langle\overline{\mathcal{B}}\rangle\psi)$. Therefore, we have shown that if $\mathcal{M}, \rho, [d, d'] \vDash \mathbf{E}\langle\overline{\mathcal{B}}\rangle\psi$, then $\mathcal{M}, [d, t] \vDash g(\mathbf{E}\langle\overline{\mathcal{B}}\rangle\psi)$. The converse holds in a similar way.

Fig. 6. Models $\mathcal{M}$ and $\mathcal{M}'$ (p is only true on intervals $[c, b]$ and $[c', b']$.)

- $\varphi = \mathbf{E}\langle\overline{\mathcal{E}}\rangle\psi$. The proof of this is similar to the case of $\mathbf{E}\langle\overline{\mathcal{B}}\rangle\psi$.

Theorem 1. *On expressivity, HS is a fragment of HS*.*

Proof. It immediately follows from Proposition 7.

In fact, Proposition 7 demonstrates a stronger result than the one stated in Theorem 1. Specifically, Proposition 7 shows that the expressivity of HS coincides with the HS*-fragment $\mathcal{L}^{\mathrm{HS}^*}_{\mathbf{E}\langle\rangle}$. It can be seen that HS is a fragment of HS*, analogous to the relationship between CTL and CTL*. Furthermore, since it is established that the logic HS is undecidable (see [11]), it follows that HS* is undecidable as well.

Next, we are going to show that HS* is strictly more expressive than HS. First, we will show the following proposition.

Proposition 8. *For each HS-formula φ, we have that $\mathcal{M}, [a, b] \models \varphi$ iff $\mathcal{M}', [a', b'] \models \varphi$, where $\mathcal{M}$ and $\mathcal{M}'$ are depicted in Fig. 6.*

Proof. First, we define a function f as follows: for each $x \in \{a, b, c\}$,

$$f(x) = x'.$$

It is obvious that f is a function from the $\mathcal{M}$-domain $S^{\mathcal{M}}$ onto the $\mathcal{M}'$-domain $S^{\mathcal{M}'}$. Moreover, it can be easily checked that f is a order-preserving function, that is, if $x \leq y$ then $f(x) \leq f(y)$ (i.e. $x' \leq y'$). Hence, it follows that if $[x, y]$ is an interval in $\mathcal{M}$, then $[x', y']$ is an interval in $\mathcal{M}'$.

Second, we show the following claim:

> For each HS-formula φ, $\mathcal{M}, [x, y] \models \varphi$ iff $\mathcal{M}', [x', y'] \models \varphi$, where $x, y \in S^{\mathcal{M}}$ and $x \leq y$.

It is obvious that Proposition 8 follows from this claim.

We prove this claim by induction on φ. The Boolean cases are straightforward. We will focus on only the cases of $\langle B \rangle\varphi$ and $\langle\overline{B}\rangle\varphi$; the cases of $\langle E \rangle\varphi$ and $\langle\overline{E}\rangle\varphi$ can be proved similarly.

- $\langle B \rangle\varphi$.
    - $x = y$. It follows that $f(x) = f(y)$ (i.e. $x' = y'$). It is obvious that $\mathcal{M}, [x, y] \models \varphi$ iff $\mathcal{M}', [x', y'] \models \varphi$.
    - $x < y$. It follows that $x \neq b$. Then, there are two cases: $x = a$ or $x = c$.

* $x = a$.

For the case $y = b$, we have the followings:

$$\mathcal{M}, [a, b] \models \langle B \rangle \varphi$$
$$\Longleftrightarrow \mathcal{M}, [a, a] \models \varphi \text{ or } \mathcal{M}, [a, c] \models \varphi \quad \text{(by Definition 4)}$$
$$\Longleftrightarrow \mathcal{M}', [a', a'] \models \varphi \text{ or } \mathcal{M}', [a', c'] \models \varphi \text{ (by IH)}$$
$$\Longleftrightarrow M', [a', b'] \models \langle B \rangle \varphi \quad \text{(by Definition 4)}$$

For the case $y = c$, we have the followings:

$$\mathcal{M}, [a, c] \models \langle B \rangle \varphi$$
$$\Longleftrightarrow \mathcal{M}, [a, a] \models \varphi \quad \text{(by Definition 4)}$$
$$\Longleftrightarrow \mathcal{M}', [a', a'] \models \varphi \quad \text{(by IH)}$$
$$\Longleftrightarrow M', [a', c'] \models \langle B \rangle \varphi \text{ (by Definition 4)}$$

* $x = c$.

Due to $x < y$, it follows that $y = b$. Then, we have the followings:

$$\mathcal{M}, [c, b] \models \langle B \rangle \varphi$$
$$\Longleftrightarrow \mathcal{M}, [c, c] \models \varphi \quad \text{(by Definition 4)}$$
$$\Longleftrightarrow \mathcal{M}', [c', c'] \models \varphi \quad \text{(by IH)}$$
$$\Longleftrightarrow M', [c', b'] \models \langle B \rangle \varphi \text{ (by Definition 4)}$$

– $\langle \overline{B} \rangle \varphi$.

- $x = a$.
  * $y = a$. We then have the followings:

$$\mathcal{M}, [a, a] \models \langle \overline{B} \rangle \varphi$$
$$\Longleftrightarrow \mathcal{M}, [a, c] \models \varphi \text{ or } \mathcal{M}, [a, b] \models \varphi \quad \text{(by Definition 4)}$$
$$\Longleftrightarrow \mathcal{M}', [a', c'] \models \varphi \text{ or } \mathcal{M}', [a', b'] \models \varphi \text{ (by IH)}$$
$$\Longleftrightarrow \mathcal{M}', [a', a'] \models \langle \overline{B} \rangle \varphi \quad \text{(by Definition 4)}$$

  * $y = c$. We then have the followings:

$$\mathcal{M}, [a, c] \models \langle \overline{B} \rangle \varphi$$
$$\Longleftrightarrow \mathcal{M}, [a, b] \models \varphi \quad \text{(by Definition 4)}$$
$$\Longleftrightarrow \mathcal{M}', [a', b'] \models \varphi \quad \text{(by IH)}$$
$$\Longleftrightarrow \mathcal{M}', [a', c'] \models \langle \overline{B} \rangle \varphi \text{ (by Definition 4)}$$

  * $y = b$. Since there is no point t in $\mathcal{M}$ satisfying $b < t$, by Definition 4, we have that $\mathcal{M}, [a, b] \not\models \langle \overline{B} \rangle \varphi$. Due to $f(y) = f(b) = b'$, similarly, we also have that $\mathcal{M}', [a', b'] \not\models \langle \overline{B} \rangle \varphi$. Therefore, we have $\mathcal{M}, [a, b] \models \langle \overline{B} \rangle \varphi$ iff $\mathcal{M}', [a', b'] \models \langle \overline{B} \rangle \varphi$.
- $x = c$. Due to $x \leq y$, there are two cases: $y = c$ or $y = b$.
  * $y = c$. We then have the followings:

$$\mathcal{M}, [c, c] \models \langle \overline{B} \rangle \varphi$$
$$\Longleftrightarrow \mathcal{M}, [c, b] \models \varphi \quad \text{(by Definition 4)}$$
$$\Longleftrightarrow \mathcal{M}', [c', b'] \models \varphi \quad \text{(by IH)}$$
$$\Longleftrightarrow \mathcal{M}', [c', c'] \models \langle \overline{B} \rangle \varphi \text{ (by Definition 4)}$$

      * $y = b$. Since there is no point t in $\mathcal{M}$ satisfying $b < t$, by Definition 4, we have that $\mathcal{M}, [c, b] \nvDash \langle \overline{B} \rangle \varphi$. Due to $f(b) = b'$, similarly, we also have that $\mathcal{M}', [c', b'] \nvDash \langle \overline{B} \rangle \varphi$. Therefore, we have $\mathcal{M}, [c, b] \vDash \langle \overline{B} \rangle \varphi$ iff $\mathcal{M}', [c', b'] \vDash \langle \overline{B} \rangle \varphi$.

- $x = b$.

  Due to $x \leq y$, it follows that $y = b$. Since there is no point t in $\mathcal{M}$ satisfying $b < t$, by Definition 4, we have that $\mathcal{M}, [b, b] \nvDash \langle \overline{B} \rangle \varphi$. Due to $f(b) = b'$, similarly, we also have that $\mathcal{M}', [b', b'] \nvDash \langle \overline{B} \rangle \varphi$. Therefore, we have $\mathcal{M}, [b, b] \vDash \langle \overline{B} \rangle \varphi$ iff $\mathcal{M}', [b', b'] \vDash \langle \overline{B} \rangle \varphi$.

Theorem 2. *HS* is strictly more expressive than HS.*

Proof. By Proposition 8, we know that there are two pointed models $\mathcal{M}, [a, b]$ and $\mathcal{M}', [a', b']$ that HS cannot distinguish them. To show that HS* is strictly more expressive than HS, we only need to show that there is a formula $\varphi \in \mathcal{L}^{\mathsf{HS}^*}$ that can distinguish them.

Firstly, we will show that $\mathcal{M}, \rho, [a, b] \vDash \mathbf{E}[E][B]p$, where ρ is a path passing through $[a, b]$. Let δ be the path $\{a, b\}$ in $\mathcal{M}$. By the semantics of HS*, we have that $\mathcal{M}, \delta, [b, b] \vDash [B]p$. Moreover, since, on path δ, the interval $[b, b]$ is the only interval that strictly ends the interval $[a, b]$, it follows that $\mathcal{M}, \delta, [a, b] \vDash [E][B]p$. Hence, $\mathcal{M}, \rho, [a, b] \vDash \mathbf{E}[E][B]p$.

Secondly, we will show that $\mathcal{M}', \rho', [a', b'] \nvDash \mathbf{E}[E][B]p$, where ρ' is the path $\{a', c', b'\}$ in $\mathcal{M}'$. By the definition of $\mathcal{M}'$, we know that $\mathcal{M}', \rho', [c', c'] \nvDash p$. By the semantics of HS*, it follows that $\mathcal{M}', \rho', [c', b'] \nvDash [B]p$. Since the interval $[c', b']$ ends the interval $[a', b']$, we then have that $\mathcal{M}', \rho', [a', b'] \nvDash [E][B]p$. Moreover, since ρ' is the only path passing through the interval $[a', b']$, it follows that $\mathcal{M}', \rho', [a', b'] \nvDash \mathbf{E}[E][B]p$.

Thus, we have shown that the HS*-formula $\mathbf{E}[E][B]p$ can distinguish the two pointed models $\mathcal{M}, [a, b]$ and $\mathcal{M}', [a', b']$.

5   Translation into MPL

It is shown in [11] that the logic HS can be translated into first-order logic. In this section, we will show that the logic HS* can be translated into Monadic Path Logic (MPL). MPL is a monadic second-order logic in which set quantification is restricted to paths (see [10]).

The MPL considered in this section is the first-order logic in [11] extended with path quantifiers. We use the individual variable symbols $x_1, x_2, \ldots$, which will designate time points, the path variable symbols $X, Y, \ldots$, which will designate sets of time points. Besides the equivalence symbol $\equiv$ and the binary relation symbol $\prec$ for partial order, the MPL also contains binary relation symbols P for each propositional letter $p \in \mathbf{P}$. Moreover, each set variable is also a unary predicate symbol.

Definition 9 (Language of MPL). *The language $\mathcal{L}^{MPL}$ is defined by the following BNF rules:*

$$\varphi ::= x_1 \equiv x_2 \mid x_1 \prec x_2 \mid P(x_1, x_2) \mid X(x_1) \mid \neg\varphi \mid (\varphi \wedge \varphi) \mid \forall x \varphi \mid \forall X \varphi$$

where x_1, x_2 are individual variables and X is a set variable.

The formula $P(x_1, x_2)$ means the interval $[x_1, x_2]$ satisfies the proposition p, and $X(x_1)$ means the time point x_1 is in the path X.

Definition 10 (Interpretation of MPL). *A* structure $\mathcal{U}$ *of MPL is defined as a pair $\mathcal{U} = \langle U, \eta \rangle$, where U is a nonempty set (the domain) and η is an assignment that associates the symbol $\prec$ with a strict partial order on the domain and associates each binary relation symbol P with a binary relation on the domain. An* interpretation *of MPL consists of a structure $\mathcal{U}$ and an assignment $\mathcal{I}$ that assigns each individual variable an element of the domain and each set variable a subset of the domain.*

Note that in each interpretation of MPL, the symbol $\prec$ is interpreted as a strict partial order on the domain. A *path* in a strict partial order is defined in Definition 7.

Definition 11 (Semantics of MPL). *The satisfaction relation $\Vdash$ between MPL-interpretations and MPL-formulas is defined as follows:*

$$
\begin{aligned}
\mathcal{U}, \mathcal{I} &\Vdash x_1 \equiv x_2 &&\Longleftrightarrow \mathcal{I}(x_1) = \mathcal{I}(x_2) \\
\mathcal{U}, \mathcal{I} &\Vdash x_1 \prec x_2 &&\Longleftrightarrow (\mathcal{I}(x_1), \mathcal{I}(x_2)) \in \eta(\prec) \\
\mathcal{U}, \mathcal{I} &\Vdash X(x_1) &&\Longleftrightarrow \mathcal{I}(x_1) \in \mathcal{I}(X) \\
\mathcal{U}, \mathcal{I} &\Vdash P(x_1, x_2) &&\Longleftrightarrow (\mathcal{I}(x_1), \mathcal{I}(x_2)) \in \eta(P) \\
\mathcal{U}, \mathcal{I} &\Vdash \neg\varphi &&\Longleftrightarrow \mathcal{U}, \mathcal{I} \nVdash \varphi \\
\mathcal{U}, \mathcal{I} &\Vdash \varphi \wedge \psi &&\Longleftrightarrow \mathcal{U}, \mathcal{I} \Vdash \varphi \text{ and } \mathcal{U}, \mathcal{I} \Vdash \psi \\
\mathcal{U}, \mathcal{I} &\Vdash \forall x \varphi &&\Longleftrightarrow \text{for all } d \in U : \mathcal{U}, \mathcal{I}(d/x) \Vdash \varphi \\
\mathcal{U}, \mathcal{I} &\Vdash \forall X \varphi &&\Longleftrightarrow \text{for all path } \rho \text{ in } \mathcal{U} : \mathcal{U}, \mathcal{I}(\rho/X) \Vdash \varphi
\end{aligned}
$$

Next, we are going to define a translation function from the language of HS* into the language of MPL and show that the translation preserves the truth value of formulas.

Definition 12. *The translation function $h : \mathcal{L}^{HS^*} \to \mathcal{L}^d MPL$ is defined as follows:*

$$
\begin{aligned}
h(p) &= P(x_1, x_2) \wedge X(x_1) \wedge X(x_2) \\
h(\neg\varphi) &= \neg h(\varphi) \\
h(\varphi \wedge \psi) &= h(\varphi) \wedge h(\psi) \\
h(\langle \mathcal{B} \rangle \varphi) &= \exists x_3 (x_1 \preceq x_3 \wedge x_3 \prec x_2 \wedge h(\varphi)[x_3/x_2]) \\
h(\langle \mathcal{E} \rangle \varphi) &= \exists x_3 (x_1 \prec x_3 \wedge x_3 \preceq x_2 \wedge h(\varphi)[x_3/x_1]) \\
h(\langle \overline{\mathcal{B}} \rangle \varphi) &= \exists x_3 (x_2 \prec x_3 \wedge h(\varphi)[x_3/x_2]) \\
h(\langle \overline{\mathcal{E}} \rangle \varphi) &= \exists x_3 (x_3 \prec x_1 \wedge h(\varphi)[x_3/x_1]) \\
h(\mathbf{A}\varphi) &= \forall Y (h(\varphi)[Y/X])
\end{aligned}
$$

where $x \preceq y$ is an abbreviation of $x \prec y \vee x \equiv y$ and $\varphi[x/y]$ is the substitution of x for all free occurrences of y.

It is easy to check that there are three free variables in $h(\varphi)$, namely, individual variables x_1, x_2 and set variable X.

Next, we will show that the translation preserves the truth value of formulas. Firstly, for each interval model of $\mathtt{HS}^*$, we define a corresponding $\mathtt{MPL}$-structure.

Definition 13. *Given an interval model $\mathcal{M} = \langle S, \mathbb{I}(S), V \rangle$ where $S = \langle D, < \rangle$, then the $\mathtt{MPL}$-structure $\mathcal{U}^{\mathcal{M}} = \langle U, \eta \rangle$ based on $\mathcal{M}$ is defined as follows: $U = D$, $\eta(\prec) = <$ and $\eta(P) = V(p)$ for each $p \in \mathbf{P}$.*

Theorem 3. *For each $\varphi \in \mathcal{L}^{\mathit{HS}^*}$, we have that $\mathcal{M}, \rho, [d, d'] \vDash \varphi$ if and only if $\mathcal{U}^{\mathcal{M}}, \mathcal{I}(d/x_1, d'/x_2, \rho/X) \Vdash h(\varphi)$.*

Proof. We prove it by induction on the structure of φ.

(1) $\varphi = p$. Let $\mathcal{M}, \rho, [d, d'] \vDash p$. By Definition 8, we know that $[d, d'] \in V(p)$. By Definition 13, we know that $\eta(P) = V(p)$ for every $p \in \mathbf{P}$. It follows that $[d, d'] \in \eta(P)$. Let $\mathcal{I}(x_1) = d, \mathcal{I}(x_2) = d', \mathcal{I}(\rho) = X$, then we have $(\mathcal{I}(x_1), \mathcal{I}(x_2)) \in \eta(P)$ and $\mathcal{I}(x_1) \in \mathcal{I}(X), \mathcal{I}(x_2) \in \mathcal{I}(X)$. By Definition 11, we have $\mathcal{U}^{\mathcal{M}}, \mathcal{I}(d/x_1, d'/x_2, \rho/X) \Vdash P(x_1, x_2) \wedge X(x_1) \wedge X(x_2)$. By Definition 12, it follows that $\mathcal{U}^{\mathcal{M}}, \mathcal{I}(d/x_1, d'/x_2, \rho/X) \Vdash h(p)$. Therefore, we have shown that if $\mathcal{M}, \rho, [d, d'] \vDash p$, then $\mathcal{U}^{\mathcal{M}}, \mathcal{I}(d/x_1, d'/x_2, \rho/X) \Vdash h(p)$. The converse holds in a similar way.

(2) $\varphi = \neg\psi$. It can be easily checked by inductive hypothesis.

(3) $\varphi = \psi_1 \wedge \psi_2$. It can be easily checked by inductive hypothesis.

(4) $\varphi = \langle \mathcal{B} \rangle \psi$. Let $\mathcal{M}, \rho, [d, d'] \vDash \langle \mathcal{B} \rangle \psi$ By Definition 8, we know that there exists $t \in \rho$ such that $d \leq t < d'$ and $\mathcal{M}, \rho, [d, t] \vDash \psi$. By inductive hypothesis, we have $\mathcal{U}^{\mathcal{M}}, \mathcal{I}(d/x_1, t/x_3, \rho/X) \Vdash h(\psi)$. Let $\mathcal{I}(x_2) = d'$. Since $d \leq t < d'$, by Definition 13, we have $x_1 \preceq x_3 \wedge x_3 \prec x_2$. It follows that $\mathcal{U}^{\mathcal{M}}, \mathcal{I}(d/x_1, d'/x_2, \rho/X) \Vdash \exists x_3 (x_1 \preceq x_3 \wedge x_3 \prec x_3 \wedge h(\psi)[x_3/x_2])$. By Definition 12, we know that $\mathcal{U}^{\mathcal{M}}, \mathcal{I}(d/x_1, d'/x_2, \rho/X) \Vdash h(\langle \mathcal{B} \rangle \psi)$. Therefore, we have shown that if $\mathcal{M}, \rho, [d, d'] \vDash \langle \mathcal{B} \rangle \psi$, then $\mathcal{U}^{\mathcal{M}}, \mathcal{I}(d/x_1, d'/x_2, \rho/X) \Vdash h(\langle \mathcal{B} \rangle \psi)$. The converse holds in a similar way.

(5) $\varphi = \langle \mathcal{E} \rangle \psi$. The proof of this is similar to the case of $\langle \mathcal{B} \rangle \psi$.

(6) $\varphi = \langle \overline{\mathcal{B}} \rangle \psi$. Let $\mathcal{M}, \rho, [d, d'] \vDash \langle \overline{\mathcal{B}} \rangle \psi$ By Definition 8, we know that there exists $t \in \rho$ such that $d' < t$ and $\mathcal{M}, \rho, [d, t] \vDash \psi$. By inductive hypothesis, we have $\mathcal{U}^{\mathcal{M}}, \mathcal{I}(d/x_1, t/x_3, \rho/X) \Vdash h(\psi)$. Let $\mathcal{I}(x_2) = d'$. Since $d' < t$, by Definition 13, we have $x_2 \prec x_3$. It follows that $\mathcal{U}^{\mathcal{M}}, \mathcal{I}(d/x_1, d'/x_2, \rho/X) \Vdash \exists x_3 (x_2 \prec x_3 \wedge h(\psi)[x_3/x_2])$. By Definition 12, we know that $\mathcal{U}^{\mathcal{M}}, \mathcal{I}(d/x_1, d'/x_2, \rho/X) \Vdash h(\langle \overline{\mathcal{B}} \rangle \psi)$. Therefore, we have shown that if $\mathcal{M}, \rho, [d, d'] \vDash \langle \overline{\mathcal{B}} \rangle \psi$, then we have $\mathcal{U}^{\mathcal{M}}, \mathcal{I}(d/x_1, d'/x_2, \rho/X) \Vdash h(\langle \overline{\mathcal{B}} \rangle \psi)$. The converse holds in a similar way.

(7) $\varphi = \langle \overline{\mathcal{E}} \rangle \psi$. The proof of this is similar to the case of $\langle \overline{\mathcal{B}} \rangle \psi$.

(8) $\varphi = \mathbf{A}\psi$. Let $\mathcal{M}, \rho, [d, d'] \vDash \mathbf{A}\psi$. By Definition 8, we know that for each path ρ' through $[d, d']$: $\mathcal{M}, \rho', [d, d'] \vDash \psi$. By inductive hypothesis, we know that for each ρ', $\mathcal{U}^{\mathcal{M}}, \mathcal{I}(d/x_1, d'/x_2, \rho'/Y) \Vdash h(\psi)$. By Definition 11, we know that $\mathcal{U}^{\mathcal{M}}, \mathcal{I}(d/x_1, d'/x_2, \rho'/Y) \Vdash \forall Y h(\psi)$. Let $\mathcal{I}(X) = \rho$. It follows that $\mathcal{U}^{\mathcal{M}}, \mathcal{I}(d/x_1, d'/x_2, \rho/X) \Vdash \forall Y (h(\psi)[Y/X]$. By Definition 12, we have that $\mathcal{U}^{\mathcal{M}}, \mathcal{I}(d/x_1, d'/x_2, \rho/X) \Vdash h(\mathbf{A}\psi)$. Therefore, we have shown if $\mathcal{M}, \rho, [d, d'] \vDash \mathbf{A}\psi$, then $\mathcal{U}^{\mathcal{M}}, \mathcal{I}(d/x_1, d'/x_2, \rho/X) \Vdash h(\mathbf{A}\psi)$. The converse holds in a similar way.

6   Conclusion

In this paper, we extend the interval temporal logic HS with path quantifiers (called HS* in this paper), which enhance its expressivity over branching interval models. Our extension of HS is analogous to the extension from CTL to CTL*. We modify the original semantics of HS by evaluating formulas on pairs of intervals and paths, rather than on intervals alone. This approach allows for the expression of interval relations without relying on the assumption of linear interval properties. Furthermore, it reveals that the temporal operators in HS represent a combined quantification of paths and time points. Consequently, similar to CTL, we cannot freely nest the quantifications of paths and time points. This limitation can be overcome in HS*. We also show that the HS* can be translated into Monadic Path Logic (MPL), a monadic second-order logic in which set quantification is restricted to paths.

Several intriguing directions for future work arise from this study. In addition to the standard research on the axiomatization and model-checking complexity of this logic, there are two natural extensions to consider. The first one is to characterize the fragment of MPL whose expressive power coincides with that of HS*. Moreover, it is shown in [3] that HS is as expressive as CTL over computation-tree-based interval models. Thus, the second one is to compare the expressivity of HS* and CTL* over computation-tree-based interval models.

Acknowledgments. The authors thank the anonymous reviewers for their detailed comments that improved the presentation of the paper. This work was funded by the Fundamental Research Funds for the Central Universities (No. 63253022).

Disclosure of Interests. The authors have no competing interests to declare that are relevant to the content of this article.

References

1. Aceto, L., Della Monica, D., Goranko, V., Ingólfsdóttir, A., Montanari, A., Sciavicco, G.: A complete classification of the expressiveness of interval logics of Allen's relations: the general and the dense cases. Acta Informatica **53**, 207–246 (2016)
2. Allen, J.F.: Maintaining knowledge about temporal intervals. Commun. ACM **26**(11), 832–843 (1983)
3. Bozzelli, L., Molinari, A., Montanari, A., Peron, A., Sala, P.: Interval vs. point temporal logic model checking: an expressiveness comparison. ACM Trans. Comput. Logic (TOCL) **20**(1), 1–31 (2018)
4. Bresolin, D., Montanari, A., et al.: A tableau-based decision procedure for a branching-time interval temporal logic. In: Proceedings of the 4th International Workshop on Methods for Modalities (M4M), pp. 38–53 (2005)
5. Chaochen, Z., Hansen, M.R.: An adequate first order interval logic. In: International Symposium on Compositionality, pp. 584–608. Springer (1997)
6. Della Monica, D., Goranko, V., Montanari, A., Sciavicco, G., et al.: Interval temporal logics: a journey. Bull. EATCS **3**(105) (2013)

7. Goranko, V., Montanari, A., Sala, P., Sciavicco, G.: A general tableau method for propositional interval temporal logics: theory and implementation. J. Appl. Log. **4**(3), 305–330 (2006)
8. Goranko, V., Montanari, A., Sciavicco, G.: Propositional interval neighborhood temporal logics. JUCS - J. Univ. Comput. Sci. **9**(9), 1137–1167 (2003). https://doi.org/10.3217/jucs-009-09-1137
9. Goranko, V., Rumberg, A.: Temporal logic. In: Zalta, E.N., Nodelman, U. (eds.) The Stanford Encyclopedia of Philosophy. Metaphysics Research Lab, Stanford University, Summer 2024 edn. (2024)
10. Gurevich, Y., Shelah, S.: The decision problem for branching time logic. J. Symbolic Logic **50**(3), 668–681 (1985). http://www.jstor.org/stable/2274321
11. Halpern, J.Y., Shoham, Y.: A propositional modal logic of time intervals. J. ACM (JACM) **38**(4), 935–962 (1991)
12. Lodaya, K.: Sharpening the undecidability of interval temporal logic. In: Jifeng, H., Sato, M. (eds.) ASIAN 2000. LNCS, vol. 1961, pp. 290–298. Springer, Heidelberg (2000). https://doi.org/10.1007/3-540-44464-5_21
13. Marx, M., Venema, Y., Marx, M., Venema, Y.: Multi-dimensional modal logic. Springer (1997)
14. Moszkowski, B.C.: Reasoning about digital circuits. Stanford University (1983)
15. Valentin Goranko, A.M., Sciavicco, G.: A road map of interval temporal logics and duration calculi. J. Appl. Non-Classical Logics **14**(1–2), 9–54 (2004). https://doi.org/10.3166/jancl.14.9-54
16. Van Benthem, J.: The logic of time: a model-theoretic investigation into the varieties of temporal ontology and temporal discourse, vol. 156. Springer Science & Business Media, 2 edn. (1991)
17. Venema, Y.: Expressiveness and completeness of an interval tense logic. Notre Dame J. Formal Logic **31**(4), 529–547 (1990)
18. Venema, Y.: A modal logic for chopping intervals. J. Log. Comput. **1**(4), 453–476 (1991)

The Modal Logic of n-State Frames

Xiaolong Liang[1], Yì N. Wáng[2(✉)], and Thomas Ågotnes[3,1]

[1] School of Philosophy, Shanxi University, Taiyuan, China
[2] School of Philosophy and Social Development, Shandong University, Jinan, China
ynw@xixilogic.org
[3] Department of Information Science and Media Studies, University of Bergen, Bergen, Norway

Abstract. Modal logics often have the finite model property. However, for many applications the state space is not only finite, but of *fixed* size n for a positive number n. An example is reasoning about a social network where the relations can vary but the set of agents (corresponding to states) is fixed. In this paper we investigate the modal logic of frames with n states, for any positive integer n, in the basic modal language. This defines a family of modal logics parameterized by n. We give complete axiomatizations of many of these logics, and characterize the computational complexity of the basic logic.

Keywords: modal logic · n-state frames · axiomatization · completeness · complexity

1 Introduction

In many contexts, reasoning about systems with a finite, and often fixed, number of states is essential. For example, the state space might be induced by a finite set of variables, with 2^n possible combinations of n Boolean variables. In social networks, states might represent individual agents, such as a committee fixed at exactly nine members. Dunbar's number suggests a theoretical upper limit of approximately 150 individuals with whom one can maintain meaningful social relationships [7,16]. Similarly, peer-to-peer networks are often designed with a fixed or maximum number of nodes, and distributed systems like blockchains may impose node limits during testing or operation.

Although these examples highlight the practical significance of finite-state systems, this paper tackles a more foundational issue: given a positive integer n, what modal logic precisely characterizes the class of all frames with exactly n states? Despite its apparent simplicity and seeming precedence, this question, to the best of our knowledge, remains unaddressed in the literature. This is not to imply that there is a lack of modal logic frameworks incorporating numerical aspects. On the contrary, several modal logics with counting mechanisms have been developed, including:[1]

---

[1] We do not delve here into first- or higher-order logics with counting quantifiers, such as logics of generalized quantifiers.

© The Author(s), under exclusive license to Springer Nature Switzerland AG 2026
V. Goranko et al. (Eds.): LORI 2025, LNCS 16010, pp. 95–109, 2026.
https://doi.org/10.1007/978-981-95-2481-5_7

- Graded modal logic [9,12], which reasons about statements with varying degrees of necessity or possibility using modalities that constrain the number of successors. The description logic $\mathcal{ALCN}$ with number restrictions [2] is a multi-modal extension, and further developments compare degrees of necessity or possibility between modal formulas [11,17].
- Logics that count the number of elements in the extension of a modal formula [1][2], extended to modalities comparing extension counts of modal formulas [3,10].
- Other approaches, such as weighted modal logic [14,20] and probabilistic modal logic [8,15,21], where numerical restrictions appear in different forms.

These frameworks are conceptually relevant to our work but distinct in their aims and methods.

The question we ask is even more fundamental. While basic modal logic (and many other modal logics) have the finite frame property, so the class of finite frames is characterized by the same axioms as the class of all frames, we are as mentioned above interested in the class of frames with a *fixed* number of states. What is the modal logic of frames with 27 states? To see that it is no longer the same, consider the most trivial example of the one-state frame on which $p \to \Box p$ is valid. Or the class of (the 16) two-state frames, on which $\left(\neg(p \vee q) \wedge \Diamond p \wedge \Diamond q\right) \to \Diamond(p \wedge q)$ holds. In this paper we give a complete axiomatization of the class of all frames with n states in the basic modal language, for any positive integer n. The result in fact carries over to a large class of completeness results; to any canonical modal logic, wrt. any class of frames that is closed under generated submodels.

Additionally, we investigate the computational complexity of the satisfiability problem for the basic modal logic over n-state frames. We show that when n is fixed, the problem is NP-complete, and when n is part of the input, it is PSPACE-complete.

The rest of the paper is structured as follows. Section 2 provides preliminary definitions, notational conventions, and a brief initial observation. In Sect. 3, we present axiomatizations, and in Sect. 4, we look at computational complexity for the basic logic. We conclude with a discussion in Sect. 5.

2   The Basics

We briefly review standard basic concepts from modal logic, and adapt them to n-state frames. For a more detailed treatment of standard concepts and theory, we refer the reader to textbooks on modal logic, such as [4].

We work within the standard language of unary modal logic, built upon a countable set of atomic propositions, denoted $\text{PROP} = \{p_1, p_2, \dots\}$. Formulas, represented by φ, are constructed according to the following grammar (where $p \in \text{PROP}$):

$$\varphi ::= p \mid \neg\varphi \mid (\varphi \wedge \varphi) \mid \Diamond\varphi$$

---

[2] Our logic is a fragment of the logic $\mathcal{MLC}$ from [1]: any formula φ in our logic of n-frames translates to $(\varphi \wedge (\top \geq n) \wedge (\top \leq n))$ in $\mathcal{MLC}$.

Standard definitions apply for other Boolean connectives and the $\Box$ operator.

A *frame* is a pair $F = (W, R)$ where W is a nonempty set of states (the *domain* of F) and R is a binary relation on W. For a positive integer n, an *n-frame* is a frame whose domain contains exactly n states. A *model* based on a frame F is a pair $M = (F, V)$ where $V : \text{PROP} \to \wp(W)$ is a valuation function mapping each atomic proposition to the set of states in which it is true. A *pointed model* is a pair (M, w) where w is a state in model M. If the model M is based on frame F, we also say that the pointed model (M, w) is based on F. Parentheses in these structures may be omitted for brevity. The *satisfaction* or *truth* of a formula in a pointed model (M, w) is defined inductively as follows:

$$
\begin{aligned}
M, w &\models p & &\Longleftrightarrow\; w \in V(p) \\
M, w &\models \neg\varphi & &\Longleftrightarrow\; \text{not } M, w \models \varphi \\
M, w &\models (\varphi \wedge \psi) & &\Longleftrightarrow\; M, w \models \varphi \text{ and } M, w \models \psi \\
M, w &\models \Diamond\varphi & &\Longleftrightarrow\; \text{there exists } u \in W \text{ such that } wRu \text{ and } M, u \models \varphi.
\end{aligned}
$$

A formula φ is *valid* (denoted $\models \varphi$) if it is satisfied in all pointed models. A formula φ is *satisfiable* (denoted $Sat(\varphi)$) if it is satisfied in at least one pointed model. Similarly, for a positive integer n, a formula φ is *n-valid* (denoted $\models_n \varphi$) if it is satisfied in all pointed models based on n-frames, and φ is *n-satisfiable* (denoted $Sat_n(\varphi)$) if it is satisfied in at least one pointed model based on an n-frame.

The following notational conventions will be useful. Let φ be a formula and Γ be a set of formulas. Given a pointed model (M, w), $M, w \models \Gamma$ means that $M, w \models \psi$ for all $\psi \in \Gamma$. $Sat_n(\Gamma)$ (where n is a positive integer) means that there is a pointed model (M, w) based on an n-frame such that $M, w \models \Gamma$.

Lemma 1. *Let n be a positive integer. For any set Γ of formulas,*

$$Sat_n(\Gamma) \;\Longleftrightarrow\; \exists k \leq n.\; Sat_k(\Gamma).$$

Proof. The direction from left to right is immediate. For the converse, assume $Sat_k(\Gamma)$ for some $k \leq n$. Let (M^k, w) be a pointed model where $M^k = (W, R, V)$, with (W, R) being a k-frame, and $M^k, w \models \Gamma$. We construct a model $M^n = (W', R', V')$ as follows: $W' = W \cup \{w_{k+1}, \ldots, w_n\}$, where $w_{k+1}, \ldots, w_n$ are distinct states not in W; $R' = R$; and $V'(p) = V(p)$ for all $p \in \text{PROP}$. It is clear that M^n is based on an n-frame, since W' contains n states. Moreover, for all formulas χ, we have $M^k, w \models \chi$ if and only if $M^n, w \models \chi$, since the newly introduced states $w_{k+1}, \ldots, w_n$ in M^n are unreachable from w. It follows that $M^n, w \models \Gamma$. $\qquad\Box$

3 Axiomatization

We introduce an axiom schema to characterize the modal logic of n-frames. For any positive integer n, consider the following axiom schema A_n, where $x_1, \ldots, x_n$ are integers satisfying $1 \leq x_i \leq n$:

$$
\left(\varphi_0 \wedge \bigwedge_{1 \leq i \leq n} \Diamond^{x_i}\varphi_i \right) \to \bigvee_{0 \leq j < k \leq n} \Diamond^{x_k}(\varphi_j \wedge \varphi_k) \tag{A_n}
$$

where $\Diamond^1 = \Diamond$ and $\Diamond^{i+1} = \Diamond\Diamond^i$ denotes the i-fold composition of $\Diamond$.

For a given $n \geq 1$, the A_n axiom says that if we can reach states where $\varphi_1, \ldots, \varphi_n$ are true, through sequences of at least one step, and φ_0 is true in the current state, then at least two of those formulas $(\varphi_0, \ldots, \varphi_n)$ must be true in the same state. This is clearly the case for n-frames. In fact, the examples discussed in the introduction follow from this schema in addition to standard modal reasoning. For $n = 1$, A_n is $(\varphi_0 \wedge \Diamond^{x_1} \varphi_1) \rightarrow \Diamond^{x_1}(\varphi_0 \wedge \varphi_1)$. From the instance $(p \wedge \Diamond \neg p) \rightarrow \Diamond \bot$ (let $\varphi_0 = p$, $\varphi_1 = \neg p$ and $x_1 = 1$), we get that $\Box\top \rightarrow (p \rightarrow \Box p)$ and thus $p \rightarrow \Box p$ by modal reasoning. For $n = 2$, let $\varphi_0 = \neg(p \vee q)$, $\varphi_1 = p$, $\varphi_2 = q$, $x_1 = x_2 = 1$, and we get $\left(\neg(p \vee q) \wedge \Diamond p \wedge \Diamond q\right) \rightarrow \Diamond(p \wedge q)$ more or less directly.

We will show formally below that A_n indeed holds on n-frames, and so the question is whether n-frames have other properties. We now show that in a precise sense the answer is "no": the axiomatic system $\mathbf{K}$ extended with the A_n axiom schema is sound and complete with respect to all n-frames. For given n, let the axiomatic system $\mathbf{K}n$ be $\mathbf{K}$ extended with the A_n schema.

Soundness of $\mathbf{K}n$, and indeed of any sound modal logic extended with A_n, follows immediately from the following.

Theorem 2 (soundness). *The schema A_n is valid on the class of n-frames.*

Proof. Let $M, w_0 \models \varphi_0 \wedge \bigwedge_{1 \leq i \leq n} \Diamond^{x_i} \varphi_i$, where M is a model based on an n-frame. Thus, for each $1 \leq i \leq n$ there is a state w_i reachable from w_0 in x_i steps, such that $M, w_i \models \varphi_i$. Since there are only n states, at least two of $w_0, w_1, \ldots, w_n$ must be the same. Say, w_j and w_k – let j be the smallest index of the two (so $k > 0$). We thus have that $M, w_k \models (\varphi_j \wedge \varphi_k)$ and $M, w \models \Diamond^{x_k}(\varphi_j \wedge \varphi_k)$. $\qquad\square$

Theorem 3 (completeness). $\mathbf{K}n$ *is strongly complete with respect to the class of n-frames: for any $\mathbf{K}n$-consistent set Γ of formulas, $Sat_n(\Gamma)$ holds.*

Proof. We recall the definition of the standard canonical model $M^c = (W^c, R^c, V^c)$:

- W^c is the set of all maximal $\mathbf{K}n$-consistent set of formulas (hereafter *MCSs*);
- For $\Gamma, \Delta \in W^c$, $\Gamma R^c \Delta$ if and only if $\varphi \in \Delta$ implies $\Diamond\varphi \in \Gamma$ for all φ;
- $V^c(p) = \{\Gamma \in W^c \mid p \in \Gamma\}$ for each atomic proposition p.

Since $\mathbf{K}n$ is a normal modal logic, the Canonical Model Theorem ([4, Theorem 4.22]) ensures that $\mathbf{K}n$ is strongly complete with respect to M^c. That is, for any $\mathbf{K}n$-consistent set Γ, there exists an MCS $\Gamma^+ \supseteq \Gamma$ such that $M^c, \Gamma^+ \models \Gamma$.

We proceed by arguing two claims.

(1) *The canonical model is based on an "n-out frame", where every state has fewer than n descendants (reachable states via R^c and not counting itself).* Suppose, for contradiction, that some MCS $\Gamma_0 \in W^c$ has at least n distinct descendants $\Gamma_1, \ldots, \Gamma_n$, each reachable from Γ_0 via R^c in k_i steps, where $1 \leq k_1 \leq \cdots \leq k_n \leq n$ (without loss of generality). Since $\Gamma_0, \Gamma_1, \ldots, \Gamma_n$ are

pairwise distinct MCSs, for each $i = 0, \ldots, n$, there exists a formula $\gamma_i \in \Gamma_i$ such that $\gamma_i \notin \Gamma_j$ for all $j \neq i$. Define:

$$P_0' = \gamma_0 \qquad P_1' = \gamma_1 \wedge \neg\gamma_0 \qquad \cdots \qquad P_n' = \gamma_n \wedge \neg\gamma_{n-1} \wedge \cdots \wedge \neg\gamma_0.$$

By construction, $P_m' \in \Gamma_m$ for $m = 0, \ldots, n$. Since for all MCSs X and Y, $X R^c Y$ implies $\Diamond\varphi \in X$ for all $\varphi \in Y$, it follows that $\Diamond^{k_i} P_i' \in \Gamma_0$ for each $i = 1, \ldots, n$, where k_i is the steps from Γ_0 to Γ_i via R^c. Thus, $P_0' \wedge \Diamond^{k_1} P_1' \wedge \cdots \wedge \Diamond^{k_n} P_n' \in \Gamma_0$. By A_n that means that $\Diamond^{k_i}(P_j' \wedge P_i') \in \Gamma_0$ for some $0 \leq j < i \leq n$. But that contradicts consistency of Γ_0.

(2) *If a set Γ of formulas is satisfied in a pointed model based on an n-out frame, then it is satisfied in a pointed model based on an n-frame.* Let (M, w) be a pointed model based on an n-out frame such that $M, w \models \Gamma$. Consider the submodel M' generated by w (see [4, Definition 2.5] for point-generated submodels). Since M is an n-out frame, M' has at most n states, and $M, w \models \Gamma$ if and only if $M', w \models \Gamma$. By Lemma 1, any set satisfied in a model with at most n states is satisfied in a model based on an n-frame. Thus, Γ is satisfied in such a model.

From (1), M^c is based on an n-out frame, and from (2), any set satisfiable in an n-out frame is satisfiable in an n-frame. Since every $\mathbf{K}n$-consistent set is satisfied in some M^c state, $\mathbf{K}n$ is strongly complete with respect to the class of n-frames. $\square$

Note that the proof in the previous theorem only relied on two properties of the underlying logic K and its associated class of frames: the logic is canonical, and the class of frames is closed under taking generated submodels. Specifically, since A_n is a Sahlqvist formula, it is complete with respect to the first-order class of frames it defines, i.e., the n-out frames, and thus complete for n-frames, provided that truth is preserved under taking generated submodels. The result immediately generalizes to other such logics and frame classes, including the following. Let $\mathbf{S4n}$ be $\mathbf{S4}$ extended with the A_n axiom schema, and so on.

Corollary 4. *The axiomatic systems $\boldsymbol{Tn}$, $\boldsymbol{KDn}$, $\boldsymbol{KBn}$, $\boldsymbol{K4n}$, $\boldsymbol{S4n}$ and $\boldsymbol{S5n}$ are all sound and complete with respect to the class of n-frames satisfying the corresponding frame properties (i.e., reflexivity; seriality; symmetry; transitivity; reflexivity and transitivity; and equivalence relations, respectively).*[3]

4   Computational Complexity

This section investigates the computational complexity of the satisfiability problem for the modal logic of n-frames. We consider two distinct scenarios:

1. *Fixed n*: given a fixed positive integer n and a formula φ, what is the worst-case time complexity, with respect to the length of φ (denoted $|\varphi|$), of determining whether φ is satisfiable in an n-frame (i.e., whether $Sat_n(\varphi)$)?

---

[3] The system $\mathbf{S4n}$ is equivalent to $\mathbf{S4}$ extended with the axiom schema $\boldsymbol{alt}_n$ (see, e.g., [5, Section 3.5] for details on $\boldsymbol{alt}_n$).

2. *Variable n, i.e., n is part of the input*: given n (encoded in $\lceil \log n \rceil$ bits) and a formula φ (of length $|\varphi|$), what is the worst-case time complexity, with respect to both $\lceil \log n \rceil$ and $|\varphi|$, of determining whether $Sat_n(\varphi)$?

These two scenarios lead to different complexity results, which we will address in the remainder of this section. As noted in the introduction, our logic is a fragment of the logic $\mathcal{MLC}$ [1]. From a complexity perspective, this fragment is simpler: the satisfiability problem for $\mathcal{MLC}$ in the variable-n scenario is EXPTIME-hard and in 2-NEXPTIME, much higher than ours.

4.1   When n is Fixed

When n is a fixed constant, the satisfiability problem for a given formula φ is NP-complete. The argument rests on the fact that the size of the models we need to consider is bounded by a constant. Specifically, every n-frame contains n states, and there are at most 2^{n^2} non-isomorphic n-frames. Since n is constant, these numbers are also constant. For each of these finitely many n-frames, we can guess a valuation for each state. Then, using an algorithm analogous to the bottom-up semantic tableau method, we can verify in polynomial time whether the formula φ is satisfied in a given state of the model. This provides a polynomial-time verifier for the truth of φ in a given model, demonstrating that the satisfiability problem is in NP, similar to the well-known result for propositional logic [6,18]. Furthermore, the problem is NP-hard because propositional logic—the satisfiability problem for which is known to be NP-complete—is a sublogic of it; any instance of propositional logic can be trivially translated into this logic with any fixed n, preserving satisfiability.

It is crucial to emphasize that this argument relies on n being a fixed constant, independent of the input formula φ. Therefore, the size of the model remains bounded and does not contribute to the input complexity. The situation changes dramatically when n becomes part of the input, as we will discuss in the following section.

4.2   When n is Part of the Input

We investigate the complexity of the $Sat_n(\varphi)$ problem, i.e., the satisfiability problem for a given formula φ and a positive integer n, which asks whether φ is satisfiable in a pointed model based on an n-frame. We show that this problem is PSPACE-complete, where the input size is defined as $|\varphi| + \lceil \log n \rceil$, reflecting the length of the formula and the logarithmic encoding of n.

PSPACE Hardness

Theorem 5. *The $Sat_n(\varphi)$ problem is PSPACE-hard.*

Proof. To establish PSPACE-hardness, we reduce the standard modal satisfiability problem, $Sat(\varphi)$ (i.e., satisfiability in the modal logic K), to $Sat_n(\varphi)$. First, we prove the equivalence: $Sat(\varphi)$ holds if and only if $Sat_{2^{|\varphi|}}(\varphi)$ holds.

- Left-to-right: by [13, Theorem 3.2], if φ is satisfiable in standard modal logic, it is satisfiable in a model with at most $2^{|\varphi|}$ worlds. Since such a model is based on a frame with size bounded by $2^{|\varphi|}$, φ is $2^{|\varphi|}$-satisfiable, i.e., $Sat_{2^{|\varphi|}}(\varphi)$ holds.
- Right-to-left: if $Sat_{2^{|\varphi|}}(\varphi)$ holds, then φ is satisfiable in a pointed model based on a $2^{|\varphi|}$-frame, which is a standard Kripke model. Thus, $Sat(\varphi)$ holds by definition.

Next, consider the reduction's complexity. The input size of $Sat_n(\varphi)$ is $|\varphi| + \lceil \log n \rceil$. Constructing the number $2^{|\varphi|}$ (in decimal) requires $O(|\varphi|)$ bits, and this can be done in polynomial time (in the length of φ). Thus, the reduction from $Sat(\varphi)$ to $Sat_{2^{|\varphi|}}(\varphi)$ is polynomial-time computable. Since the satisfiability problem for the modal logic **K** (the $Sat(\varphi)$ problem) is PSPACE-hard [19], and there exists an efficient (polynomial-time) reduction from $Sat(\varphi)$ to the satisfiability problem for the logic of n-frames (the $Sat_n(\varphi)$ problem), it follows that $Sat_n(\varphi)$ is PSPACE-hard. $\qquad\square$

A PSPACE Decision Procedure Using the Tableau Method. We adapt the tableau method, as described in Sect. 6.3 of [13], to develop a PSPACE algorithm for deciding $Sat_n(\varphi)$, i.e., whether a formula φ is satisfied in a pointed model based on an n-frame.

A *propositional tableau* is a set T of formulas that satisfies the following conditions:

1. If $\neg\neg\varphi \in T$, then $\varphi \in T$;
2. If $\varphi \wedge \psi \in T$, then both $\varphi \in T$ and $\psi \in T$;
3. If $\neg(\varphi \wedge \psi) \in T$, then either $\neg\varphi \in T$ or $\neg\psi \in T$;
4. For no formula φ are both φ and $\neg\varphi$ in T.

A *modal tableau* is a tuple $T = (W, R, L)$, where (W, R) is a frame and L is a labeling function assigning to each $w \in W$ a set $L(w)$ of formulas such that:

1. $L(w)$ is a propositional tableau;[4]
2. If $\Diamond\varphi \in L(w)$, then there exists $w' \in W$ such that wRw' and $\varphi \in L(w')$;
3. If $\neg\Diamond\varphi \in L(w)$ and wRw', then $\neg\varphi \in L(w')$.

A *modal n-tableau* is a modal tableau with exactly n states. We say that T is a *modal (n-)tableau for* φ if it is a modal $(n$-$)$tableau and $\varphi \in L(w)$ for some state $w \in W$.

Our tableau construction builds on the approach in Sect. 6.3 of [13], with key adaptations to limit the number of states to n. This restriction introduces two requirements: (i) we must monitor the total number of states in the tableau, and (ii) when applying rules for diamond formulas, we must reuse existing states

---

[4] Note that w qualifies as a state only when $L(w)$ forms a propositional tableau. During the tableau construction, we may introduce "internal nodes" with labels that do not satisfy this requirement. Thus, all states are nodes in the construction process, but the converse does not always hold.

rather than only create new ones, ensuring the tableau remains within the n-state bound.

The construction of the modal n-tableau for φ follows, with terminology such as "witness," "blatantly inconsistent," and "pre-tableau" as defined in [13].

Step 1. Start with a tree consisting of a single node s_0 (the root), labeled $L(s_0) = \{\varphi\}$. Initialize a counter $C = 1$ to track the number of nodes.

Step 2. Repeat (a)–(c) below until none of them apply:

(a) *Forming a propositional tableau.* If s is a leaf node, not marked "stop," with $L(s)$ neither blatantly inconsistent nor a propositional tableau, and ψ is the least witness to this, then:

   (i) If ψ is of the form $\neg\neg\psi'$, create a successor node s' of s with $L(s') = L(s) \cup \{\psi'\}$;

   (ii) If ψ is of the form $\psi_1 \wedge \psi_2$, create a successor node s' of s with $L(s') = L(s) \cup \{\psi_1, \psi_2\}$;

   (iii) If ψ is of the form $\neg(\psi_1 \wedge \psi_2)$, create two successors s_1 and s_2 of s (each on a separate branch), with $L(s_1) = L(s) \cup \{\neg\psi_1\}$ and $L(s_2) = L(s) \cup \{\neg\psi_2\}$.

(b) *Handling the $\Diamond$-operator.* Suppose s is a leaf node, $C \leq n$, $L(s)$ is a propositional tableau (making s a state) and not blatantly inconsistent. Let m be the number of states in the branch containing s (internal nodes are not counted), and let $L(s)$ contain k formulas of the form $\Diamond\psi$, denoted $\Diamond\psi_1, \ldots, \Diamond\psi_k$. For each $i = 1, \ldots, k$, apply one of the following actions:

   (i) *Reuse an unpicked state.* Select an existing state t from the current branch not yet chosen for any $j < i$. Create a duplicate t' as a successor of s, and set $L(t') = L(t) \cup \{\neg\chi \mid \neg\Diamond\chi \in L(s)\} \cup \{\psi_i\}$.

     – For efficiency, if t already has a duplicate t'' in the branch, use t'' instead of t.

     – If $L(t') = L(t)$ (i.e., no new formulas are added), mark t' "stop." Since $L(t')$ may not be a propositional tableau, apply 2(a) immediately to generate a descendant state t'_x. If t' is marked "stop," mark t'_x "stop" as well. If t'_x is not "stop," proceed to Step 2(b) on t'_x.

     – For each successor state u of t (excluding the new t'), create a successor u' of t'_x with $L(u') = L(u) \cup \{\neg\chi \mid \neg\Diamond\chi \in L(t'_x)\}$. If $L(u') = L(u)$, mark u' "stop." If $L(u')$ is not a propositional tableau, apply Step 2(a) to produce a descendant state u'_y, marking it "stop" if u' is "stop." Proceed to Step 2(b) on u'_y if it is not "stop."

     – Repeat this process to replicate the structure of t's descendants under s.

   (ii) *Reuse a picked state.* Select a state t already chosen for some $j < i$, where t' is its most recent duplicate successor of s. Update $L(t') = L(t') \cup \{\neg\chi \mid \neg\Diamond\chi \in L(s)\} \cup \{\psi_i\}$. If $L(t')$ changes, remove the "stop" mark from t' and all its successors, then apply Step 2(b) to t' as if it were a leaf. If $L(t')$ is unchanged, retain the "stop" mark.

   (iii) *Create a new state.* If $C < n$, create a new state t as a successor of s with $L(t) = \{\neg\chi \mid \neg\Diamond\chi \in L(s)\} \cup \{\psi_i\}$, and increment C by 1.

Each combination of choices for the k formulas generates a distinct branch from s. The total number of possible branches is at most $A^k_{m+k} = \frac{m!}{k!}$, as this counts the ways to assign the k $\Diamond$-formulas to m existing states or new ones within the n-bound.

(c) *Marking nodes as satisfiable.* For brevity, we use the symbol "$\checkmark$" in place of "satisfiable" where space is limited (e.g., in Figs. 1 and 2). If a node s is not yet marked "satisfiable," mark it "satisfiable" if one of the following holds:

   (i) $L(s)$ is not a propositional tableau, and at least one successor s' of s is marked "satisfiable;"

   (ii) $L(s)$ is a propositional tableau, contains no formulas of the form $\Diamond\psi$, and is not blatantly inconsistent;

   (iii) $L(s)$ is a propositional tableau, s has successors (possibly across different branches), and there exists a branch where all successors of s on that branch are marked "satisfiable."

Step 3. Check the root node of the tree. If it is marked "satisfiable," return "φ is n-satisfiable"; otherwise, return "φ is not n-satisfiable."

To illustrate, we present an example of tableau construction for the formula $\varphi = \big(p \wedge \neg(p \wedge q)\big) \wedge \Big(\big(\neg\Diamond p \wedge \Diamond(r \wedge q)\big) \wedge \Diamond\Diamond\neg q\Big)$, adapted from [13, Fig. 3]. Here, φ is slightly modified to enrich the example, providing greater clarity for the discussion that follows. Figure 1 depicts a classical modal tableau without restrictions on the number of states; for a detailed explanation, see [13]. In the following, we focus on constructing a modal n-tableau for the same formula, incorporating the n-state constraint.

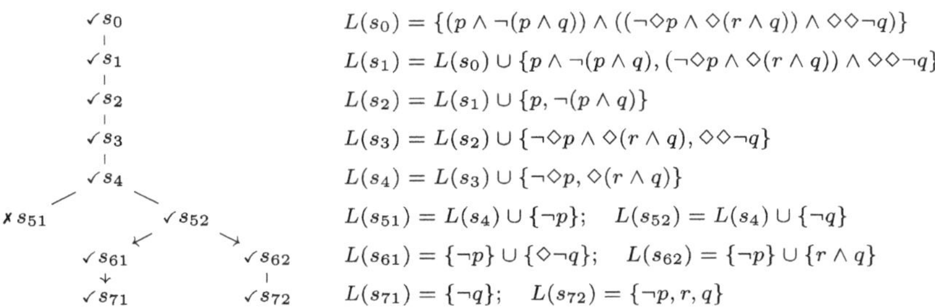

Fig. 1. Classical tableau construction for $\big(p \wedge \neg(p \wedge q)\big) \wedge \Big(\big(\neg\Diamond p \wedge \Diamond(r \wedge q)\big) \wedge \Diamond\Diamond\neg q\Big)$.

Figure 2 illustrates the modal n-tableau construction for φ. We begin by explaining the naming conventions. Subscripts indicate the order of node generation within the propositional tableau construction. When a branch occurs, we assign a number to the edge (or a group of modal successor edges), and the superscript of a node records its branch path. For instance, s_5^{251} denotes the fifth iteration of the root node s_0 after applying propositional rules, at which point it becomes a state, located in the first subbranch of the fifth subbranch of the second main branch. All nodes labeled with s within the same branch (e.g., "251") represent a single state (e.g., s_0–s_4, s_5^2 and s_5^{251}), though only the node with the lowest height is fully expanded (e.g., s_5^{251})—others contain only a subset of the state's formulas. In contrast, t_0^{251}, marked with a "t" prefix, represents a distinct state in the same "251" branch, in its initial version. "$\checkmark$" marks satisfiable nodes and "$\boldsymbol{X}$" unsatisfiable nodes, supplementing the notation introduced earlier.

The state s_5^2 includes two diamond formulas, $\Diamond\Diamond\neg q$ and $\Diamond(r \wedge q)$, and one box formula, $\neg\Diamond p$, giving rise to five subbranches. The first subbranch, "21," represents the case where s_5^2 itself witnesses both diamond formulas: by resolving the leftmost $\Diamond$, the remaining subformulas persist in s_5^2. Subbranches "22" and "23" correspond to scenarios where one diamond formula is witnessed by s_5^2 and the other by a new state, t_0^{22} or t_0^{23}, respectively. Subbranches "24" and "25" depict cases where both diamond formulas are witnessed by new states: in "24," they are satisfied by a single new state, t_0^{24}, while in "25," they are witnessed by two distinct new states, t_0^{25} and u_0^{25}. The box formula $\neg\Diamond p$ propagates $\neg p$ to all successors of s_5^2. This leads to blatant inconsistency in nodes s_5^{21}, s_5^{22} and s_5^{23} (e.g., contradicting p from the original formula), marking them "$\boldsymbol{X}$" to indicate unsatisfiability.

The duplicate nodes s_5^{241} (from s_5^2) and t_1^{241} (from t_1^{24}) remain unchanged and are thus marked "stop" to prevent looping. Similarly, the duplicates s_5^{251} (from s_5^2), t_0^{251} (from t_0^{25}), and u_0^{251} (from u_0^{25}) are marked "stop" for the same reason.

The tableau in Fig. 2 contains at most four states (e.g., in branch "254"), qualifying it as a modal n-tableau for $n \geq 4$. To accommodate a model with even fewer states, we can adapt the tableau, as shown in Fig. 3.

Lemma 6. *For any formula φ and positive integer n, the construction of the modal n-tableau for φ terminates. Furthermore, it does so using space polynomial in $|\varphi| + \lceil \log n \rceil$.*

Proof. Let $|\varphi| = m$. For any node s in the tableau tree, $L(s)$ contains only subformulas of φ or their negations, so $|L(s)| \leq 2m$. Consider Step 2(a), which resolves Boolean operators. Since each application of Step 2(a) processes a witness formula in $L(s)$, reducing the number of unresolved Boolean operators, it can be applied at most m times before $L(s)$ either becomes a propositional tableau (making s a state) or becomes blatantly inconsistent, halting further expansion at s. Next, consider Step 2(b), driven by $\Diamond$ operators. When s' is a successor state of s created by a $\Diamond\psi$ rule, the modal depth of ψ (the maximum nesting of modal operators in any formula of $L(s')$) is less than that of $\Diamond\psi$ in $L(s)$. For duplicated states (as in Step 2(b)(i)), the $\Diamond$ formula triggering the duplication is "condensed" (i.e., replaced by ψ_i). Existing $\Diamond$ formulas in copied states persist until processed, at which point they are either condensed or the node is marked "stop" to prevent infinite loops. Thus, the modal depth decreases along each path, ensuring the tree's depth is finite. The tree's finiteness follows: Step 2(a) bounds Boolean expansions, Step 2(b) bounds modal expansions via decreasing depth and the n-state limit, and Steps 2(c) and 3 terminate once the tree is fully constructed.

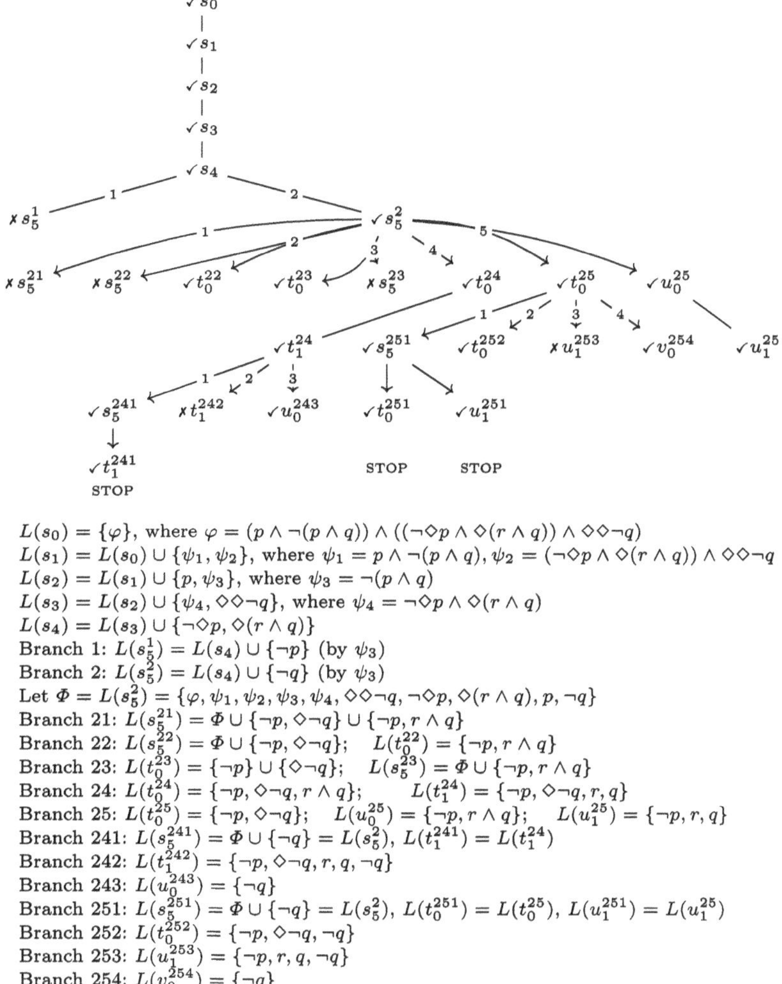

$L(s_0) = \{\varphi\}$, where $\varphi = (p \wedge \neg(p \wedge q)) \wedge ((\neg\Diamond p \wedge \Diamond(r \wedge q)) \wedge \Diamond\Diamond\neg q)$
$L(s_1) = L(s_0) \cup \{\psi_1, \psi_2\}$, where $\psi_1 = p \wedge \neg(p \wedge q)$, $\psi_2 = (\neg\Diamond p \wedge \Diamond(r \wedge q)) \wedge \Diamond\Diamond\neg q$
$L(s_2) = L(s_1) \cup \{p, \psi_3\}$, where $\psi_3 = \neg(p \wedge q)$
$L(s_3) = L(s_2) \cup \{\psi_4, \Diamond\Diamond\neg q\}$, where $\psi_4 = \neg\Diamond p \wedge \Diamond(r \wedge q)$
$L(s_4) = L(s_3) \cup \{\neg\Diamond p, \Diamond(r \wedge q)\}$
Branch 1: $L(s_5^1) = L(s_4) \cup \{\neg p\}$ (by ψ_3)
Branch 2: $L(s_5^2) = L(s_4) \cup \{\neg q\}$ (by ψ_3)
Let $\Phi = L(s_5^2) = \{\varphi, \psi_1, \psi_2, \psi_3, \psi_4, \Diamond\Diamond\neg q, \neg\Diamond p, \Diamond(r \wedge q), p, \neg q\}$
Branch 21: $L(s_5^{21}) = \Phi \cup \{\neg p, \Diamond\neg q\} \cup \{\neg p, r \wedge q\}$
Branch 22: $L(s_5^{22}) = \Phi \cup \{\neg p, \Diamond\neg q\}$; $L(t_0^{22}) = \{\neg p, r \wedge q\}$
Branch 23: $L(t_0^{23}) = \{\neg p\} \cup \{\Diamond\neg q\}$; $L(s_5^{23}) = \Phi \cup \{\neg p, r \wedge q\}$
Branch 24: $L(t_0^{24}) = \{\neg p, \Diamond\neg q, r \wedge q\}$; $L(t_1^{24}) = \{\neg p, \Diamond\neg q, r, q\}$
Branch 25: $L(t_0^{25}) = \{\neg p, \Diamond\neg q\}$; $L(u_0^{25}) = \{\neg p, r \wedge q\}$; $L(u_1^{25}) = \{\neg p, r, q\}$
Branch 241: $L(s_5^{241}) = \Phi \cup \{\neg q\} = L(s_5^2)$, $L(t_1^{241}) = L(t_1^{24})$
Branch 242: $L(t_1^{242}) = \{\neg p, \Diamond\neg q, r, q, \neg q\}$
Branch 243: $L(u_0^{243}) = \{\neg q\}$
Branch 251: $L(s_5^{251}) = \Phi \cup \{\neg q\} = L(s_5^2)$, $L(t_0^{251}) = L(t_0^{25})$, $L(u_1^{251}) = L(u_1^{25})$
Branch 252: $L(t_0^{252}) = \{\neg p, \Diamond\neg q, \neg q\}$
Branch 253: $L(u_1^{253}) = \{\neg p, r, q, \neg q\}$
Branch 254: $L(v_0^{254}) = \{\neg q\}$

Fig. 2. Modal n-tableau $(n \geq 4)$ for $(p \wedge \neg(p \wedge q)) \wedge \big((\neg\Diamond p \wedge \Diamond(r \wedge q)) \wedge \Diamond\Diamond\neg q\big)$.

On to space complexity. The parameter n is compared against the counter C, requiring only $O(\log n)$ bits. For Boolean operators, Step 2(a) generates at most m nodes per branch to form a propositional tableau. Modal rules in Step 2(b) complicate this:

- If each $\Diamond$ formula created a new state (up to m times), the branch length would be $O(m^2)$, as there are at most m modal formulas, each spawning a subtree of depth $O(m)$ via Step 2(a).
- Reusing states (Steps 2(b)(i) and (ii)) increases this: duplicating a state t copies its subtree, but only the latest duplicate is used, and the number of states in a branch is capped at m. Each $\Diamond$ formula triggers at most m dupli-

cations (bounded by subformulas), and each duplication spawns a subtree of size $O(m)$, yielding a branch size of $O(m^3)$.

– The number of branches, however, may grow exponentially (e.g., A^k_{m+k} from Step 2(b)), where $k \leq m$ is the number of $\Diamond$ formulas.

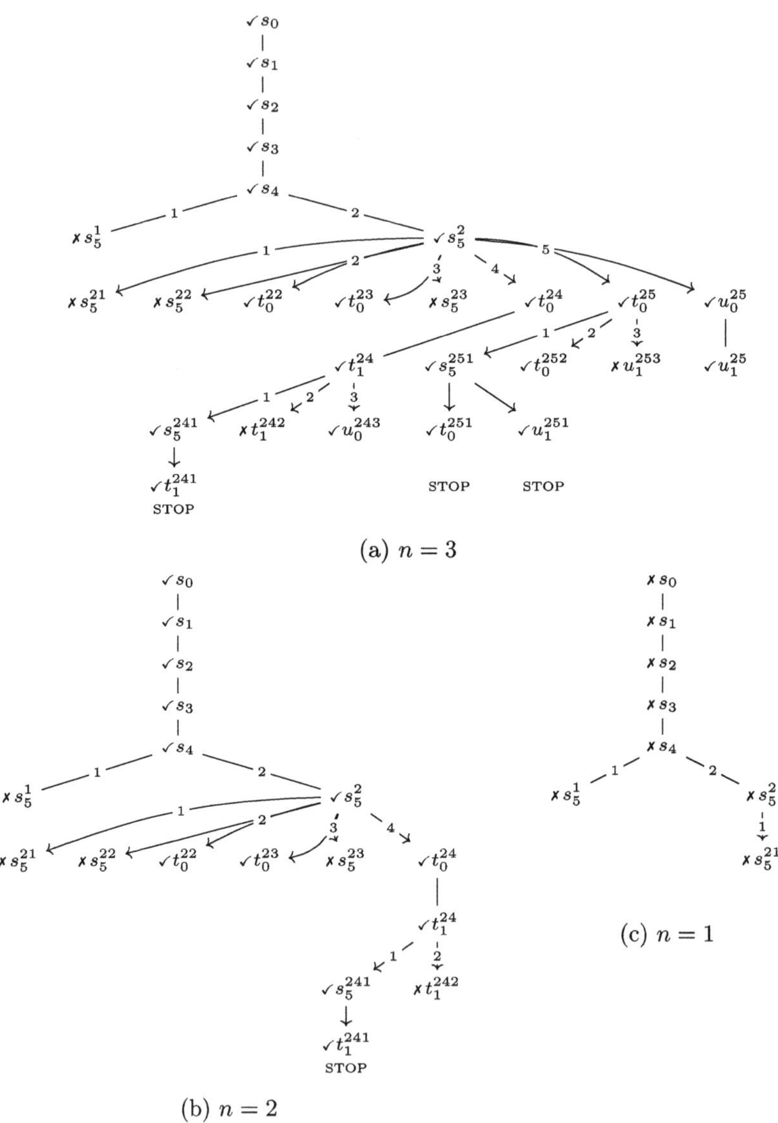

(a) $n = 3$

(b) $n = 2$

(c) $n = 1$

Fig. 3. Modal n-tableau ($n = 1, 2, 3$) for $\left(p \wedge \neg(p \wedge q)\right) \wedge \left(\left(\neg\Diamond p \wedge \Diamond(r \wedge q)\right) \wedge \Diamond\Diamond\neg q\right)$.

Despite exponential branching, a polynomial-space algorithm exists by exploring branches nondeterministically, reusing space. Adapting the proof of [13, Theorem 6.11], for PSPACE decidability of modal satisfiability, our construction fits in $O(m^4)$ space: $O(m^3)$ for a branch's nodes, with $O(m)$ additional bits to track state labels and counters, processed in a depth-first manner. Thus, total space is polynomial in $m + \log n$. $\qquad\square$

Lemma 7. *A formula φ is n-satisfiable if and only if the construction of the modal n-tableau for φ returns "φ is satisfiable."*

Proof. Suppose the construction terminates with the root s_0 marked "satisfiable" (Step 3). Consider a branch where all states are marked "satisfiable." The states in this branch (excluding internal nodes) form the domain W of a model $M = (W, R, V)$, where R is defined by successor relations from the tableau (as depicted by arrows in the figures), and $V(p) = \{w \in W \mid p \in L(w)\}$ for each atomic proposition p.[5] Let s' be a state (i.e., $L(s')$ is a propositional tableau) reachable from s_0 in this branch. Since $\varphi \in L(s_0)$ and s_0 is marked "satisfiable," the tableau rules—reflecting satisfaction conditions—ensure $M, s' \models \varphi$. For example, if $\Diamond\psi \in L(w)$, a successor w' exists with $\psi \in L(w')$, satisfying the modal semantics. The number of states $|W| = k \leq n$ by the n-bound in Step 2(b). Thus, φ is k-satisfiable, and since $k \leq n$, it is n-satisfiable (by Lemma 1).

For the converse, we prove the contrapositive: if the root s_0 is not marked "satisfiable," then φ is not n-satisfiable. We show by induction on the height of a node s in the tree that if s is not marked "satisfiable," then $L(s)$ is inconsistent in $\mathbf{K}n$, implying φ is not n-satisfiable by the soundness.

- Base case: s is a leaf. By Step 2(c)(ii), s is not marked "satisfiable" if and only if $L(s)$ is blatantly inconsistent (e.g., contains both ψ and $\neg\psi$). Thus, $L(s)$ is inconsistent.
- Inductive step: If $L(s)$ is not a propositional tableau (an internal node), Step 2(c)(i) implies s is not "satisfiable" only if no successor s' is "satisfiable." By the induction hypothesis, $L(s')$ is inconsistent for each s'. Step 2(a) expands $L(s)$ into $L(s')$ via Boolean rules (e.g., $\neg(\psi_1 \wedge \psi_2)$ splits to $\neg\psi_1$ or $\neg\psi_2$). Since all successor options are inconsistent, propositional logic implies $L(s)$ is inconsistent. If $L(s)$ is a propositional tableau (i.e., a state fully expanded for propositional rules), Step 2(c)(iii) implies that s is not marked "satisfiable" if, in every branch, at least one successor s' is not marked "satisfiable." Suppose $\Diamond\psi \in L(s)$, and s' is created by Step 2(b) with $L(s') = \{\neg\chi \mid \neg\Diamond\chi \in L(s)\} \cup \{\psi\}$. Let $\{\neg\Diamond\chi_1, \ldots, \neg\Diamond\chi_m, \Diamond\psi\} \subseteq L(s)$ and $L(s') = \{\neg\chi_1, \ldots, \neg\chi_m, \psi\}$. By the induction hypothesis, $L(s')$ is inconsistent, so $\vdash \neg\chi_1 \wedge \cdots \wedge \neg\chi_m \rightarrow \neg\psi$. This implies $\vdash \neg\Diamond\chi_1 \wedge \cdots \wedge \neg\Diamond\chi_m \rightarrow \neg\Diamond\psi$ (by the K axiom and necessitation). Since $\{\neg\Diamond\chi_1, \ldots, \neg\Diamond\chi_m, \Diamond\psi\} \subseteq L(s)$, $L(s)$ is inconsistent. $\qquad\square$

From Lemmas 6 and 7, we obtain the following result:

Theorem 8. *The $Sat_n(\varphi)$ problem—determining whether a modal formula φ is satisfied in a pointed model with n states—is in PSPACE.* $\qquad\square$

---

[5] Multiple valuations may replace this condition, as $L(w)$ need not include either p or $\neg p$ for every p. Any such V suffices for the proof.

5   Discussion

We showed that the logic of n-frames is characterized by the A_n axiom schema. To be more precise, A_n is canonical for what we called the n-out property. It follows that completeness results for any canonical modal logic can be extended to completeness of the logic with the A_n axiom wrt. the same model class restricted to n-out frames. If the model class is closed under generated submodels (the typical case), then we also get completeness wrt. n-frames.

Note that depending on the underlying logic, A_n can be further simplified. E.g., for any logic with the **4** axiom, $(\varphi_0 \wedge \bigwedge_{1 \leq i \leq n} \Diamond \varphi_i) \rightarrow \bigvee_{0 \leq j < k \leq n} \Diamond(\varphi_j \wedge \varphi_k)$ is sufficient.

The complexity of the satisfiability problem over n-frames is non-trivial. It turns out to be PSPACE-complete for $\mathbf{K}n$, like for the underlying logic $\mathbf{K}$, assuming binary representation of n, the standard representation of numbers in complexity theory. Note that this means that the model we are looking for is potentially exponentially larger than the input, which goes some way towards an intuitive understanding of the result. This would of course not be the case if n is represented in unary. See [1] for an exploration of both types of encodings in the context of modal logic. We leave the details of the unary case to future work. We only looked at complexity in the $\mathbf{K}n$ case, and characterizations of computational complexity for other logics over n-frames are open problems.

There is plenty of other opportunities for future work. More general modalities (similarity types) like binary modalities and languages with more than one modality like multi-agent languages come to mind. In the multi-agent setting, adding common knowledge would maybe be particularly interesting. A Kleene-star modality would allow a similar simplification as in the case with the **4** axiom above. It is also worth noting in this context that the A_n axioms (together with taking generated submodels) is a way to get finite models without using filtration or selection.

Acknowledgments. We express our gratitude to the anonymous reviewers for their valuable and insightful comments.

Disclosure of Interests. Yì N. Wáng and Xiaolong Liang have received funding support from the Project of Humanities and Social Sciences from Chinese Ministry of Education (No. 24YJA72040002).

References

1. Areces, C., Hoffmann, G., Denis, A.: Modal logics with counting. In: Dawar, A., de Queiroz, R. (eds.) Logic, Language, Information and Computation. Lecture Notes in Artificial Intelligence, vol. 6188, pp. 98–109. Springer, Heidelberg (2010). https://doi.org/10.1007/978-3-642-13824-9_9
2. Baader, F., Calvanese, D., McGuinness, D.L., Nardi, D., Patel-Schneider, P.F.: The Description Logic Handbook: Theory, Implementation, and Applications. Cambridge University Press, 2 edn. (2007). https://doi.org/10.1017/CBO9780511711787

3. van Benthem, J., Icard, T.: Interleaving logic and counting. Bull. Symb. Log. **29**(4), 503–587 (2023). https://doi.org/10.1017/bsl.2023.30
4. Blackburn, P., de Rijke, M., Venema, Y.: Modal Logic, Cambridge Tracts in Theoretical Computer Science, vol. 53. Cambridge University Press (2001). https://doi.org/10.1017/CBO9781107050884
5. Chagrov, A., Zakharyaschev, M.: Modal Logic. Oxford University Press (1997)
6. Cook, S.A.: The complexity of theorem-proving procedures. In: STOC '71: Proceedings of the Third Annual ACM Symposium on Theory of Computing. pp. 151–158. Association for Computing Machinery, New York (1971). https://doi.org/10.1145/800157.805047
7. Dunbar, R.I.M.: Neocortex size as a constraint on group size in primates. J. Hum. Evol. **22**(6), 469–493 (1992). https://doi.org/10.1016/0047-2484(92)90081-J
8. Fagin, R., Halpern, J.Y.: Reasoning about knowledge and probability. J. ACM **41**(2), 340–367 (1994). https://doi.org/10.1145/174652.174658
9. Fine, K.: In so many possible worlds. Notre Dame J. Formal Logic **13**(4), 516–520 (1972). https://doi.org/10.1305/ndjfl/1093890715
10. Fu, X., Zhao, Z.: Axiomatization of modal logic with counting. Logic Journal of the IGPL, p. jzae079, June 2024. https://doi.org/10.1093/jigpal/jzae079
11. Fu, X., Zhao, Z.: Modal logic with "most". Stud. Logica. (2024). https://doi.org/10.1007/s11225-024-10159-5
12. Goble, L.F.: Grades of modality. Logique et Anal. (N.S.) **13**, 323–334 (1970)
13. Halpern, J.Y., Moses, Y.: A guide to completeness and complexity for modal logics of knowledge and belief. Artif. Intell. **54**(3), 319–379 (1992). https://doi.org/10.1016/0004-3702(92)90049-4
14. Hansen, M., Larsen, K.G., Mardare, R., Pedersen, M.R.: Reasoning about bounds in weighted transition systems. Logical Methods in Computer Science **14**(4), 1–32 (2018). https://doi.org/10.23638/LMCS-14(4:19)2018
15. Heifetz, A., Mongin, P.: The modal logic of probability. In: Gilboa, I. (ed.) Proceedings of the 7th Conference on Theoretical Aspects of Rationality and Knowledge (TARK-98), pp. 175–185. Morgan Kaufmann (1998)
16. Hernando, A., Villuendas, D., Vesperinas, C., Abad, M., Plastino, A.: Unravelling the size distribution of social groups with information theory in complex networks. Europ. Phys. J. B **76**(1), 87–97 (2010). https://doi.org/10.1140/epjb/e2010-00216-1
17. Van der Hoek, W.: Qualitative modalities. Internat. J. Uncertain. Fuzziness Knowl.-Based Syst. **4**(1), 45–60 (1996). https://doi.org/10.1142/S0218488596000044
18. Karp, R.M.: Reducibility among combinatorial problems. In: Miller, R.E., Thatcher, J.W., Bohlinger, J.D. (eds.) Complexity of Computer Computations. pp. 85–103. The IBM Research Symposia Series, Springer, Boston, MA (1972). https://doi.org/10.1007/978-1-4684-2001-2_9
19. Ladner, R.E.: The computational complexity of provability in systems of modal propositional logic. SIAM J. Comput. **6**(3), 467–480 (1977). https://doi.org/10.1137/0206033
20. Larsen, K.G., Mardare, R.: Complete proof systems for weighted modal logic. Theoret. Comput. Sci. **546**(12), 164–175 (2014). https://doi.org/10.1016/j.tcs.2014.03.007
21. Larsen, K.G., Skou, A.: Bisimulation through probabilistic testing. Inf. Comput. **94**(1), 1–28 (1991). https://doi.org/10.1016/0890-5401(91)90030-6

Evidence Diffusion in Social Networks: a Topological Perspective

Aybüke Özgün[(✉)], Sonja Smets, and Teodor-Ştefan Zotescu

ILLC, University of Amsterdam, Amsterdam, The Netherlands
{a.ozgun,s.j.l.smets}@uva.nl

Abstract. This paper explores the effect of social structures on the beliefs and knowledge of agents who reason in an evidence-based manner. We introduce Evidence Diffusion Models for the formal analysis of multi-agent evidence-based reasoning, focusing on the context of a social network where evidence pieces are being communicated between agents via threshold-limited diffusion. Our models bridge the gap between multi-agent Topological Evidence Models (Partitional Models) and Threshold Models for Diffusion. Firstly, we show that, in our setting, network structures are expressible and known both defeasibly and infallibly by the agents. Then, we prove that defeasible knowledge and (defeasible) 'evidence-based distributed knowledge' are easily lost under the diffusion of pieces of evidence in a network, whereas 'group knowledge with distributed evidence' is strongly robust. Finally, we obtain so-called Cluster Theorems characterising the evidential and network conditions for evidence cascades to form, and for individuals and groups to obtain knowledge in the diffusion process.

Keywords: Dynamic Epistemic Logic · Topological Evidence Models · Logics for Social Networks · Distributed Knowledge · Formal Epistemology

1 Introduction

We position this paper within the recent trend of using logic for the formal analysis of social phenomena [2,7,10,14,19–21,24,27]. More specifically we adopt Dynamic Epistemic Logic [3,4,9] to model evidence-based reasoning in social networks. Here we focus on agents in a multi-agent system whose epistemic and doxastic attitudes are directly tied to their available evidence (aligned with earlier work in [3,18]). These agents are able to share their evidence with their neighbors in the network, allowing for evidence to spread in this network via a process of diffusion. Our leading example is the spread of evidence within a scientific community. In a diffusion round, individual agents (e.g. scientists) share evidence, which will then have effects on the epistemic states of other agents and thereby on the epistemic states of groups of agents in the network. In this paper, we are particularly interested in the stability of these epistemic attitudes

© The Author(s), under exclusive license to Springer Nature Switzerland AG 2026
V. Goranko et al. (Eds.): LORI 2025, LNCS 16010, pp. 110–123, 2026.
https://doi.org/10.1007/978-981-95-2481-5_8

under the diffusion process. In the context of evidence-based reasoning, we will distinguish different group attitudes in this paper. We investigate the notions of 'Group Knowledge with Distributed Evidence' and 'Evidence-based Distributed Knowledge', which, as we shall see, behave quite differently. We provide the definitions of these attitudes and analyze their stability or robustness under the process of evidence diffusion. Further, we zoom in on so-called Cluster Theorems, which characterize the evidential and network conditions for evidence cascades to form, and for individuals and groups to obtain a certain epistemic state in the diffusion process.

This paper is organized as follows. In Sect. 2 we provide the preliminary topological notions for evidence-based knowledge, introducing our logic for evidence diffusion (both the language and semantics). In Sect. 3 we continue with an analysis of the dynamics of our models to describe the processes of diffusion, distinguishing evidence diffusion and law distribution. In Sect. 4 we analyze the robustness of our epistemic notions under the diffusion process and state our main cluster theorems. We end with a brief conclusion in Sect. 5.[1]

2   Evidence Models for Diffusion

We introduce now the preliminary topological notions we will be using throughout the paper. For a more in-depth introduction to topology, we refer the reader to a standard topology textbook, e.g. [17].

Topological Preliminaries: A *topological space* is a pair $\mathcal{X} = (X, \tau)$, where X is a non-empty set and $\tau \subseteq \mathcal{P}(X)$ such that (1) $X, \emptyset \in \tau$, (2) for all finite $\mathcal{U} \subseteq \tau$ $\bigcap \mathcal{U} \in \tau$, and (3) for all $\mathcal{U} \subseteq \tau$, $\bigcup \mathcal{U} \in \tau$. Elements of τ are called *open sets* (or *opens*) and their complements are called *closed sets*.

Given a family $\mathcal{E} \subseteq \mathcal{P}(X)$ of subsets of X, there exists a unique, smallest topology $\tau_{\mathcal{E}}$ with $\mathcal{E} \subseteq \tau_{\mathcal{E}}$. The family $\tau_{\mathcal{E}}$ consists of $\emptyset$, X, all finite intersections of the elements of $\mathcal{E}$, and all arbitrary unions of these finite intersections. $\mathcal{E}$ is called a *subbasis* for $\tau_{\mathcal{E}}$, and $\tau_{\mathcal{E}}$ is said to be *generated* by $\mathcal{E}$. The set $\overline{\mathcal{E}}$ of finite intersections of members of $\mathcal{E}$ forms a basis for $\tau_{\mathcal{E}}$.

Given a topological space $\mathcal{X} = (X, \tau)$ and $A \subseteq X$, we define its interior as $\mathrm{Int}(A) = \bigcup \{U \subseteq A \mid U \in \tau\}$, i.e. the greatest open contained in A. The closure of A is $\mathrm{Cl}(A) = \bigcap \{C \supseteq A \mid C = X \setminus U$ for some $U \in \tau\}$, i.e. the smallest closed set containing A. We call a set $A \subseteq X$ dense if $\mathrm{Cl}(A) = X$, or equivalently if for all $U \in \tau \setminus \{\emptyset\}$, $A \cap U \neq \emptyset$.

We now introduce evidence diffusion models. Throughout the paper, we work with a non-empty, finite set of agents Ag.

Definition 1 (Evidence Diffusion Model). *An* evidence diffusion model *is a tuple* $\mathfrak{M} = \langle X, \{\mathcal{E}_i\}_{i \in Ag}, \{\Pi_i\}_{i \in Ag}, \mathcal{N}, \llbracket \cdot \rrbracket, \theta \rangle$, *where X is a non-empty set of*

---

[1] The results in this paper are based on the work in [27]. Due to the page-limit of this paper, several proofs are omitted from the text and can be found in the appendix that accompanies this paper, which is available here.

possible worlds; $\mathcal{E}_i$ is agent i's set of basic pieces of evidence such that $X \in \mathcal{E}_i$ and $\emptyset \notin \mathcal{E}_i$; Π_i is a partition of X, representing agent i's hard information/infallible knowledge, such that $\Pi_i \subseteq \mathcal{E}_i$; $[\![\cdot]\!] : \mathbb{P} \to X$ is a standardly defined valuation function; and $\theta \in [0,1]$ is a diffusion threshold (fixed for all agents and pieces of evidence). Finally, $\mathcal{N} : Ag \mapsto \wp(Ag)$ is a neighbor function, satisfying: irreflexivity ($a \notin \mathcal{N}(a)$), symmetry ($a \in \mathcal{N}(b)$ implies $b \in \mathcal{N}(a)$) and connectedness (for every $i \neq j$ there exist $k_0, ..., k_n$ agents such that $k_0 = i, k_n = j$ and for every $\ell \in \{1, ..., n\}$, $k_\ell \in \mathcal{N}(k_{\ell-1})$).

This paper fits within the approach to epistemic logic which investigates knowledge not as a primitive notion, but as derived from other pieces of information, called broadly *evidence* [23]. We say broadly because we include in our notion of evidence false or misleading pieces of information, as well as priors, biases and *a priori* knowledge [3,18]. Topological evidence models [3,18], which our setting is based on, replace the usual accessibility relation of Kripke models, which represents epistemic indistinguishability, with a set of basic pieces of evidence for each agent, $\mathcal{E}_i$, the elements of which represent *directly observable* pieces of evidence. Topologically, this is a subbasis, and the elements of the topology generated by this subbasis, τ_i, represent simply evidence, which we understand as any piece of information derived by taking finite conjunctions (intersection) and arbitrary disjunctions (union) of basic evidence. We will also make use of $\overline{\mathcal{E}}_i$, the basis generated by closing $\mathcal{E}_i$ under finite intersections.

We take the same approach to multi-agent topological evidence models as [12], who introduces partitions Π_i of the set of possible worlds for each agent. These partitions represent the agents' hard information, which might not be the same for each agent. Given this, however, we need to *localize* the usual topological and epistemic notions [3,18] to the agents' hard information. Effectively, we introduce a notion of density restricted to the subspace induced by a specific partition cell, which will represent what the agent defeasibly knows, given what she infallibly knows. Let x be a world in X. Then $\Pi_i(x)$ is the unique $\pi \in \Pi_i$ such that $x \in \pi$. We denote by $\tau|_{\Pi_i(x)} = \{U \cap \Pi_i(x) \mid U \in \tau\}$, the subspace topology obtained by restricting τ to $\Pi_i(x)$. A set $U \subseteq X$ is i-locally dense in $\Pi_i(x)$ if and only if U is dense in $\tau|_{\Pi_i(x)}$.

Finally, we impose three constraints on the network, all motivated by the modeled phenomenon: (1) symmetry, because scientific connections are mutual; (2) connectedness, because excluding isolated agents ensures a proper network; and (3) irreflexivity, so self-connections are excluded and the threshold θ reflects genuine scientific influence.

The *language $\mathcal{L}$ of evidence diffusion for group knowledge* is given by the following grammar, where $p \in \mathbb{P}$, a countable set of propositional atoms, and $i, j \in Ag$:

$$\varphi := p \mid N_{ij} \mid \neg\varphi \mid (\varphi \wedge \varphi) \mid E_i\varphi \mid [\Pi_i]\varphi \mid K_i\varphi \mid \text{ƎD}\varphi \mid \text{ꓘD}\varphi$$

We read N_{ij} as 'j is i's neighbor' and $E_i\varphi$ as 'φ is a basic piece of evidence of agent i'. For the individual knowledge operators, we have $[\Pi_i]\varphi$ read as 'i

infallibly knows that φ', $K_i\varphi$ read as 'i defeasibly knows that φ'. Finally, we read the group knowledge operators as follows: $\exists\mathbf{D}\varphi$ is read as 'there is group knowledge with distributed evidence that φ', and $\forall\mathbf{D}\varphi$ as 'there is evidence-based distributed knowledge that φ'. In the following, $[\![\varphi]\!]_{\mathfrak{M}} = \{x \in X \mid \mathfrak{M}, x \vDash \varphi\}$ is the *truth set* of φ in a model $\mathfrak{M}$. We omit the subscript $\mathfrak{M}$ when the model is contextually clear. We follow the usual rules for elimination of the parenthesis in the language.

Definition 2 (Semantics for $\mathcal{L}$). *Given an evidence diffusion model* $\mathfrak{M} = \langle X, \{\mathcal{E}_i\}_{i\in Ag}, \{\Pi_i\}_{i\in Ag}, \mathcal{N}, [\![\cdot]\!], \theta\rangle$ *and* $x \in X$, *we interpret* $\mathcal{L}$ *recursively as:*

$$\begin{aligned}
&\mathfrak{M}, x \vDash p && \textit{iff } x \in [\![p]\!] \\
&\mathfrak{M}, x \vDash N_{ij} && \textit{iff } j \in \mathcal{N}(i) \\
&\mathfrak{M}, x \vDash \neg\varphi && \textit{iff not } \mathfrak{M}, x \vDash \varphi \\
&\mathfrak{M}, x \vDash \varphi \wedge \psi && \textit{iff } \mathfrak{M}, x \vDash \varphi \textit{ and } \mathfrak{M}, x \vDash \psi \\
&\mathfrak{M}, x \vDash E_i\varphi && \textit{iff } [\![\varphi]\!] \in \mathcal{E}_i \\
&\mathfrak{M}, x \vDash [\Pi_i]\varphi && \textit{iff } \Pi_i(x) \subseteq [\![\varphi]\!] \\
&\mathfrak{M}, x \vDash K_i\varphi && \textit{iff there exists } U \in \tau_i \textit{ such that } x \in U \subseteq [\![\varphi]\!] \\
&&& \textit{and } U \textit{ is } i\textit{-locally dense in } \Pi_i(x)
\end{aligned}$$

Note that our E_i operator is different from the usual E_i operator on topological evidence models [3,12,18], where it means 'agent i has a basic piece of evidence *for* φ', which is interpreted as $\exists e \in \mathcal{E}_i(e \subseteq [\![\varphi]\!])$.

For the distributed operators we write the definitions separately, as they are more complex. We begin with a topological concept which is necessary for group knowledge with distributed evidence, the join topology, which is the smallest topology containing all the agents' topologies.

Definition 3 (Join Topology). *Let* X *be a set of possible worlds and let* $Ag = \{1,...,n\}$ *and for each* $i \in Ag$, *let* τ_i *be a topology on* X. *Then the join topology, denoted by* $\bigvee_{i\in Ag}\tau_i$ *is the smallest topology containing* τ_i *for each* $i \in Ag$. *Formally,* $\bigcup_{i\in Ag}\tau_i$ *constitutes a subbasis and* $\bigvee_{i\in Ag}\tau_i$ *is the topology generated by this subbasis.*

Definition 4 (Group Knowledge with Distributed Evidence). $\mathfrak{M}, x \vDash$ $\exists\mathbf{D}\varphi$ *if and only if for every* $i \in Ag$, *there exists* $U_i \in \tau_i$ *such that:*

1. $Cl_{\vee\tau_\cap}(\bigcap_{i\in Ag} U_i) = \bigcap_{i\in Ag} \Pi_i(x)$, *where* $\vee\tau_\cap = \bigvee_{i\in Ag}\tau_i\big|_{\bigcap_{i\in Ag}\Pi_i(x)}$;
   (If the agents communicated all their information, both hard and soft, they would have a justification...)
2. $x \in \bigcap_{i\in Ag} U_i$;
   (...which is factive...)
3. $\bigcap_{i\in Ag} U_i \subseteq [\![\varphi]\!]$;
   (... and which entails φ.)

What this notion encodes is the knowledge the agents would come to possess if they put all of their evidence and infallible knowledge together and reasoned as one agent. It was formally defined by [1] and we concur with [12] that it is not the appropriate notion of evidence-based distributed knowledge. For a more detailed analysis of why this is, consult [12,27].

In our framework, since we encode inter-agent communication, we show in Proposition 5 that this coincides with the knowledge of an agent given certain communication conditions. As [18] points out, this notion of knowledge is not monotonic, in the sense that some proposition φ might be individual knowledge (there is an agent i such that $K_i\varphi$ holds at the actual world) but not group knowledge with distributed evidence (the actual world doesn't make $\exists D\varphi$ true).

Definition 5 (Evidence-based Distributed Knowledge). $\mathfrak{M}, x \vDash \exists D\varphi$ *if and only if for every* $i \in Ag$, *there exists* $U_i \in \tau_i$ *such that:*

1. U_i *is* $\Pi_i(x)$*-locally dense;*
   (All agents have justifications...)
2. $x \in \bigcap_{i \in Ag} U_i$*;*
   (... which are factive ...)
3. $\bigcap_{i \in Ag} U_i \subseteq [\![\varphi]\!]$.
   (... and when put together, these justifications entail φ.)

This notion is more akin to the regular understanding of distributed knowledge [11,13], coinciding with the one on Kripke models in the sense that distributed knowledge is that which is entailed by what the agents know individually. Despite its conceptual appeal, as shown in Proposition 7, this notion is not robust: it is very easily lost under evidence diffusion.

The definition of *logical consequence* is standard [5]: for any $\Gamma \subseteq \mathcal{L}$ and $\varphi \in \mathcal{L}$, Γ *logically entails* φ, $\Gamma \vDash \varphi$, iff for all models $\mathfrak{M}$ and all $x \in X$: if $\mathfrak{M}, x \vDash \psi$ for all $\psi \in \Gamma$, then $\mathfrak{M}, x \vDash \varphi$. We write $\mathfrak{M} \vDash \varphi$ if for every world $x \in X$, $\mathfrak{M}, x \vDash \varphi$; and $\vDash \varphi$ means that $\mathfrak{M} \vDash \varphi$ for every evidence diffusion model $\mathfrak{M}$, that is, φ is *valid*.

3   Evidential Dynamics

We now go on to define two dynamic operators that capture, on the one hand the sharing of soft information, or evidence, and on the other hand the sharing of hard information, or certain knowledge. We call the former Evidence Diffusion and the latter Law Distribution.

We model the spread of evidence across the network as a communication process which is *social, local* [22], and *threshold-limited* [10]. An agent i will adopt a piece of evidence e if out of her network of scientific contacts, $\mathcal{N}(i)$, the proportion of agents who have this piece of evidence in their evidence sets to all the agents in $\mathcal{N}(i)$ is over the threshold θ. We restrict attention to the basic pieces of evidence because the combined pieces of evidence and the open sets in the topology of evidence represent the result of agents' reasoning with the basic pieces of evidence, and we only want the 'empirical data' to be transmitted, so agents on the receiving end can do their own logical reasoning on the basis of it.

Definition 6 (Evidence Diffusion). *The result of applying evidence diffusion to the evidence diffusion model* $\mathfrak{M} = \langle X, \{\mathcal{E}_i\}_{i \in Ag}, \{\Pi_i\}_{i \in Ag}, \mathcal{N}, \llbracket \cdot \rrbracket, \theta \rangle$ *is the evidence diffusion model* $\mathfrak{M}^{\sim} = \langle X, \{\mathcal{E}_i^{\sim}\}_{i \in Ag}, \{\Pi_i\}_{i \in Ag}, \mathcal{N}, \llbracket \cdot \rrbracket, \theta \rangle$, *where for any* $i \in Ag$,

$$\mathcal{E}_i^{\sim} = \mathcal{E}_i \cup \{e \in \bigcup_{j \in Ag} \mathcal{E}_j \mid \frac{|\{j \mid j \in \mathcal{N}(i) \ and \ e \in \mathcal{E}_j\}|}{|\{j \mid j \in \mathcal{N}(i)\}|} \geq \theta\}.$$

Note that evidence diffusion does not change the threshold. When an agent receives a new piece of evidence through diffusion, we say that agent 'adopted' that piece of evidence. The diffusion rule could easily be adapted so that combined evidence or opens get transmitted as well. Unlike other models of threshold-limited influence [6, 7, 10, 22], what is subjected to diffusion in our model is not the belief directly, but the 'data' for belief. We believe that this strikes a middle ground between agents who are completely swayed by those around them and those whose epistemic states are completely independent of the evidence of other agents in their network; acknowledging both the effects of social influence on belief and the rational, evidence-based nature of belief.

We call a countably infinite sequence of applications of this transformation a diffusion sequence.

Definition 7 (Diffusion Sequence). *Given an evidence diffusion model* $\mathfrak{M} = \langle X, \{\mathcal{E}_i\}_{i \in Ag}, \{\Pi_i\}_{i \in Ag}, \mathcal{N}, \llbracket \cdot \rrbracket, \theta \rangle$, *we define its diffusion sequence to be the tuple* $\mathcal{S}_{\mathfrak{M}} = \langle \mathfrak{M}_0, \mathfrak{M}_1, ... \rangle$, *where* $\mathfrak{M}_0 = \mathfrak{M}$ *and for any* $n \in \mathbb{N}$, $\mathfrak{M}_{n+1} = \mathfrak{M}_n^{\sim}$.

Given a diffusion sequence $\mathcal{S}_{\mathfrak{M}} = \langle \mathfrak{M}_0, \mathfrak{M}_1, ... \rangle$, we denote the basic evidence set of agent i in the model $\mathfrak{M}_k$ for $k \in \mathbb{N}$ by $\mathcal{E}_i^k$. Similarly $\overline{\mathcal{E}}_i^k$ denotes the associated combined evidence set, τ_i^k their topology of evidence, and $\llbracket \varphi \rrbracket^k$ the truth set of proposition φ at diffusion step k.

The sharing of hard information is captured using the partitions that encode each agent's infallible knowledge. When all agents combine their infallible knowledge, they arrive at a common partition of possible worlds: the coarsest common refinement of their individual partitions. This is the least refined partition that incorporates all their hard information. Note that the change to the agents' basic evidence sets is required to make $\mathfrak{M}^{\wedge}$ an evidence diffusion model.

Definition 8 (Coarsest Common Refinement). *Given finitely many partitions* $\Pi_1, ..., \Pi_n$ *of* X, *the* coarsest common refinement, $\bigwedge_{i=1}^n \Pi_i$, *of these partitions is defined as:*

$$\bigwedge_{i=1}^n \Pi_i = \{\bigcap_{i=1}^n \pi_i \mid \pi_i \in \Pi_i \ and \ \bigcap_{i=1}^n \pi_i \neq \emptyset\}.$$

Definition 9 (Law Establishment). *Given an evidence diffusion model* $\mathfrak{M} = \langle X, \{\mathcal{E}_i\}_{i \in Ag}, \{\Pi_i\}_{i \in Ag}, \mathcal{N}, \llbracket \cdot \rrbracket, \theta \rangle$, *the result of applying law establishment to* $\mathfrak{M}$ *is the model* $\mathfrak{M}^{\wedge} = \langle X, \{\mathcal{E}_i^{\wedge}\}_{i \in Ag}, \{\Pi_i^{\wedge}\}_{i \in Ag}, \mathcal{N}, \llbracket \cdot \rrbracket, \theta \rangle$, *where for every* $i \in Ag$, $\Pi_i^{\wedge} = \bigwedge_{j \in Ag} \Pi_j$ *and* $\mathcal{E}_i^{\wedge} = \mathcal{E}_i \cup \Pi_i^{\wedge}$.

Having defined these model transformations, we extend our language with corresponding dynamic modalities and obtain natural semantic clauses for them.

The *dynamic language* $\mathcal{L}^{\square}$ *of evidence diffusion for group knowledge* is obtained by extending the static language $\mathcal{L}$ with the dynamic modalities $[\sim]\varphi$ (read 'adopt φ') and $[\bigwedge]\varphi$ (read 'laws φ') for evidence diffusion and law establishment, respectively.

Given an evidence diffusion model $\mathfrak{M} = \langle X, \{\mathcal{E}_i\}_{i \in Ag}, \{\Pi_i\}_{i \in Ag}, \mathcal{N}, [\![\cdot]\!], \theta \rangle$ and $x \in X$, we interpret the new dynamic operators as follows:

$$\mathfrak{M}, x \vDash [\sim]\varphi \ \text{ iff } \ \mathfrak{M}^{\sim}, x \vDash \varphi$$
$$\mathfrak{M}, x \vDash [\textstyle\bigwedge]\varphi \ \text{ iff } \ \mathfrak{M}^{\bigwedge}, x \vDash \varphi.$$

4 Robustness Under Diffusion and Cluster Theorems

One characteristic of this framework, highlighted by the validity in Proposition 1, is that the network structure is transparent to all agents, that is, all agents infallibly (and also defeasibly), know who is connected to whom in the network.

Proposition 1. *For every* $i, j, k \in Ag$, $\vDash N_{jk} \leftrightarrow [\Pi_i]N_{jk}$ *and* $\vDash N_{jk} \leftrightarrow K_i N_{jk}$.

Proof. Follows from the fact that the semantic clause for N_{jk} is world-independent, that is $[\![N_{jk}]\!] = X$ or $[\![N_{jk}]\!] = \emptyset$. Moreover, $x \in \Pi_i(x) \subseteq Cl_{\tau_i}(\Pi_i(x)) \subseteq X$.

Even though we defined diffusion sequences as countably infinite, the following proposition establishes that a fixpoint is always reached at some step within the sequence. In other words, diffusion always 'ends' at some point and the model no longer changes.

Notation 1. *For an evidence diffusion model* $\mathfrak{M} = \langle X, \{\mathcal{E}_i\}_{i \in Ag}, \{\Pi_i\}_{i \in Ag}, \mathcal{N}, [\![\cdot]\!], \theta \rangle$, *we let* $Ag_e = \{i \in Ag \mid e \in \mathcal{E}_i\}$, the set of all agents who have e as a basic piece of evidence. *Given the diffusion sequence* $\mathcal{S}_{\mathfrak{M}} = \langle \mathfrak{M}, \mathfrak{M}_1, ... \rangle$ *and* $k \in \mathbb{N}$, *we let* $Ag_e^k = \{i \in Ag \mid e \in \mathcal{E}_i^k\}$, the set of all agents who have e as a basic piece of evidence at the kth diffusion.

Proposition 2. *For any evidence diffusion model* $\mathfrak{M}$, *there exists some* $n \in \mathbb{N} \leq |Ag| - 1$ *where the diffusion sequence* $\mathcal{S}_{\mathfrak{M}}$ *reaches a fixed point, that is, for any* $\mathfrak{M}_m \in \mathcal{S}_{\mathfrak{M}}$ *with* $m \geq n$, $\mathfrak{M}_m = \mathfrak{M}_n$. *For clarity, we will often denote the upper bound* $|Ag| - 1$ *by* fix *(for fixpoint).*

Proof. Let $\mathfrak{M} = \langle X, \{\mathcal{E}_i\}_{i \in Ag}, \{\Pi_i\}_{i \in Ag}, \mathcal{N}, [\![\cdot]\!], \theta \rangle$ be an evidence diffusion model. By the definition of $\mathcal{S}_{\mathfrak{M}}$, for any $i \in Ag$ and $k \in \mathbb{N}$, $\mathcal{E}_i^k \subseteq \mathcal{E}_i^{k+1}$. Thus, for all $e \in \bigcup_{i \in Ag} \mathcal{E}_i$, $Ag_e^k \subseteq Ag_e^{k+1}$. Note that the set of agents Ag is finite, and so for every $e \in \bigcup_{i \in Ag} \mathcal{E}_i$, there exists $n \in \mathbb{N}$ such that $Ag_e^n = Ag_e^{n+1}$. Clearly, the slowest diffusion scenario for some $e \in \bigcup_{i \in Ag} \mathcal{E}_i$ is that in which $|Ag_e| = 1$ and for every $k \in \mathbb{N}$, $|Ag_e^{k+1}| = |Ag_e^k| + 1$, which is clearly bounded by $|Ag| - 1$.

Using the dynamic language of evidence diffusion for group knowledge, we can capture whether or not an agent receives a piece of evidence at the next diffusion round.

Definition 10 (Free Evidence for agent i). *Given an evidence diffusion model $\mathfrak{M} = \langle X, \{\mathcal{E}_i\}_{i \in Ag}, \{\Pi_i\}_{i \in Ag}, \mathcal{N}, [\![\cdot]\!], \theta \rangle$, we call a piece of evidence $e \in \bigcup_{i \in Ag} \mathcal{E}_i$ free for i in $\mathfrak{M}$ if and only if there exists $\mathcal{G} \subseteq \mathcal{N}(i)$ such that $e \in \mathcal{E}_j$ for every $j \in \mathcal{G}$ and $\frac{|\mathcal{G}|}{|\mathcal{N}(i)|} \geq \theta$.*

Abbreviation 1. *We define the formula $[\sim]^n \varphi$ as the abbreviation defined recursively by $[\sim]^0 \varphi := \varphi$ and $[\sim]^{n+1} \varphi := [\sim][\sim]^n \varphi$.*

Making use of this abbreviation, we can express that some piece of evidence φ is free for some agent i.

$$\Phi_i \varphi := \bigvee_{\{G \subseteq F \subseteq Ag \mid \frac{|G|}{|F|} \geq \theta\}} (\bigwedge_{j \in F} N_{ij} \wedge \bigwedge_{j \notin F} \neg N_{ij} \wedge \bigwedge_{j \in G} E_j \varphi).$$

Before we proceed with our analysis of robustness results, we want to restrict attention to Boolean formulas. Our agents share their direct observations and infallible first-order knowledge, not what they know about other agents' evidence or knowledge, so it stands to reason that we focus on first-order group and individual knowledge and evidence. The following proposition establishes that the truth sets of Boolean formulas does not change under our model transformations. We denote by $\mathcal{L}_{CPL}$ the language of classical propositional logic.

Proposition 3. *Given an evidence diffusion model $\mathfrak{M} = \langle X, \{\mathcal{E}_i\}_{i \in Ag}, \{\Pi_i\}_{i \in Ag}, \mathcal{N}, [\![\cdot]\!], \theta \rangle$ and $\varphi \in \mathcal{L}_{CPL}$, $[\![\varphi]\!]_{\mathfrak{M}} = [\![\varphi]\!]_{\mathfrak{M}\sim} = [\![\varphi]\!]_{\mathfrak{M}\wedge}$.*

The following validity describes that the only way an agent i comes to possess evidence after diffusion is if they already had that piece of evidence, or if it was free for i in the initial model.

Proposition 4. *For any $i \in Ag$ and any $\varphi \in \mathcal{L}_{CPL}$, $\models [\sim]E_i \varphi \leftrightarrow \Phi_i \varphi \vee E_i \varphi$.*

The following proposition shows that defeasible evidence is the knowledge of a hypothetical agent that has all of the hard knowledge ('laws') and soft knowledge (evidence) of all of the agents combined. The definition of group knowledge with distributed evidence (Definition 4) already hinted at this fact.

Proposition 5. *Given an evidence diffusion model $\mathfrak{M} = \langle X, \{\mathcal{E}_i\}_{i \in Ag}, \{\Pi_i\}_{i \in Ag}, \mathcal{N}, [\![\cdot]\!], \theta \rangle$, if for every $e \in \bigcup_{j \in Ag} \mathcal{E}_j$, e is free for i in some model in $\mathcal{S}_{\mathfrak{M}}$, then for any $\varphi \in \mathcal{L}_{CPL}$, $\mathfrak{M} \models \boxplus D\varphi \leftrightarrow [\wedge][\sim]^{fix} K_i \varphi$.*

Since distributed evidence is the knowledge of this *hypothetical* agent, it stands to reason that the *actual* diffusion pattern of a model does not affect it—distributed evidence is robust (in fact, invariant) under diffusion.

118   A. Özgün et al.

Proposition 6. *For any φ in $\mathcal{L}_{CPL}$, $\models \exists D\varphi \leftrightarrow [\sim]\exists D\varphi$*

In stark contrast, our notion of evidence-based distributed knowledge is not robust under evidence diffusion. It is so easily lost under evidence diffusion that we cannot assure its preservation even if a proposition is known both distributively and individually by all the agents before evidence diffusion.

Proposition 7. $\not\models (\mathcal{X}D\varphi \wedge \bigwedge_{i \in Ag} K_i\varphi) \rightarrow [\sim]\mathcal{X}D\varphi$

Proof. Consider the evidence diffusion model $\mathfrak{M} = \langle\{1,2,3,4\},\{\mathcal{E}_{\text{Alice}},$ $\mathcal{E}_{\text{Bob}}\},\{\Pi_{Alice},\Pi_{Bob}\},\mathcal{N},[\![\cdot]\!],0.5\rangle$ for two agents $Ag = \{Alice, Bob\}$, where $[\![p]\!] = P = \{1,2\}$ (green in Fig. 1) and $[\![q]\!] = Q = \{1,3\}$ (blue in Fig. 1); $\mathcal{E}_{\text{Alice}} = \{X,P,Q\}$ and $\mathcal{E}_{\text{Bob}} = \{X,P,\neg(P \leftrightarrow Q)\}$; $\Pi_{\text{Alice}} = \Pi_{\text{Bob}} = \{X\}$, and Alice and Bob are connected (as depicted by the arrows in Fig. 1). We stipulate that 1 is the actual world. Since the partitions of both agents is the whole space, local density with respect to the agents' partitions and global density coincide.

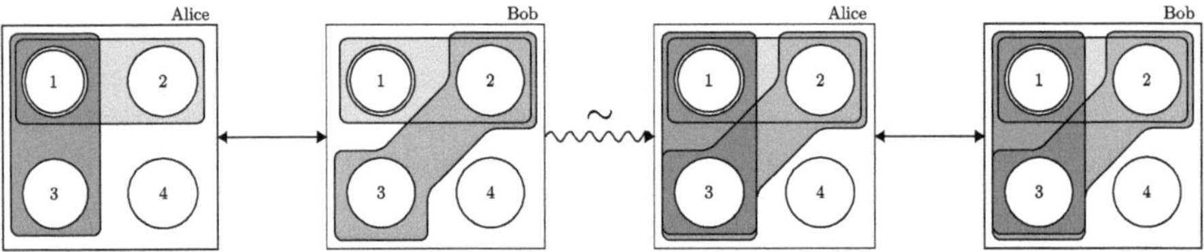

Fig. 1. Distributed knowledge loss

$\mathfrak{M},1 \models \mathcal{X}Dp$ since $[\![p]\!] = P$ itself is a dense open set both in τ_{Alice} and τ_{Bob} and $1 \in P$. For the same reasons, we also have $\mathfrak{M},1 \models K_{\text{Alice}}p$ and $\mathfrak{M},1 \models K_{\text{Bob}}p$, thus, $\mathfrak{M},1 \models \bigwedge_{i \in Ag} K_i p$. However, $\mathfrak{M},1 \not\models [\sim]\mathcal{X}Dp$: after evidence diffusion, we obtain that $\{3\} \in \tau_{\widetilde{Alice}}$ and $\{3\} \in \tau_{\widetilde{Bob}}$. So, any dense open set $U_1 \in \tau_{\widetilde{Alice}}$ and $U_2 \in \tau_{\widetilde{Bob}}$ is such that $3 \in U_1 \cap U_2 \not\subseteq P$.

One condition that is sufficient for the preservation of evidence-based distributed knowledge under evidence diffusion is that no agent receives so-called misleading evidence (defined in [3,18]) through the diffusion process.

Definition 11 (Misleading evidence). *We call a piece of evidence e i-misleading at $x \in X$ if and only if there is some $e' \in \overline{\mathcal{E}}_i$ such that $x \notin (e \cap e')$ and $(e \cap e') \notin \overline{\mathcal{E}}_i \cup \{\emptyset\}$.*

Proposition 8. *Given an evidence diffusion model $\mathfrak{M} = \langle X, \{\mathcal{E}_i\}_{i \in Ag},$ $\{\Pi_i\}_{i \in Ag}, \mathcal{N}, [\![\cdot]\!], \theta\rangle$ such that for all i, there is no i-misleading piece of evidence that is free for i, we have that for any $\varphi \in \mathcal{L}_{CPL}$, $\mathfrak{M} \models \mathcal{X}D\varphi \rightarrow [\sim]\mathcal{X}D\varphi$.*

The condition that no agent i receives an i-misleading piece of evidence is, however, not a necessary condition for the preservation of evidence-based distributed knowledge. For a simple example, consider a variation $\mathfrak{M}'$ of the model presented in the proof of Proposition 7, where $\mathcal{E}'_{\text{Alice}} = \{X, P\}$ and $\mathcal{E}'_{\text{Bob}} = \{X, \{2\}\}$. In this case, $\mathfrak{M}', 1 \vDash \rtimes\!\mathsf{D}p$, and $\{2\} \in \mathcal{E}_{\text{Bob}}$ is misleading for Alice. Yet, $\mathfrak{M}', 1 \vDash [\sim]\rtimes\!\mathsf{D}p$.

The following theorem specifies the conditions under which a so-called *cascade* [6, 10] is formed for a piece of evidence, i.e. when it propagates to all agents. It makes use of the notion of clusters, which we introduce now, following [6, 10].

Definition 12 (Cluster of Density d). *Given a set of agents Ag and network $\langle Ag, \mathcal{N} \rangle$ defined on Ag, a subset of the set of agents $C \subseteq Ag$ is called a* cluster *of density d if and only if for all $i \in C$, $\frac{|\mathcal{N}(i) \cap C|}{|\mathcal{N}(i)|} \geq d$.*

Theorem 1 (Evidence Cluster Theorem). *For every model $\mathfrak{M} = \langle X, \{\mathcal{E}_i\}_{i \in Ag}, \{\Pi_i\}_{i \in Ag}, \mathcal{N}, \llbracket \cdot \rrbracket, \theta \rangle$ with $\theta \neq 0$ and $e \in \bigcup_{i \in Ag} \mathcal{E}_i$, all agents will eventually adopt e if and only if there does not exist a cluster C of density greater than $1 - \theta$ such that $C \subseteq Ag \setminus Ag_e$.*

Corollary 1. *For every model $\mathfrak{M} = \langle X, \{\mathcal{E}_i\}_{i \in Ag}, \{\Pi_i\}_{i \in Ag}, \mathcal{N}, \llbracket \cdot \rrbracket, \theta \rangle$ with $\theta \neq 0$, $e \in \bigcup_{i \in Ag} \mathcal{E}_i$, and $i \in Ag$, $e \in \mathcal{E}_i^{fix}$ if and only if for every cluster C of density greater than $1 - \theta$ with $i \in C$, there exists an agent $j \in C$ such that $e \in \mathcal{E}_j$.*

From this corollary, an agent will end up adopting a piece of evidence if and only if there is an 'early adopter' [10, Ch. 19] of that piece of evidence in every cluster of density greater than $1 - \theta$ that she is a part of. If we take cluster density as a measure of group cohesion, we can rephrase this as saying: agents refrain from adopting pieces of evidence if they have any contact group that is 'close enough' in which none of the members have adopted that piece of evidence. Put in these terms, it certainly seems like the process of adopting pieces of evidence from diffusion is not a rational process, as we would like to think that scientists have the ability to individually and independently evaluate groundbreaking science when they come across it. In many real cases of collective failure though, such as the case of Ignaz Semmelweis [27, Ch. 2], it certainly seems like precisely this kind of 'groupthink' is at play.

Corollary 1 not only provides the essential ingredient to establishing the necessary and sufficient conditions for evidence-based individual and distributed knowledge to be preserved in diffusion, but also to be attained at the end of a diffusion sequence. We now go on to prove theorems to this extent, the first pertaining to individual knowledge.

Theorem 2 (Knowledge Cluster Theorem). *For any model $\mathfrak{M} = \langle X, \{\mathcal{E}_i\}_{i \in Ag}, \{\Pi_i\}_{i \in Ag}, \mathcal{N}, \llbracket \cdot \rrbracket, \theta \rangle$ with $\theta \neq 0$, $x \in X$ and $\varphi \in \mathcal{L}_{CPL}$, $\mathfrak{M}, x \vDash [\sim]^{fix} K_i \varphi$ if and only if there exists $E \subseteq \bigcup_{j \in Ag} \mathcal{E}_j$ such that:*

1. There exists $U \in \tau_i^{+E}$, the topology generated from $\mathcal{E}_i \cup E$, such that $x \in U \cap \Pi_i(x) \subseteq \llbracket \varphi \rrbracket$; and

2. For every cluster C of density greater than $1 - \theta$ containing i:

    (a) for every $e \in E$ there is some $j \in C$ such that $e \in \mathcal{E}_j$; and

(b) *for all $E' \subseteq \bigcup_{j \in Ag} \mathcal{E}_j$ and $\bigcap E' \neq \emptyset$, if there exists $J \subseteq C$ such that for every $e \in E'$ there is $j \in J$ with $e \in \mathcal{E}_j$, then $\bigcap E' \cap (U \cap \Pi_i(x)) \neq \emptyset$.*

Before presenting the proof, we outline the theorem's key conditions. It consists of two parts: **receiving** and **blocking**.

In the receiving part, an agent attains individual knowledge of φ at the end of the diffusion sequence if and only if there exists a subset of the total evidence satisfying certain conditions. This subset functions as an 'information package' the agent must acquire to know φ. First, adding this package to the agent's basic evidence must ensure the existence of a set U in the resulting topology such that $U \cap \Pi_i(x) \subseteq [\![\varphi]\!]$, meaning the evidence *supports* φ. Second, by Theorem 1, all elements of this package must become part of the agent's evidence by the end of the diffusion sequence. Thus, the package both exists and is received.

The blocking part ensures that no subset of evidence the agent already has or will receive disrupts the density of U. Again, this follows from the Evidence Cluster Theorem (Theorem 1). Together, these conditions establish a dense subset of $[\![\varphi]\!]$, ensuring that the agent defeasibly knows φ. We now proceed to the proof.

Proof. Let $\mathfrak{M} = \langle X, \{\mathcal{E}_i\}_{i \in Ag}, \{\Pi_i\}_{i \in Ag}, \mathcal{N}, [\![\cdot]\!], \theta \rangle$ be an evidence diffusion model with $\theta \neq 0$ and $x \in X$.

($\Rightarrow$) Suppose that $\mathfrak{M}, x \vDash [\sim]^{fix} K_i \varphi$. This means, by the definition of $[\sim]$, that $\mathfrak{M}_{fix}, x \vDash K_i \varphi$, that is, there exists $U \in \tau_i^{fix}$, the topology of evidence generated from $\mathcal{E}_i^{fix}$, such that $U \cap \Pi_i(x) \subseteq [\![\varphi]\!]^{fix} = [\![\varphi]\!]$ (by Proposition 3) and U is i-locally dense in $\Pi_i(x)$. Now, for the set E required by the consequent, take $E = \mathcal{E}_i^{fix} \setminus \mathcal{E}_i$. For the subset of φ required, we take U mentioned above. The topology generated by $\mathcal{E}_i \cup E = \mathcal{E}_i \cup (\mathcal{E}_i^{fix} \setminus \mathcal{E}_i) = \mathcal{E}_i^{fix}$ is exactly τ_i^{fix}, so we know by assumption that $U \in \tau_i^{fix}$ and $x \in U \cap \Pi_i(x) \subseteq [\![\varphi]\!]$, so (1) is proven.

For (2), let C be a cluster of density greater than $1 - \theta$ such that $i \in C$. To prove (2a), suppose further, toward contradiction, that there is $e \in E$ such that $e \notin \mathcal{E}_j$ for all $j \in C$. Then, by Corollary 1, $e \notin \mathcal{E}_i^{fix}$, contradicting $e \in E \subseteq \mathcal{E}_i^{fix}$.

To prove (2b), take an arbitrary $E' \subseteq \bigcup_{j \in Ag} \mathcal{E}_j$ and suppose there exists $J \subseteq C$ such that for every $e \in E'$ there is $j \in J$ with $e \in \mathcal{E}_j$. By by Corollary 1, we have that for every $e \in E'$, $e \in \mathcal{E}_i^{fix}$, so that $\bigcap E' \in \tau_i^{fix}$. But note that, by assumption, U is i-locally dense in $\Pi_i(x)$ with respect to τ_i^{fix}. Thus, we cannot have that $\bigcap E' \cap (U \cap \Pi_i(x)) = \emptyset$. And since E' was picked arbitrarily, the second requirement is met as well.

($\Leftarrow$) Now suppose that there exists $E \subseteq \bigcup_{j \in Ag} \mathcal{E}_j$ that satisfies items (1) and (2) and prove that $\mathfrak{M}, x \vDash [\sim]^{fix} K_i \varphi$, that is, there is $U \in \tau_i^{fix}$ such that $x \in U$ and U is i-locally dense in $\Pi_i(x)$ with respect to τ_i^{fix}.

By the assumption, we know that there exists a set $E \subseteq \bigcup_{j \in Ag} \mathcal{E}_j$ such that there exists $U \in \tau_i^{+E}$ with $x \in U \cap \Pi_i(x) \subseteq [\![\varphi]\!]$. By assumption, for every cluster C of density greater than $1 - \theta$ containing i and for every $e \in E$, there is some

$j \in C$ such that $e \in \mathcal{E}_j$. Thus, by Corollary 1, for every $e \in E$, $e \in \mathcal{E}_i^{fix}$, and indeed $\tau_i^{+E} \subseteq \tau_i^{fix}$, entailing $U \in \tau_i^{fix}$.

Now to show that U is i-locally dense in $\Pi_i(x)$ with respect to τ_i^{fix}, suppose the contrary. This means that there exists $A \in \tau_i^{fix}$ such that $(U \cap \Pi_i(x)) \cap (A \cap \Pi_i(x)) = \emptyset$. That is, $A \cap (U \cap \Pi_i(x)) = \emptyset$. From this we can infer that there exists $e \in \overline{\mathcal{E}}_i^{fix}$ such that $e \cap (U \cap \Pi_i(x)) = \emptyset$. Moreover, there exists $E' \subseteq \mathcal{E}_i^{fix}$ such that $\bigcap E' \cap (U \cap \Pi_i(x)) = \emptyset$. But now note that for every $e' \in E'$, $e' \in \mathcal{E}_i^{fix}$, and so by the evidence cluster theorem we have that for every cluster C of density greater than $1 - \theta$ there must exist $j \in C$ such that $e' \in \mathcal{E}_j$. So we have found $E' \subseteq \bigcup_{j \in Ag} \mathcal{E}_j$ such that there exists $J \subseteq C$ with every $e' \in E'$ having a corresponding $j \in J$ with $e' \in \mathcal{E}_j$, but $\bigcap E' \cap (U \cap \Pi_i(x)) = \emptyset$, contradiction. So U is i-locally dense in $\Pi_i(x)$ with respect to τ_i^{fix}.

The following corollary establishes the conditions for what is known in the literature as 'mutual knowledge' [11,16], which amounts to 'everybody knows'.

Corollary 2. *For any evidence diffusion model* $\mathfrak{M} = \langle X, \{\mathcal{E}_i\}_{i \in Ag}, \{\Pi_i\}_{i \in Ag}, \mathcal{N}, [\![\cdot]\!], \theta \rangle$ *with* $\theta \neq 0$, $x \in X$ *and* $\varphi \in \mathcal{L}_{CPL}$, $\mathfrak{M}, x \vDash [\sim]^{fix} \bigwedge_{i \in Ag} K_i \varphi$ *if and only if the conditions in the previous theorem are met for every agent.*

For evidence-based distributed knowledge, the conditions are similar, but they, of course, have to mimic the definition of evidence-based distributed knowledge instead of individual knowledge. What we mean by this is that instead of requiring a factive dense open subset of φ, we require that each agent has an 'information package' which gives them a factive dense set, and that all of these sets taken together entail φ (i.e. their intersection is a subset of $[\![\varphi]\!]$). The proof is similar to the previous one.

Theorem 3 (Evidence-Based Distributed Knowledge Cluster Theorem). *For any evidence diffusion model* $\mathfrak{M} = \langle X, \{\mathcal{E}_i\}_{i \in Ag}, \{\Pi_i\}_{i \in Ag}, \mathcal{N}, [\![\cdot]\!], \theta \rangle$ *with* $\theta \neq 0$, $x \in X$ *and* $\varphi \in \mathcal{L}_{CPL}$, $\mathfrak{M}, x \vDash [\sim]^{fix} \mathcal{D}\varphi$ *if and only if for every* $i \in Ag$ *there exists* $E \subseteq \bigcup_{j \in Ag} \mathcal{E}_j$ *such that there exists* $U_i \in \tau_i^{+E}$, *the topology generated by* $\mathcal{E}_i \cup E$, *with* $x \in U_i$ *such that:*

1. *for every cluster* C *of density greater than* $1 - \theta$ *containing* i, *we have:*
   (a) *for every* $e \in E$, *there exists* $j \in C$ *such that* $e \in \mathcal{E}_j$; *and*
   (b) *for every* $E' \subseteq \bigcup_{j \in Ag} \mathcal{E}_j$, *if there exists* $J \subseteq C$ *such that for all* $j \in J$, $e \in \mathcal{E}_j$, *then* $\bigcap E' \cap (U_i \cap \Pi_i(x)) \neq \emptyset$;
2. $\bigcap_{i \in Ag} U_i \subseteq [\![\varphi]\!]$.

5   Further Work

There are several directions in which this research could be taken further, which includes a continued logical investigation of our formal setting as well as the study of its implications and relevance for debates in philosophy of science, social epistemology and related philosophical studies. Axiomatizing evidence diffusion models is a natural next step for this line of research. Computer implementations would enable simulations, allowing for quantitative analyses of evidence diffusion on specific models, and in turn more direct comparisons with related work, specifically that of Zollman [25,26]. Another avenue is to explore communication protocols that differ from evidence diffusion, especially those that allow for various social-epistemic factors at play in more fine-grained representations of scientific networks. For instance, one could imagine a diffusion protocol which more closely mimics the replication of results in scientific practice. That is to say, before an agent adopts a piece of evidence from a neighbor, the agent 're-runs the experiment', and only adopts that evidence if the experiment is validated. Whether or not this evidence gets validated can be modeled probabilistically: if the evidence is true (contains the real world), then it replicates with, say, 99% probability, the remaining 1% representing false negatives; and if the evidence is false, it replicates with 5% probability, representing false negatives. Another possible communication protocol worth exploring would be one involving deliberation[2], where agents have to persuade their neighbors to adopt a piece of evidence, similar to [8].

Beyond these refinements, extending the model to allow agent-specific thresholds could better capture variations in skepticism, while evidence-specific thresholds could reflect differences in plausibility. Further, incorporating qualitative expert deference principles—where non-expert agents adopt beliefs based on expert testimony—would offer an alternative to purely quantitative diffusion mechanisms.

Finally, investigating common knowledge and common evidence within the diffusion framework, alongside a systematic exploration of the Independence Thesis [15] in this model, could provide deeper insights into the epistemic trade-offs between network structure and collective reasoning.

Acknowledgement. The research priority area Human(e) AI at the University of Amsterdam has contributed to sponsoring the authors' participation at the conference. We thank the anonymous reviewers of LORI 2025 for their valuable feedback.

References

1. Abarca, A.I.R.: Topological models for group knowledge and belief. MSc Thesis, University of Amsterdam, Amsterdam (2015)
2. Baltag, A., Bezhanishvili, N., Özgün, A., Smets, S.: A topological approach to full belief. J. Philos. Log. **48**(2), 205–244 (2019)
3. Baltag, A., Bezhanishvili, N., Özgün, A., Smets, S.: Justified belief, knowledge, and the topology of evidence. Synthese **200**(6), 512 (2022)
4. Baltag, A., Renne, B.: Dynamic Epistemic Logic. In: Stanford Encyclopedia of Philosophy, Winter 2016 edn. Metaphysics Research Lab, Stanford University (2016)

---

[2] We thank one of our anonymous reviewers for this suggestion.

5. Blackburn, P., de Rijke, M., Venema, Y.: Modal Logic. Cambridge University Press (2010)
6. Christoff, Z.: Dynamic logics of networks: information flow and the spread of opinion. Ph.D. thesis, University of Amsterdam (2016)
7. Christoff, Z., Hansen, J.U., Proietti, C.: Reflecting on social influence in networks. J. Log. Lang. Inf. **25**(3–4), 299–333 (2016)
8. Ding, H., Pivato, M.: Deliberation and epistemic democracy. J. Econ. Behav. Organ. **185**, 138–167 (2021)
9. van Ditmarsch, H., van der Hoek, W., Kooi, B.P.: Dynamic Epistemic Logic. Synthese Library, vol. 337. Springer, Dordrecht (2007)
10. Easley, D., Kleinberg, J.: Networks, Crowds, and Markets: Reasoning about a Highly Connected World. Cambridge University Press, Cambridge (2010)
11. Fagin, R., Halpern, J.Y., Moses, Y., Vardi, M.: Reasoning About Knowledge. The MIT Press (1995)
12. Fernández González, S.: Generic models for topological evidence logics. MSc thesis, University of Amsterdam (2018)
13. Halpern, J.Y., Moses, Y.: Knowledge and common knowledge in a distributed environment. J. ACM **37**(3), 549–587 (1990)
14. Liu, F., Seligman, J., Girard, P.: Logical dynamics of belief change in the community. Synthese **191**(11), 2403–2431 (2014). https://doi.org/10.1007/s11229-014-0432-3
15. Mayo-Wilson, C., Zollman, K.J.S., Danks, D.: The independence thesis: when individual and social epistemology diverge. Philos. Sci. **78**(4), 653–677 (2011)
16. Meyer, J.J.C., van der Hoek, W.: Epistemic Logic for AI and Computer Science, vol. 41. Cambridge University Press, Cambridge (2004)
17. Munkres, J.R.: Topology, 2nd edn. Prentice Hall, Upper Saddle River (2000)
18. Özgün, A.: Evidence in epistemic logic: a topological perspective. Ph.D. thesis, University of Amsterdam, Amsterdam (2017)
19. Seligman, J., Liu, F., Girard, P.: Logic in the community. In: Banerjee, M., Seth, A. (eds.) ICLA 2011. LNCS (LNAI), vol. 6521, pp. 178–188. Springer, Heidelberg (2011). https://doi.org/10.1007/978-3-642-18026-2_15
20. Smets, S., Velázquez-Quesada, F.R.: A logical study of group-size based social network creation. J. Log. Alegbr. Methods Program **106**, 117–140 (2019)
21. Smets, S., Velázquez-Quesada, F.R.: A closeness- and priority-based logical study of social network creation. J. Logic Lang. Inform. **29**(1), 21–51 (2020). https://doi.org/10.1007/s10849-019-09311-5
22. Valente, T.W.: Social network thresholds in the diffusion of innovations. Soc. Netw. **18**(1), 69–89 (1996)
23. van Benthem, J., Fernández-Duque, D., Pacuit, E.: Evidence and plausibility in neighborhood structures. Ann. Pure Appl. Logic **165**(1), 106–133 (2014)
24. Zhen, L., Seligman, J.: A logical model of the dynamics of peer pressure. Electron. Notes Theor. Comput. Sci. **278**, 275–288 (2011)
25. Zollman, K.J.S.: The communication structure of epistemic communities. Philos. Sci. **74**(5), 574–587 (2007)
26. Zollman, K.J.: Network epistemology: communication in epistemic communities. Philos Compass **8**(1), 15–27 (2013)
27. Zotescu, T.Ş.: Multi-agent topological models for evidence diffusion. MSc thesis, University of Amsterdam (2024)

Functional Dependence in Uniform Dependence Model

Chenwei Shi and Wenfei Ouyang[(✉)]

Tsinghua University – University of Amsterdam Joint Research Centre for Logic,
Department of Philosophy, Tsinghua University, Beijing, China
`scw@tsinghua.edu.cn, oywf@mails.tsinghua.edu.cn`

Abstract. This paper studies the representation theorem of functional dependence relation in uniform dependence models, where local and global functional dependence relations coincide. We simplify the construction in the proof provided by Baltag and van Benthem [2] and take a deeper look into the simplified construction. Moreover, we demonstrate the impossibility of representing some disjunctive functional dependence in a uniform dependence model.

Keywords: Functional dependence relation · Representation theorem · Uniform dependence model

1 Introduction

Functional dependence is a basic concept in relational databases [5]. Within relational database theory, functional dependence can be interpreted as follows: an attribute y *functionally depends* on an attribute x means that for any two assignments f and g, $f(x) = g(x) \implies f(y) = g(y)$. More generally, Y *functionally depends* on a set of attributes X means that for any two assignments f and g, $\forall x \in X(f(x) = g(x)) \implies \forall y \in Y(f(y) = g(y))$. A simple example is shown in the following relational database:

This table provides data about 5 Pokémons by listing their attributes: Power(P), Magic(M), and Speed(S). In terms of functional dependence, Speed functionally depends on Power, because for each Pokémon, if the attribute Power has the same point, then the attribute Speed has the same point. However, Power does not functionally depend on Speed, because Psyduck has the same Speed points as Bulbasaur but different Power points from it. Using $D_X Y$ to denote the expression "set Y functionally depends on a set X" (the brackets will be omitted for singletons), it is the case that $D_P S$ but not the case that $D_S P$.

In 1974, Armstrong showed that functional dependence could be characterized by the following axioms [1]. For a fixed set of variables V, $x, y \in V$ and $X, X', Y, Z \subseteq V$:

C. Shi—granted by the Fund for Basic Research in Humanities of Tsinghua University.

V. Goranko et al. (Eds.): LORI 2025, LNCS 16010, pp. 124–135, 2026.
https://doi.org/10.1007/978-981-95-2481-5_9

Table 1. A table of the data of Pokémons' attributes.

Pokémon	Attributes		
	Power(P)	Magic(M)	Speed(S)
Pikachu	A	B	A
Squirtle	A	C	A
Bulbasaur	B	A	B
Psyduck	C	A	B
Magikarp	B	C	B

1. $D_x x$;
2. $D_X Y, D_Y Z \implies D_X Z$;
3. $D_X y, X \subseteq Z \implies D_Z y$.

Based on this result, dependency theory is developed in database theory, where first-order logic is used to characterize all kinds of (in)dependency (cf. [7]).

Several logical approaches to functional dependence proposed in recent years have witnessed a revival of interest in functional dependence. For example, team semantics of dependence logic [4,8], Chap. 8 in [6] and the logic of functional dependence (LFD) [2].

In LFD, local functional dependence is primitive, in contrast to global dependence, that is the main focus of database theory and team semantics. Its semantics similarly relies on an underlying relational database (seen as a first-order model restricted to a subset of admissible assignments), but dependence relations are evaluated at an assignment locally. From the modal perspective, each admissible assignment is viewed as a "possible world" in Kripke semantics. Two worlds are connected with respect to a set of variables when they assign the same values to these variables.

In [2], it is proved that the original axiomatic characterization of Armstrong is also complete with respect to a dependence model where the local dependence and the global dependence coincide. We call such a dependence model a "uniform dependence model". One of the main contributions of our paper is simplification of the construction used in the proof of the representation theorem in [2].

In addition, we consider another type of dependence – disjunctive dependence. In Table 1, Power or Magic disjunctively depends on Speed, in the sense that if the value of Speed is fixed, then either the value of Power or the value of Magic is also fixed. While the basic functional dependence in uniform dependence models can be completely axiomatized, this paper shows that this is not the case for disjunctive dependence in uniform dependence models.

The paper is structured as follows. Section 2 briefly reviews the original representation theorem, especially the key construction provided in [2]. In Sect. 3, we present our new way of constructing uniform dependence models. Section 4 introduces two variants of our construction and the corresponding relations they represent. In Sect. 5, we turn to disjunctive dependence in uniform dependence models and discuss a negative result concerning its axiomatization.

2   Representation Theorem of Uniform Model

In this section, we introduce the representation theorem in [2, Proposition 2.6.2] for closure relations. We start with formal definitions for the key concepts.

Definition 1. *A dependence model M is a pair (O, A), where O is a set of values and $A \subseteq O^V$ is a set of assignments which assign to each variable in V a value in O.*

An attribute or variable y locally depends on X at an assignment s (denoted by $D_X^s y$) in a dependence model $M = (O, A)$ if and only if:

$$\text{for all assignments } t \text{ in } M, \text{ if } s =_X t, \text{ then } s =_y t$$

where $s =_X t$ denotes "$s(x) = t(x)$ for all $x \in X$". Global dependence in a dependence model M (denoted by $D_X^M y$) is $D_X^s y$ for every assignment s in M.

A dependence model M is a uniform dependence model if for any $s \in A$ and $X, Y \subseteq V$, $D_X^s Y$ if and only if $D_X^M Y$.

Example 1. In Table 2, even though Power does not functionally depend on Speed globally, Power functionally depends on Speed on Pikachu locally. A trivial example of uniform model $M = (O, A)$ is when A is a singleton. We will see a non-trivial one in the following section.

Definition 2 (Closure relation). *Let $R \subseteq \wp(V) \times V$ be a relation between subsets of variables and single variables. We write $R_X y$ when $(X, y) \in R$ and write $R_X Y$ if $(X, y) \in R$ for all $y \in Y$. We call R a closure relation if it satisfies the following properties:*

- *Reflexivity: For every $x \in V$, we have $R_x x$;*[1]
- *Transitivity: If $R_X Y$ and $R_Y Z$, then $R_X Z$;*
- *Monotonicity: If $R_X y$ and $X \subseteq Z$, then $R_Z y$.*

Given a set of variables $X \subseteq V$, we define its R-closure as $\tilde{X} = \{y \in V : R_X y\}$. A subset Y is called R-closed if $Y = \tilde{Y}$. We denote by Γ the family of all R-closed subsets of V.

The following observation explains why we call a relation satisfying the three Armstrong axioms "closure relation".

Remark 1.   – *Γ is a topped intersection structure, that is to say,*
    1. *Γ is closed under arbitrary intersection;*
    2. *$V \in \Gamma$.*
– *The operator $\tilde{\ }$ is a closure operator on $\wp(V)$, that is, it satisfies:*
    1. *$X \subseteq \tilde{X}$*
    2. *$X \subseteq Y \Rightarrow \tilde{X} \subseteq \tilde{Y}$*
    3. *$\tilde{\tilde{X}} = \tilde{X}$*

---

[1] Instead of $R_{\{x\}}\{y\}$, we write $R_x y$ for brevity.

- $(\Gamma, \wedge, \vee)$ is a complete lattice ordered by $\subseteq$ where $\bigwedge_{i \in I} A_i = \bigcap_{i \in I} A_i$ and $\bigvee_{i \in I} A_i = \bigcap \{B \in \Gamma \mid \bigcup_{i \in I} A_i \subseteq B\}$.

For more details on the relationship between topped intersection structures, closure operators and complete lattices, see [3, Section 2 and 7]. We only note that they are different forms of the same structure. In fact, closure relations are another form of the same structure. We have seen how to define a closure operator from a closure relation. It is not hard to verify that from a closure operator c we can define a closure relation as follows: $R_X y$ iff $y \in c(X)$.

Since there is one-one correspondence between closure relations and closure operators, the following representation theorem ([2, Proposition 2.6.2]) can also be viewed as a representation theorem for topped intersection structures, closure operators and complete lattices.

Theorem 1 (Representation Theorem of Uniform Model). *For every closure relation R, there is a uniform dependence model $\mathbf{M}_R$ where $R = D^s$ for all assignments $s \in A$. Moreover, if V is finite, then $\mathbf{M}_R$ is finite as well, of size bounded by $2^{2^{|V|}}$.*

In the proof of the above theorem provided in [2], the definition of the uniform dependence model relies on the following equivalence relations on subsets of Γ:

$$\mathcal{A} \sim_x \mathcal{B} \quad \text{iff} \quad x \in \bigcap(\mathcal{A} \triangle \mathcal{B}),$$

where $\mathcal{A} \triangle \mathcal{B}$ denotes the symmetric difference of the two families.

Let $[\mathcal{A}]_x := \{\mathcal{B} \mid \mathcal{A} \sim_x \mathcal{B}\}$ and for each $\mathcal{A} \subseteq \Gamma$ define a function $s^{\mathcal{A}} : V \to O := \{[\mathcal{A}]_x \mid x \in V, \mathcal{A} \subseteq \Gamma\}$ as

$$s^{\mathcal{A}} : x \mapsto [\mathcal{A}]_x.$$

The dependence model is given by $\mathbf{M}_R = (O, A)$, where $A = \{s^{\mathcal{A}} \mid \mathcal{A} \subseteq \Gamma\}$.

We illustrate the above construction using the following example, which also gives a taste of its cumbersomeness:

Example 2. Let $V = \{x, y, z\}$ and $\Gamma = \{\{x, y, z\}, \{y, z\}, \{z\}\}$. We spell out the assignment $s^{\mathcal{A}}$ where $\mathcal{A} = \{\{x, y, z\}, \{y, z\}\}$.

$s^{\mathcal{A}}(x) = [\mathcal{A}]_x = \{\{\{x, y, z\}, \{y, z\}\}, \{\{y, z\}\}\}$

$s^{\mathcal{A}}(y) = [\mathcal{A}]_y = \{\{\{x, y, z\}, \{y, z\}\}, \{\{x, y, z\}\}, \{\{y, z\}\}, \emptyset\}$

$s^{\mathcal{A}}(z) = [\mathcal{A}]_z = \{\{\{x, y, z\}, \{y, z\}\}, \{\{x, y, z\}, \{y, z\}, \{z\}\},$

$\qquad\qquad \{\{y, z\}, \{z\}\}, \{\{x, y, z\}, \{z\}\}, \emptyset,$

$\qquad\qquad \{\{x, y, z\}\}, \{\{y, z\}, \{z\}\}, \{\{z\}\}\}$

The example makes it clear that while there is a one-one correspondence between the set of assignments A and the set $\wp(\Gamma)$, the values each variable can take belong to $\wp\wp(\Gamma)$. In the next section, we will show that it is enough to take $\wp(\Gamma)$ as the set of values O.

3   Proof Simplified

We propose a lighter construction to prove the representation theorem as follows:

Definition 3 (Armstrong dependence model). *An Armstrong dependence model[2] for a closure relation R, $M_R = (O_R, A_R)$, is defined by*

- $O_R = \wp(\Gamma)$;
- $A_R = \{s^{\mathcal{A}} : \mathcal{A} \subseteq \Gamma\}$ *where* $s^{\mathcal{A}} : V \to O$ *is defined by*

$$s^{\mathcal{A}}(x) = \mathcal{A}_x^- := \{c \in \mathcal{A} : x \notin c\}$$

The following example illustrates the new construction.

Example 3. To facilitate comparison, we take the same $V = \{x, y, z\}$ and $\Gamma = \{\{x, y, z\}, \{y, z\}, \{z\}\}$ to Example 2. The Armstrong dependence model (O_R, A_R) is illustrated in Table 2. $s^{\{\{x,y,z\},\{y,z\}\}}$ is much simpler than that in Example 2.

Table 2. Representation model

A_R	x	y	z
$s^{\{\{x,y,z\},\{y,z\},\{z\}\}}$	$\{\{y, z\}, \{z\}\}$	$\{\{z\}\}$	$\emptyset$
$s^{\{\{x,y,z\},\{y,z\}\}}$	$\{\{y, z\}\}$	$\emptyset$	$\emptyset$
$s^{\{\{y,z\},\{z\}\}}$	$\{\{y, z\}, \{z\}\}$	$\{\{z\}\}$	$\emptyset$
$s^{\{\{x,y,z\},\{z\}\}}$	$\{\{z\}\}$	$\{\{z\}\}$	$\emptyset$
$s^{\{\{x,y,z\}\}}$	$\emptyset$	$\emptyset$	$\emptyset$
$s^{\{\{y,z\}\}}$	$\{\{y, z\}\}$	$\emptyset$	$\emptyset$
$s^{\{\{z\}\}}$	$\{\{z\}\}$	$\{\{z\}\}$	$\emptyset$
$s^{\emptyset}$	$\emptyset$	$\emptyset$	$\emptyset$

We use the new construction to prove the representation theorem of uniform models.

Proposition 1. *In the dependence model (O_R, A_R) defined above, $R = D^s$ for all $s \in A_R$.*

---

[2] The reason for calling this construction "Armstrong dependence model" is that it is first used in Armstrong's proof of his representation theorem for his axiomatic characterization of functional dependency [1]. We independently discover this construction, and later on find that Armstrong used the same construction, although he did not put any emphasis on the uniformity of the dependence model.

Proof. Take an arbitrary $\mathcal{A} \in 2^{\Gamma}$.

Assume that $R_X y$ holds, we need to prove that $D^{s^{\mathcal{A}}}_X y$ (which we will subsequently abbreviate to $D^{\mathcal{A}}_X y$) Take any $\mathcal{B} \in \Gamma$ satisfying $s^{\mathcal{A}} =_X s^{\mathcal{B}}$. We need to prove that $s^{\mathcal{A}} =_y s^{\mathcal{B}}$, that is, $\mathcal{A}^-_y = \mathcal{B}^-_y$. Since $R_X y$, for all $g \in \Gamma$, if $y \notin g$, then $x \notin g$ for some $x \in X$. Take any $g \in \mathcal{A}^-_y$. Then there is some $x \in X$ such that $g \in \mathcal{A}^-_x$. Together with $\mathcal{A}^-_x = \mathcal{B}^-_x$, it implies that $g \in \mathcal{B}^-_x$. So $g \in \mathcal{B}$. Together with $y \notin g$, it implis that $g \in \mathcal{B}^-_y$. So we have proved that $\mathcal{A}^-_y \subseteq \mathcal{B}^-_y$. The other direction is symmetric and thus it follows that $\mathcal{A}^-_y = \mathcal{B}^-_y$.

For the other direction, we assume that $\neg R_X y$ and need to prove that $\neg D^{\mathcal{A}}_X y$. There must be $g \in \Gamma$ such that $X \subseteq g$ (by existence of $\tilde{X}$, where *Reflexivity* and *Monotonicity* are used) and $y \notin g$ (by $\tilde{X}$ is R-closure and does not contain y, where *Transitivity* and *Monotonicity* are used). Take such a R-closed set g arbitrarily. Consider two cases. Case one: $g \in \mathcal{A}$. Consider $s^{\mathcal{A} \setminus \{g\}}$ and $s^{\mathcal{A}}$. By $X \subseteq g$, $(\mathcal{A} \setminus \{g\})^-_x = \mathcal{A}^-_x$ for all $x \in X$. By $y \notin g$, $(\mathcal{A} \setminus \{g\})^-_y \neq \mathcal{A}^-_y$. So $\neg D^{\mathcal{A}}_X y$. Case two: $g \notin \mathcal{A}$. Consider $s^{\mathcal{A} \cup \{g\}}$ and $s^{\mathcal{A}}$. By $X \subseteq g$, $(\mathcal{A} \cup \{g\})^-_x = \mathcal{A}^-_x$ for all $x \in X$. By $y \notin g$, $(\mathcal{A} \cup \{g\})^-_y \neq \mathcal{A}^-_y$. So $\neg D^{\mathcal{A}}_X y$.

$\square$

Remark 2. In the new construction, if we take the following subset of A_R

$$\{s^{\mathcal{A}} \in A_R : \mathcal{A} \text{ is a singleton}\}$$

then the resulting model is in essence the construction used for the proof of Proposition 2.6.1 in [2], which is not uniform.

4   Two Variants of Armstrong Dependence Model

In this section, we study two variants of the new construction and the relations they represent.

4.1   Plus Dependence Model

Let $\mathcal{A}^+_x := \{c \in \mathcal{A} : x \in c\}$ We change the definition of A_R in $\mathbf{M}_R$ to $A^+_R = \{s^{\mathcal{A}} : \mathcal{A} \subseteq \Gamma\}$ where

$$s^{\mathcal{A}}(x) := \mathcal{A}^+_x$$

and denote the new model by $\mathbf{M}^+_R$. We call this model "plus dependence model".

What kind of relations do the dependence relations in the plus dependence model represent?

Let $S \subseteq \wp(V) \times V$ be a relation defined as follows:

$$S_X y \text{ iff } \{g \in \Gamma : X \subseteq V \setminus g\} \subseteq \{g \in \Gamma : y \in V \setminus g\} \ .$$

Recall that $\Gamma = \{X \subseteq V \mid X = \tilde{X}\}$ and compare the definition of $S_X y$ with the following condition:

$$\{g \in \Gamma : X \subseteq g\} \subseteq \{g \in \Gamma : y \in g\}$$

which is equivalent to $R_X y$. This analogy leads us to the following theorem, which answers the above question.

130   C. Shi and W. Ouyang

Theorem 2 (Representation theorem for S). *For every closure relation R, there is a uniform dependence model $\mathbf{M}_R^+$ where S as defined above equals D^s for all $s \in A_R^+$. Therefore, $S = D^{\mathbf{M}_R^+}$. Moreover, if V is finite, then $\mathbf{M}_R^+$ is finite as well, of size bounded by $2^{2^{|V|}}$.*

The proof is almost the same as that in Sect. 3. We only need to change R to S and make corresponding modification.

Although the relation S is represented as the basic functional dependence relation in the plus dependence model, in the Armstrong dependence model $\mathbf{M}_R$, the relation S is represented in the following way, which is a special form of disjunctive dependence.

Proposition 2. *In the Armstrong dependence model $\mathbf{M}_R$ for a given closure relation R,*

$$S_X y \quad \textit{iff} \quad \exists x \in X : R_y x \quad \textit{iff} \quad \exists x \in X : D_y^s x$$

The proposition follows from the observation that $S_X y$ iff $\widetilde{\{y\}} \cap X \neq \emptyset$, recalling that $\widetilde{\{y\}}$ denotes the R-closure of $\{y\}$

We will come back to the issue of disjunctive dependence in Sect. 5.

4.2   Transposing Armstrong Dependence Model

In this section, we study the transposition of the Armstrong dependence model $\mathbf{M}_R$ for a closure relation R, denoted by $\mathbf{M}_R^\top$. To be more precise, in $\mathbf{M}_R^\top$,

- $V = \wp(\Gamma)$
- $O_R^\top = \wp(\Gamma)$;
- $A_R^\top = \{s^x : V \to O \mid x \in V\}$ where $s^x(\mathcal{A}) = \mathcal{A}_x^-$.

In this setting, two assignments s^x and s^y agree on $\mathcal{A} \in \wp(\Gamma)$ as a variable, written as $s^x =_\mathcal{A} s^y$, if and only if, $\mathcal{A}_x^- = \mathcal{A}_y^-$. Note that $\mathcal{A}_x^- = \mathcal{A}_y^-$ if and only if $\mathcal{A}_x^+ = \mathcal{A}_y^+$. Hence the local dependence relation D^x in $\mathbf{M}_R^\top$ at s^x and the local dependence relation D^{+x} in the transposition of the plus model $\mathbf{M}_R^{+\top}$ at s^x is the same.

Table 3. Transposed model

	$\{\{x,y,z\},\{z\}\}$	$\{\{x,y,z\},\{y,z\}\}$	$\{\{y,z\},\{z\}\}$	$\{\{x,y,z\},\{y,z\},\{z\}\}$	$\{\{x,y,z\}\}$	$\{\{y,z\}\}$	$\{\{z\}\}$	$\emptyset$
s^x	$\{\{z\}\}$	$\{\{y,z\}\}$	$\{\{y,z\},\{z\}\}$	$\{\{y,z\},\{z\}\}$	$\emptyset$	$\{\{y,z\}\}$	$\{\{z\}\}$	$\emptyset$
s^y	$\{\{z\}\}$	$\emptyset$	$\{\{z\}\}$	$\{\{z\}\}$	$\emptyset$	$\emptyset$	$\{\{z\}\}$	$\emptyset$
s^z	$\emptyset$	$\emptyset$	$\emptyset$	$\emptyset$	$\emptyset$	$\emptyset$	$\emptyset$	$\emptyset$

Example 4. The transposition of the Armstrong dependence model in Example 3 is illustrated in Table 3.

In this transposed dependence model,

$$D^{s^x}_{\{\{y,z\}\}}\{\{z\}\} \quad \text{but} \quad \text{not } D^{s^y}_{\{\{y,z\}\}}\{\{z\}\}$$

So, local dependence relations do not always coincide with global dependence relation after transposing an Armstrong dependence model.

Given $\mathbf{M}^\top_R$, let $C_x(\mathfrak{G}) := \{\mathcal{A} \in \wp(\Gamma) \mid D^{s^x}_{\mathfrak{G}}\mathcal{A}\}$ and $\Gamma^\top_x := \{\mathfrak{G} \subseteq \wp(\Gamma) \mid C_x(\mathfrak{G}) = \mathfrak{G}\}$. What kind of relation the local dependence relation at s^x in $\mathbf{M}^\top_R$ represent? We answer this question by characterizing $\Gamma^\top_x$ in terms of R and S.

Recall that $\tilde{X} = \{y \in V \mid R_X y\}$ and let $\hat{X} := \{y \in V \mid S_X y\}$. Moreover, let $X^\sharp := \tilde{X} \cap \hat{X}$, $\Gamma^\sharp := \{X \subseteq V : X = X^\sharp\}$, and $\Gamma^\sharp_x := \{X \in \Gamma^\sharp : x \in X\}$. It is not difficult to verify that $\hat{}$ and $\sharp$ are closure operators just as $\tilde{}$.

The following theorem answers the main question of this subsection, the proof of which can be found in the appendix.

Theorem 3. *The function $\varphi_x : \Gamma^\sharp_x \to \Gamma^\top_x$ defined by*

$$\varphi_x(X) = \wp(\{G \in \Gamma : X \subseteq G\} \cup \{G \in \Gamma : X \cap G = \emptyset\})$$

is a bijection.

5   Disjunctive Dependence, to be Uniform or Not?

The main question we aim to address in this section is whether any axiomatic characterization of disjunctive dependence can be complete with respect to uniform dependence models.

First, we define disjunctive dependence precisely.

Definition 4. *Given a dependence model $M = (O, A)$, $X \subseteq V$ and $\Theta \subseteq \wp(V)$,*

$$D^s_X\Theta \text{ (holds)} \quad \text{iff} \quad \text{for all } t \in A, s =_X t \text{ implies } s =_Y t \text{ for some } Y \in \Theta$$

and

$$D^M_X\Theta \text{ (holds)} \quad \text{iff} \quad D^s_X\Theta \text{ holds for all } s \in A$$

We call D^s "local disjunctive dependence relation" at s and D^M "global disjunctive dependence relation" in M. We will leave out the brackets of Θ in these two notations when it is a singleton.

Note that functional dependence relation is a special case of disjunctive dependence relation where Θ is a singleton.

In Sect. 2, we proved that every closure relation R can be represented as the dependence relation in its Armstrong dependence model. However, there is local/global dependence relation, which is impossible to be represented in any uniform dependence model.

Proposition 3. *For any $R \subseteq \wp(V) \times \wp\wp(V)$ and $X, Y, Z \in \wp(V)$, if $(X, \{Y, Z\}) \in R$ but $(X, \{Y\}) \notin R$ and $(X, \{Z\}) \notin R$, then there is no uniform model whose disjunctive dependence relation coincides with R.*

Proof. Suppose that there is a uniform model $M = (O, A)$ and an assignment $s \in A$ satisfying the relation R. Since $(X, Y) \notin R$ and $(X, Z) \notin R$, it follows that $M \not\models D_X^s Y$ and $M \not\models D_X^s Z$. By the semantics, there exist assignments $r, t \in A$ such that $s =_X r =_X t$, $s \neq_Y r$, and $s \neq_Z t$. However, because $(X, \{Y, Z\}) \in R$, we have $M \models D_X^s \{Y, Z\}$; by definition of disjunctive dependence, since $M \models D_X^s \{Y, Z\}$, each assignment that coincides with s on X must coincide on at least one of Y or Z. Given assignments r, t, it follows that $s =_Z r$ and $s =_Y t$. The following graph illustrate the relevant part of M. Two assignments are connected by an edge labeled $X, Y, \ldots$ if they agree on $X, Y, \ldots$ [3]

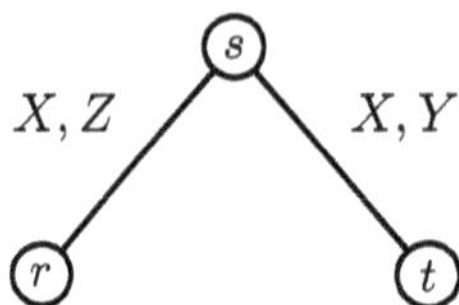

Since M is a uniform model, we also have $M \models D_X^r \{Y, Z\}$ and $M \models D_X^t \{Y, Z\}$. Moreover, given that $r =_X t$, it follows that either $r =_Y t$ or $r =_Z t$.

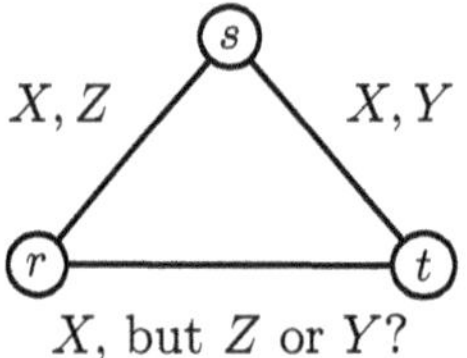

X, but Z or Y?

However, if $r =_Y t$, then, since $s =_Y t$, we would have $s =_Y r$, contradicting the earlier observation that $s \neq_Y t$; a similar contradiction arises if $r =_Z t$.   □

The argument used in the proof of the above proposition cannot be generalized, for example, to the case where $n = 3$, as shown in the following graph. It illustrates a uniform model where $D_X^M \{Y_1, Y_2, Y_3\}$ is satisfied while $D_X^M Y_1$, $D_X^M Y_2$, or $D_X^M Y_3$ is not. Assignments are represented by nodes in the graph. Two assignments are connected by an edge labeled n if they agree on Y_n. All assignments agree on X. To keep the graph clear, we omit edges labeled by X.

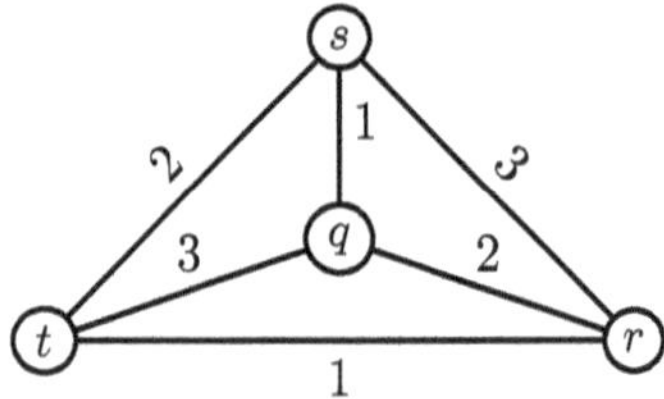

---

[3] Note that this part of M itself can serve as a dependence model in which D^s satisfy the presupposed conditions on R. It is also not hard to modify this model to have a global dependence relation satisfying the same conditions.

To the case where $n = 4$, the argument does not apply either. The next graph illustrates a uniform model where $D_X^M\{Y_1, Y_2, Y_3, Y_4\}$ is satisfied, but not $D_X^M Y_1$, $D_X^M Y_2$, $D_X^M Y_3$, or $D_X^M Y_4$.

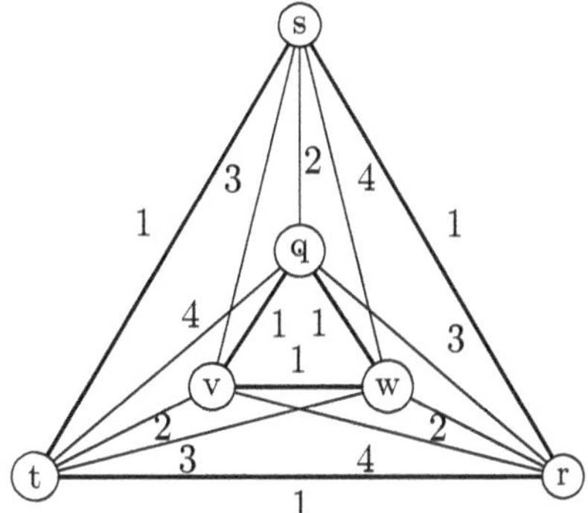

These observations lead to the following conjecture. With respect to a special class of uniform dependence models where $D_X^M Y_1$ or $D_X^M Y_2$ is satisfied whenever $D_X^M\{Y_1, Y_2\}$ is satisfied, disjunctive dependence model might have a sound and complete axiomatic characterization.

6 Future Work

To resolve the conjecture, it is helpful to know whether there can be a sound and complete axiomatic characterization of global disjunctive dependency in dependence models. We will present such a representation theorem in a future extension of this paper.

Acknowledgments. This paper was partly presented at the Tsing Ch'a session, held at the Tsinghua University - University of Amsterdam Joint Research Centre for Logic. The authors thank Johan van Benthem for his valuable feedback on this research, the audience and organizers of Tsing Ch'a session for their engagement, and the three anonymous referees for their high-quality reviews.

Disclosure of Interests. The authors have no competing interests to declare that are relevant to the content of this article.

A Proof of Theorem 3

We begin this proof with some basic properties of $D_{\mathfrak{G}}^{s^x} \mathcal{A}$ where $\mathfrak{G} \subseteq \wp(\Gamma)$ and $\mathcal{A} \in \wp(\Gamma)$. To avoid clumsy notation, we'll write D^x for D^{s^x}.

Proposition 4. *In* $\mathbf{M}_R^\top$*, for any* $s^x \in A_R^\top$*,*

1. *if* $D_{\mathfrak{G}}^x \mathcal{A}$ *and* $\mathcal{A}' \subseteq \mathcal{A}$*, then* $D_{\mathfrak{G}}^x \mathcal{A}'$*;*
2. *if* $D_{\mathfrak{G}}^x \mathcal{A}_i$ *for* $i \in I$*, then* $D_{\mathfrak{G}}^x \bigcup_{i \in I} \mathcal{A}_i$*;*

3. *for any $\mathfrak{G} \in \Gamma_x^\top$, $\bigcup \mathfrak{G} \in \mathfrak{G}$;*
4. *$D_\mathfrak{G}^x \mathcal{A}$ if and only if $D_{\bigcup \mathfrak{G}}^x \mathcal{A}$;*
5. *for any $\mathfrak{G} \in \Gamma_x^\top$, $\mathfrak{G} = C_x(\{\bigcup \mathfrak{G}\}) = \wp(\bigcup \mathfrak{G})$.*

Proof. The first property relies on the following fact.

Lemma 1. *If $\mathcal{A}' \subseteq \mathcal{A}$, then $\mathcal{A}_x^+ = \mathcal{A}_y^+$ implies that $(\mathcal{A}')_x^+ = (\mathcal{A}')_y^+$.*

The second property relies on the following fact.

Lemma 2. *If $(\mathcal{A}_i)_x^+ = (\mathcal{A}_i)_y^+$ for all $i \in I$, then $(\bigcup_{i \in I} \mathcal{A}_i)_x^+ = (\bigcup_{i \in I} \mathcal{A}_i)_y^+$.*

The third property follows from (2) and $D_\mathfrak{G}^x \mathcal{A}$, where $\mathcal{A} \in \mathfrak{G}$. The fourth property follows from the above two lemmas. The fifth property follows from the third and the fourth properties.

Next, we present several critical lemmas which lead us to a characterization of $\Gamma_x^\top$ in terms of a function $T : \wp(V) \to \wp(\Gamma)$ defined by

$$T(X) = \{G \in \Gamma : X \subseteq G\} \cup \{G \in \Gamma : X \cap G = \emptyset\} \ .$$

Let $[x]_\mathcal{A} := \{y \in V : s^y =_\mathcal{A} s^x\}$.

Lemma 3. *In $\mathbf{M}_R^\top$, for any $s^x \in A_R^\top$ and any $\mathcal{A} \in \wp(\Gamma)$, $\mathcal{A} \subseteq T([x]_\mathcal{A})$.*

Lemma 4. *For any $\mathfrak{G} \in \Gamma_x^\top$, $\bigcup \mathfrak{G} \supseteq T([x]_{\bigcup \mathfrak{G}})$.*

Proof. Take any $G \notin \bigcup \mathfrak{G}$. It follows that $\neg D_{\bigcup \mathfrak{G}}^x \{G\}$, otherwise $\mathfrak{G}$ is not closed by Proposition 4. This means that neither $[x]_{\bigcup \mathfrak{G}} \subseteq G$ nor $[x]_{\bigcup \mathfrak{G}} \cap G = \emptyset$. So it follows from the definition of T that $G \notin T([x]_{\bigcup \mathfrak{G}})$.

Lemma 3 and 4 together imply the following proposition.

Proposition 5. *For any $\mathfrak{G} \in \Gamma_x^\top$, $\bigcup \mathfrak{G} = T([x]_{\bigcup \mathfrak{G}})$. So, $\mathfrak{G} = \wp(T([x]_{\bigcup \mathfrak{G}}))$*

Proposition 6. *For any $X \subseteq V$ and any $x \in X$, $\wp(T(X)) \in \Gamma_x^\top$.*

Proof. Take an arbitrary $X \subseteq V$ and $x \in X$. It is sufficient to show that $\wp(T(X))$ is closed on x. Assume that $D_{T(X)}^x \mathcal{A}$. By Proposition 4, it is sufficient to show that $\mathcal{A} \subset T(X)$. It follows from the assumption that for all $G \in \mathcal{A}$, either $[x]_{T(X)} \subseteq G$ or $[x]_{T(X)} \cap G = \emptyset$. Together with the fact that $X \subseteq [x]_{T(X)}$, which follows from $x \in X$, it implies that $G \in T(X)$ for all $G \in \mathcal{A}$. So $\mathcal{A} \subseteq T(X)$ and by Proposition 4, $\wp(T(X)) = C_x(\{T(X)\}) \in \Gamma_x^\top \ .$

The above two propositions yield the following characterization of $\Gamma_x^\top$.

Corollary 1. *For any $x \in X$, $\Gamma_x^\top = \{\wp(T(X)) : x \in X \subseteq V\}$*

Lastly, we characterize $\Gamma_x^\top$ by $\Gamma_x^\sharp$.

Lemma 5. *For any $X \subseteq V$, $T(X) = T(X^\sharp) \ .$*

Proof. $X \subseteq X^{\sharp} \subseteq \tilde{X}$ and $\{g \in \Gamma : X \subseteq g\} = \{g \in \Gamma : \tilde{X} \subseteq g\}$ imply that $\{g \in \Gamma : X \subseteq g\} = \{g \in \Gamma : X^{\sharp} \subseteq g\}$

$X \subseteq X^{\sharp} \subseteq \hat{X}$ and $\{g \in \Gamma : X \cap g = \emptyset\} = \{g \in \Gamma : \hat{X} \cap g = \emptyset\}$ imply that $\{g \in \Gamma : X \cap g = \emptyset\} = \{g \in \Gamma : X^{\sharp} \cap g = \emptyset\}$

Together with Corollary 1, Lemma 5 implies the following proposition.

Proposition 7. $\Gamma_x^{\top} = \{\wp(T(X)) : x \in X \in \Gamma^{\sharp}\}$

Lemma 6. *For any $x \in X \in \Gamma^{\sharp}$, $X = [x]_{T(X)}$.*

Proof. $X \subseteq [x]_{T(X)}$ is obvious. For the other direction, $z \in [x]_{T(X)}$ implies that $z \in \tilde{X}$ and $z \in \hat{X}$. So $z \in X^{\sharp} = X$.

Finally, we can prove Theorem 3.

Proof. Surjectivity follows from Proposition 7. Injectivity follows from Lemma 6: for any two $X, Y \in \Gamma_x^{\sharp}$, if $\varphi_x(X) = \varphi_y(Y)$, then $T(X) = T(Y)$, which implies that $[x]_{T(X)} = [x]_{T(Y)}$. So by Lemma 6, $X = Y$.

References

1. Armstrong, W.W.: Dependency structures of data base relationships. In: IFIP Congress, vol. 74, pp. 580–583. Geneva, Switzerland (1974)
2. Baltag, A., van Benthem, J.: A simple logic of functional dependence. J. Philos. Log. **50**, 939–1005 (2021)
3. Davey, B.: Introduction to Lattices and Order. Cambridge University Press (2002)
4. Kontinen, J., Yang, F.: Complete logics for elementary team properties. J. Symb. Log. **88**(2), 579–619 (2023)
5. Maier, D.: The Theory of Relational Databases, vol. 11. Computer Science Press, Rockville (1983)
6. Väänänen, J.: Dependence Logic: A New Approach to Independence Friendly Logic, vol. 70. Cambridge University Press (2007)
7. Vardi, M.Y.: Fundamentals of dependency theory. IBM Thomas J. Watson Research Division (1985)
8. Yang, F., Väänänen, J.: Propositional logics of dependence. Ann. Pure Appl. Logic **167**(7), 557–589 (2016)

Measurement-Theoretic Foundations of Logic of Qualitative Conditional Probability

Satoru Suzuki[✉]

Faculty of Arts and Sciences, Komazawa University, 1-23-1, Komazawa, Setagaya-ku,
Tokyo 154-8525, Japan
`bxs05253@nifty.com`

Abstract. Koopman [14–16] introduces such a quaternary relation $\succsim$ that $(A, B) \succsim (C, D)$ is interpreted to mean that A given B is at least as probable as C given D. Recently Mundici [24] and Ibeling et al. [13] respectively have investigated Koopman's qualitative conditional probability from a logical point of view. However, the complete axiomatization of logic of qualitative conditional probability has been an open problem. The aim of this paper is to propose a new version of complete logic–Logic of Qualitative Conditional Probability (LQCP) by proving a theorem that can bridge the gap between the semantics and the axiomatization of LQCP.

Keywords: completeness · filtration · finite cancellation · Koopman · propositional logic · qualitative conditional probability relation · measurement theory · representation theorem

1 Motivation

Kratzer [18] provides comparative epistemic modals such as "at least as likely (probable) as" with their models in terms of a qualitative ordering on propositions derived from a qualitative ordering on possible worlds. Yalcin [35] shows that Kratzer's model does not validate some intuitively valid inference schemata and validates some intuitively invalid ones. He adopts a model based directly on a probability measure for comparative epistemic modals. His model does not cause this problem. However, as Kratzer [19, p. 25] says, "Our semantic knowledge alone does not give us the precise quantitative notions of probability and desirability that mathematicians and scientists work with", Yalcin's model seems to be unnatural as a model of comparative epistemic modals. Segerberg [30] proposes a complete logic PK of qualitative probability. Gärdenfors [8] simplifies the model of PK. Holliday and Icard [11] prove that not only a probability measure model but also a qualitatively additive measure model and a revised version of Kratzer's model do not cause Yalcin's problem. PK has a binary sentential operator as a qualitative probability operator, whereas Delgrande et al. [3] propose

V. Goranko et al. (Eds.): LORI 2025, LNCS 16010, pp. 136–150, 2026.
https://doi.org/10.1007/978-981-95-2481-5_10

a complete logic of qualitative probability with (m, n)-ary sentential operator with a finite model. We [34] propose a complete logic of qualitative probability without the limitation of the size of domain. Although, generally speaking, it is needless to say that the notion of logic of qualitative conditional probability itself is important, here we would like to present a new argument about a concrete example with which logics of qualitative absolute probability cannot deal but logics of qualitative conditional probability can deal.

(1) Probably, φ.

According to Yalcin [35], (1) is analyzed into

(2) φ is more probable than not φ.

For example, in PK (2) is expressed as

(3) $\varphi \gg \neg\varphi$,

where $\varphi \gg \psi$ is a binary sentential operator that is interpreted to mean that φ is more probable than ψ. Then, what is a logical form of (4)?:

(4) Probably, if φ, then ψ.[1]

If we tried to express (4) in PK, we would have (5):

(5) $(\varphi \supset \psi) \gg \neg(\varphi \supset \psi)$,

where $\supset$ is a material-implicational connective. (5) is equivalent to (6):

(6) $\neg(\varphi \wedge \neg\psi) \gg (\varphi \wedge \neg\psi)$.

Intuitively, (4) should be considered to imply at least (7):

(7) $(\varphi \wedge \psi) \gg (\varphi \wedge \neg\psi)$.[2]

However, if we assume the semantics of $\varphi \geqslant \psi$ in PK that is interpreted to mean that φ is at least as probable as ψ, then (8) holds:

(8) $\neg(\varphi \wedge \neg\psi) \geqslant (\varphi \wedge \psi)$.

So (6) does not always imply (7). So (5) is not suitable as a logical form of (4). Then, when $\rightarrow$ is introduced as an indicative-conditional connective, is (9) suitable as a logical form of (4)?:

(9) $(\varphi \rightarrow \psi) \gg \neg(\varphi \rightarrow \psi)$.

When $\varphi \approx \psi$ is interpreted to mean that φ is as probable as ψ, in PK (10) should hold:

---

[1] One should notice that "if φ, then ψ" in (4) is not a subjunctive (counterfactual) but indicative (factual) conditional.

[2] Counterfactual conditionals do not have to satisfy (7).

(10) $(\varphi \rightarrow \psi) \approx *$,

where $*$ is $\varphi \wedge \psi, \varphi \wedge \neg\psi, \neg\varphi \wedge \psi$ or $\neg\varphi \wedge \neg\psi$. Neither of them works. When $\psi|\varphi$ is interpreted to mean that ψ given φ and its semantics is given, we might introduce (11) as an axiom:

(11) $(\varphi \rightarrow \psi) \approx \psi|\varphi$.

(11) is considered to be a qualitative variant of *Stalnaker's Hypothesis* (Stalnaker [31]). So (11) can invite such bad results as qualitative variants of *Lewis* [20, 21]' *triviality results*. What should we do? We can get over the conception of logic of qualitative absolute probability and appeal to the conception of logic of qualitative conditional probability. When $(\varphi, \psi) \gg (\chi, \tau)$ is a quaternary sentential operator that is interpreted to mean that φ given ψ is more probable than χ given τ and we assume a suitable semantics of $\gg$, we can have (12):

(12) $(\psi, \varphi) \gg (\neg\psi, \varphi)$.

We can have (12) as a logical form of (4). If we assume the semantics of $\gg$ in our LQCP below, then (12) implies (13):

(13) $(\varphi \wedge \psi, \top) \gg (\varphi \wedge \neg\psi, \top)$,

which is a quaternary-relational variant of (7). Moreover (2) can be expressed as (14):

(14) $(\varphi, \top) \gg (\neg\varphi, \top)$.

Then, how can we provide $\gg$ with its semantics? Koopman [14–16] introduces such a quaternary relation $\succsim$ that $(A, B) \succsim (C, D)$ is interpreted to mean that A given B is at least as probable as C given D. He tries to give sufficient conditions for the existence of such conditional probability measure that

$$(15)\ (A, B) \succsim (C, D) \text{ iff } P(A, B) \geq P(C, D).$$

However, according to Fine [6, p. 185], Koopman's conditions can only assure that

$$\text{If } (A, B) \succsim (C, D) \text{ then } P(A, B) \geq P(C, D).$$

Koopman's conditions do not assure us of the usual property of conditional probability:

$$P(A, B \cap C) = \frac{P(A \cap C, B)}{P(C, B)}$$

Luce [22] gives sufficient conditions for (15) from a measurement-theoretic point of view. Krantz et al. [17] modify Luce's conditions. Domotor [4] gives necessary and sufficient conditions for (15) when the sample space is *finite*. Suppes and Zanotti [32] give necessary and sufficient conditions when the sample space has no limitation of its size. Suppes and Zanotti's conditions include **Archimedeanity**. No variant of **Archimedeanity** is expressible not only in the language of propositional logic but also even in that of first-order logic. So we adopt Domotor's conditions as those for the model of our propositional logic LQCP. Recently Mundici [24] and Ibeling et al. [13] respectively have investigated Koopman's qualitative conditional probability from a logical point of view. However, the complete axiomatization of logic of qualitative conditional probability has been an open problem. The aim of this paper is to propose a new version of complete logic–Logic of Qualitative Conditional Probability (LQCP) by proving a theorem that can bridge the gap between the semantics and the axiomatization of LQCP.

The structure of this paper is as follows. In Sect. 2, we give a brief explanation of a representation theorem in measurement theory. In Sect. 3, we state a hypothesis about representation of a qualitative conditional probability relation given by Domotor [4]. In Sect. 4, we propose LQCP. In Subsect. 4.1, we define the language $\mathscr{L}_{\mathsf{LQCP}}$ of LQCP. In Subsect. 4.2, we define a model $\mathfrak{M}$ of LQCP, and provide LQCP with a truth definition and a validity definition. In Subsect. 4.3, we prove a theorem that can bridge the gap between the semantics and the axiomatization of LQCP. In Subsect. 4.4, we provide LQCP with its proof system. In Subsect. 4.5, we prove the soundness and completeness of LQCP. In Sect. 5, we finish with brief concluding remarks.

2   Representation Theorem in Measurement Theory

Measurement plays an essential role not only in science but also in our daily lives. Measurement theory[3] investigates not concrete methods for measurement but conditions for *measurability*. The mathematical foundation of measurement had not been studied before Hölder [10] developed his axiomatization for the measurement of mass. Krantz et al. [17], Suppes et al. [33] and Luce et al. [23] are seen as milestones in the history of measurement theory. There are two main problems in measurement theory:

1. the *representation problem*: Given a *numerical* relational structure $\mathfrak{V}$, find conditions on an *observed* relational structure $\mathfrak{U}$ (necessary and) sufficient for the *existence* of a homomorphism f from $\mathfrak{U}$ to $\mathfrak{V}$ that preserves all the relations and operations in $\mathfrak{U}$.
2. the *uniqueness problem*: Find the type of transformation of the homomorphism f under which all the relations and operations in $\mathfrak{U}$ are preserved.

The theorem stating conditions on $\mathfrak{U}$ are (necessary and) sufficient for the existence of f is called a *representation theorem* that can furnish a solution to the representation problem. Holman [12] classifies representations of, for example, $\mathfrak{U} := (A, \succsim)$ into the following two types:

1. **Strong Representation** is defined as the construction of such a nonconstant function f on A that, for any $a, b \in A$,

$$a \succsim b \quad \text{iff} \quad f(a) \geq f(b).$$

---

[3] Roberts [28] gives a comprehensive survey of measurement theory.

2. **Weak Representation**[4] is defined as the construction of such a nonconstant function f on A that, for any $a, b \in A$,

$$\text{if} \quad a \gtrsim b, \quad \text{then} \quad f(a) \geq f(b).$$

On the other hand, the theorem stating the type of transformation up to which f above is unique is called a *uniqueness theorem*, which can furnish a solution to the uniqueness problem.

3   Hypothesis About Strong Representation of Qualitative Conditional Probability Relation

Domotor [4] tries to prove the following proposition (Theorem 10 in [4, p. 85]). The proof of Theorem 10 depends on that of Theorem 7 in [4, p. 66]. However, we agree to Ibeling et al. [13, pp. 30–32]'s argument that the proof of Theorem 7 "very briefly appeals to an *unstated* result in geometry of webs [2]" ([13, p. 31]). As far as we know, the (algebraic) verification of the proof has been carried out only in the case of $|\Omega| = 2$ (Ibeling et al. [13, pp. 42–43]). In order to regard the following proposition as a theorem (fact), the verification of the proof in the case of $|\Omega| = n \geq 1$ must be carried out. So in this paper, for the sake of argument, we would like to regard the following proposition as a *hypothesis*. Indeed because Corollary 1 (Truth Condition by Conditional Probability Measure) depends directly on this hypothesis, its proof is conditional. But since all the other technical results, including Theorem 3 (Completeness), in this paper than Corollary 1 do not directly depend on this hypothesis, their proofs are not conditional.[5][6]

Hypothesis 1 (Strong Representation). *Suppose that Ω is a nonempty finite set of elementary events, and that $\mathscr{F}$ is the Boolean algebra of subsets of Ω, and that $\gtrsim_\omega$ is a* quaternary *relation relative to $\omega \in \Omega$ on $\mathscr{F}$ (binary relation relative to ω on $\mathscr{F} \times \mathscr{F}$), and that $(A, B) >_\omega (C, D)$ is defined as $(C, D) \not\gtrsim_\omega (A, B)$, and that $(A, B) \sim_\omega (C, D)$ is defined as $(A, B) \gtrsim_\omega (C, D)$ and $(C, D) \gtrsim_\omega (A, B)$, and that $\mathscr{F}_0 := \{A : A \in \mathscr{F} \text{ and } (A, \Omega) \not\sim_\omega (\varnothing, \Omega)\}$. Then there exists a finitely additive conditional probability measure $P_\omega : \mathscr{F} \times \mathscr{F}_0 \to \mathbb{R}$ relative to $\omega \in \Omega$ satisfying, for any $A, C \in \mathscr{F}$ and any $B, D \in \mathscr{F}_0$,*[7]

$$(A, B) \gtrsim_\omega (C, D) \textit{iff } P_\omega(A, B) \geq P_\omega(C, D),$$

iff *the following conditions are met:*

---

[4] **Weak Representation** itself is not demeritorious. About the merits of weak representation, refer to Mundy [25].

[5] In Remark 4 below, we explain the relationship between Theorem 3 (Completeness) and Hypothesis 1.

[6] Of course, conditional proofs are never of no cognitive value; for example, Granville and Tucker [9]'s very short proof of Fermat's Last Theorem (which is different from Wiles' proof) in the case of $n \geq 6$ based on the *abc* conjecture.

[7] One should notice that B constitutes the condition part of $P_\omega(A, B)$.

- **Nontriviality**: $(\Omega, \Omega) >_\omega (\varnothing, \Omega)$.
- **Nonnegativity**: $(A, B) \succsim_\omega (\varnothing, C)$, for any $A \in \mathscr{F}$ and any $B, C \in \mathscr{F}_0$.
- **Absorption**: $(A \cap B, B) \succsim_\omega (A, B)$ for any $A \in \mathscr{F}$ and any $B \in \mathscr{F}_0$.
- **Connectedness**: $(A, B) \succsim_\omega (C, D)$ or $(C, D) \succsim_\omega (A, B)$, for any $A, C \in \mathscr{F}$ and any $B, D \in \mathscr{F}_0$.
- **Finite Cancellation (Addition)**: For any $A_1, \ldots, A_n, C_1, \ldots, C_n \in \mathscr{F}$ and any $B_1, \ldots, B_n, D_1, \ldots, D_n \in \mathscr{F}_0$, if, for any $i < n$, $(A_i, B_i) \succsim_\omega (C_i, D_i)$, then $(C_n, D_n) \succsim_\omega (A_n, B_n)$, given that

$$\sum_{i \leq n} f_{(A_i, B_i)} = \sum_{i \leq n} f_{(C_i, D_i)},$$

where $f_{(A,B)}$ is the characteristic function of (A, B) in the following sense:

$$f_{(A,B)}(\omega) := \begin{cases} 1 & if\ \omega \in A \cap B, \\ 0 & if\ \omega \in \overline{A} \cap B, \\ undefined\ if\ \omega \in \overline{B}. \end{cases}$$

- **Finite Cancellation (Multiplication)**: For any $A_0, \ldots, A_n, B_0, \ldots, B_n \in \mathscr{F}_0$, if, for any k with $0 < k \leq n$, $(A_k, \bigcap_{0 \leq i < k} A_i) \succsim_\omega (B_{\beta(k)}, \bigcap_{0 \leq i < \beta(k)} B_i)$, then $(\bigcap_{0 < i \leq n} A_i, A_0) \succsim_\omega (\bigcap_{0 < i \leq n} B_{\beta(i)}, B_0)$, where β is a permutation on $\{1, \ldots, n\}$.

The relation $\succsim_\omega$ satisfying the above conditions is called a qualitative conditional probability relation.

Remark 1 (Finite Cancellations (Addition) and (Multiplication)) *Finite Cancellation (Addition)* and *Finite Cancellation (Multiplication)* are the qualitative versions of the addition and multiplication laws of conditional probability, respectively.

Remark 2 (De Finetti's Tri-Event) *The notion of characteristic function $f_{(A,B)}$ of (A, B) originates in that of* tri-event. *De Finetti [7, p. 13] defines the tri-event as follows:*

> *One can consider, then, the "conditional events" (or "tri-events"), which are the events of a three valued logic: this "tri-event", "E' conditioned E''", $E'|E''$, is the logical entity capable of having three values:* true *if E'' and E' are true;* false *if E'' is true and E' false;* void *if E'' is false.*

4   Logic of Qualitative Conditional Probability (**LQCP**)

4.1   Language

We define the language $\mathscr{L}_{\mathsf{LQCP}}$ of LQCP as follows:

Definition 1 (Language).

- *Let $\mathscr{S}$ denote a set of sentential variables and a qualitative conditional probability relation symbol $\succeq$ a quaternary sentential operator.*

- *The language $\mathscr{L}_{\mathsf{LQCP}}$ of* LQCP *is given by the following BNF grammar:*

$$\varphi ::= s \mid \top \mid \neg\varphi \mid \varphi \wedge \psi \mid (\varphi, \psi) \geqslant (\chi, \tau)$$

such that $s \in \mathscr{S}$.
- *$\bot, \vee, \rightarrow$ and $\leftrightarrow$ are introduced by the standard definitions.*
- *$(\varphi, \psi) \gg (\chi, \tau) := \neg((\chi, \tau) \geqslant (\varphi, \psi))$.*
- *$(\varphi, \psi) \approx (\chi, \tau) := ((\varphi, \psi) \geqslant (\chi, \tau)) \wedge ((\chi, \tau) \geqslant (\varphi, \psi))$.*
- *The set of all well-formed formulae of $\mathscr{L}_{\mathsf{LQCP}}$ is denoted by $\Phi_{\mathscr{L}_{\mathsf{LQCP}}}$.*

4.2   Semantics

We define a structured model $\mathfrak{M}$ of LQCP as follows:

Definition 2 (Model). *$\mathfrak{M}$ is a triple (Ω, ρ, V) in which*

- *Ω is a nonempty finite set of elementary events.*
- *ρ is a qualitative conditional probability space assignment that assigns to each $\omega \in \Omega$ a qualitative conditional probability space $(\Omega, \mathscr{F}, \succsim_\omega)$ in which $\succsim_\omega$ is a qualitative conditional probability relation relative to $\omega \in \Omega$ on $\mathscr{F}$ that satisfies all of **Nontriviality, Nonnegativity, Absorption, Connectedness, Finite Cancellation (Addition),** and **Finite Cancellation (Multiplication)** of Hypothesis 1, and in which $\mathscr{F}$ contains higher-order qualitative conditional probability events (e.g., $\{\omega : (\{\omega_1 : (A_1, B_1) \succsim_{\omega_1} (C_1, D_1)\}, \{\omega_2 : (A_2, B_2) \succsim_{\omega_2} (C_2, D_2)\}) \succsim_\omega (\{\omega_3 : (A_3, B_3) \succsim_{\omega_3} (C_3, D_3)\}, \{\omega_4 : (A_4, B_4) \succsim_{\omega_4} (C_4, D_4)\})\}$, where $A_1, \ldots, A_4, C_1, \ldots, C_4 \in \mathscr{F}$ and $B_1, \ldots, B_4, D_1, \ldots, D_4 \in \mathscr{F}_0$).*
- *$\mathfrak{F} := (\Omega, \rho)$ is called a frame of* LQCP.
- *V is a truth assignment to each $s \in \mathscr{S}$ for each $\omega \in \Omega$.*

We provide LQCP with the following truth definition at $\omega \in \Omega$ in $\mathfrak{M}$, define the truth in $\mathfrak{M}$, and then define validity as follows:

Definition 3 (Truth and Validity).

- *The notion of $\varphi \in \Phi_{\mathscr{L}_{\mathsf{LQCP}}}$ being true at $\omega \in \Omega$ in $\mathfrak{M}$, in symbols $(\mathfrak{M}, \omega) \vDash_{\mathsf{LQCP}} \varphi$, is inductively defined as follows:*
  - *$(\mathfrak{M}, \omega) \vDash_{\mathsf{LQCP}} s \quad iff \quad V(\omega)(s) = true$.*
  - *$(\mathfrak{M}, \omega) \vDash_{\mathsf{LQCP}} \top$.*
  - *$(\mathfrak{M}, \omega) \vDash_{\mathsf{LQCP}} \neg\varphi \quad iff \quad (\mathfrak{M}, \omega) \nvDash_{\mathsf{LQCP}} \varphi$.*
  - *$(\mathfrak{M}, \omega) \vDash_{\mathsf{LQCP}} \varphi \wedge \psi \quad iff \quad (\mathfrak{M}, \omega) \vDash_{\mathsf{LQCP}} \varphi$ and $(\mathfrak{M}, \omega) \vDash_{\mathsf{LQCP}} \psi$.*
  - *$(\mathfrak{M}, \omega) \vDash_{\mathsf{LQCP}} (\varphi, \psi) \geqslant (\chi, \tau) \quad iff \quad (\llbracket\varphi\rrbracket^{\mathfrak{M}}, \llbracket\psi\rrbracket^{\mathfrak{M}}) \succsim_\omega (\llbracket\chi\rrbracket^{\mathfrak{M}}, \llbracket\tau\rrbracket^{\mathfrak{M}})$, where the valuation $\llbracket\varphi\rrbracket^{\mathfrak{M}} := \{\omega' \in \Omega : (\mathfrak{M}, \omega') \vDash_{\mathsf{LQCP}} \varphi\}$ and $\llbracket\varphi\rrbracket^{\mathfrak{M}}, \llbracket\chi\rrbracket^{\mathfrak{M}} \in \mathscr{F}$ and $\llbracket\psi\rrbracket^{\mathfrak{M}}, \llbracket\tau\rrbracket^{\mathfrak{M}} \in \mathscr{F}_0$.*
- *If $(\mathfrak{M}, \omega) \vDash_{\mathsf{LQCP}} \varphi$ for any $\omega \in \Omega$, we write $\mathfrak{M} \vDash_{\mathsf{LQCP}} \varphi$ and say that φ is true in $\mathfrak{M}$.*

- *If φ is true in any model $\mathfrak{M}$ on $\mathfrak{F}$, we write $\mathfrak{F} \models_{\mathsf{LQCP}} \varphi$ and say that φ is valid in $\mathfrak{F}$.*

Then the next corollary follows from Definition 3 an Hypothesis 1.

Corollary 1 (Truth Condition by Conditional Probability Measure).
There exists $P_\omega : \mathscr{F} \times \mathscr{F}_0 \to \mathbb{R}$ satisfying $(\mathfrak{M}, \omega) \models_{\mathsf{LQCP}} (\varphi, \psi) \geqslant (\chi, \tau)$ iff $([\![\varphi]\!]^{\mathfrak{M}}, [\![\psi]\!]^{\mathfrak{M}}) \succsim_\omega ([\![\chi]\!]^{\mathfrak{M}}, [\![\tau]\!]^{\mathfrak{M}})$ iff $P_\omega([\![\varphi]\!]^{\mathfrak{M}}, [\![\psi]\!]^{\mathfrak{M}}) \geq P_\omega([\![\chi]\!]^{\mathfrak{M}}, [\![\tau]\!]^{\mathfrak{M}})$.

4.3   Bridge Between Semantics and Axiomatization

With the help of Domotor [5, p. 204], we prove the following theorem that can bridge the gap the semantics and the axiomatization of LQCP:

Theorem 1 (Equivalence Between Characteristic Functions and Boolean Operations).

$$\sum_{1 \leq i \leq n} f_{(A_i, B_i)} = \sum_{1 \leq i \leq n} f_{(C_i, D_i)}$$

$$\textit{iff} \quad \bigcup_{1 \leq i_1 < \cdots < i_k \leq n} ((A_{i_1} \cap B_{i_1}) \cap \cdots \cap (A_{i_k} \cap B_{i_k})) = \bigcup_{1 \leq i_1 < \cdots < i_k \leq n} ((C_{i_1} \cap D_{i_1}) \cap \cdots \cap (C_{i_k} \cap D_{i_k}))$$

holds for any k with $1 \leq k \leq n$.

Proof. Judging from the form of the definition of $f_{(A_i, B_i)}$ of Hypothesis 1, whether $\sum_{1 \leq i \leq n} f_{(A_i, B_i)} = \sum_{1 \leq i \leq n} f_{(C_i, D_i)}$ holds or not is determined by both the number of such ω's as $\omega \in A_i \cap B_i$ and that of such ω's as $\omega \in C_i \cap D_i$. So neither the number of such ω's as $\omega \in \overline{A_i} \cap B_i$ as well as $\omega \in \overline{C_i} \cap D_i$ nor that of such ω's as $\omega \in \overline{B_i}$ as well as $\omega \in \overline{D_i}$ is relevant to whether it holds or not. So in order to prove this theorem, we have only to show by mathematical induction on n that for any $\omega \in \Omega$,

$$\sum_{1 \leq i \leq n} f_{(A_i, B_i)}(\omega) = k \text{ iff } \omega \in \bigcup_{1 \leq i_1 < \cdots < i_k \leq n} ((A_{i_1} \cap B_{i_1}) \cap \cdots \cap (A_{i_k} \cap B_{i_k})). \quad \blacksquare$$

4.4   Axiomatization

On the basis of Theorem 1, we axiomatize both **Finite Cancellation (Addition)** and **(Multiplication)**. The proof system of LQCP consists of the following:

Definition 4 (Proof System).

- *All tautologies of classical propositional logic.*
- *Axiomatic Counterpart of **Nontriviality**: $(\top, \top) \gg (\bot, \top)$.*

- *Axiomatic Counterpart of **Nonnegativity***:
  $$(((\psi, \top) \nleftrightarrow (\bot, \top)) \wedge ((\chi, \top) \nleftrightarrow (\bot, \top))) \to ((\varphi, \psi) \geqslant (\varnothing, \chi)).$$
- *Axiomatic Counterpart of **Absorption***:
  $$((\psi, \top) \nleftrightarrow (\bot, \top)) \to ((\varphi \wedge \psi, \psi) \geqslant (\varphi, \psi)).$$
- *Axiomatic Counterpart of **Connectdness***:
  $$(((\psi, \top) \nleftrightarrow (\bot, \top)) \wedge ((\tau, \top) \nleftrightarrow (\bot, \top))) \to (((\varphi, \psi) \geqslant (\chi, \tau)) \vee ((\chi, \tau) \geqslant (\varphi, \psi))).$$
- *Axiomatic Counterpart of **Finite Cancellation (Addition)***:

$$\wedge_{1 \leq i \leq n}\Big(((\psi_i, \top) \nleftrightarrow (\bot, \top)) \wedge ((\tau_i, \top) \nleftrightarrow (\bot, \top)) \Big)$$
$$\to \Big(\wedge_{1 \leq k \leq n}\Big(\vee_{1 \leq i_1 < \cdots < i_k \leq n}((\varphi_{i_1} \wedge \psi_{i_1}) \wedge \cdots \wedge (\varphi_{i_k} \wedge \psi_{i_k})) \leftrightarrow \vee_{1 \leq i_1 < \cdots < i_k \leq n}((\chi_{i_1} \wedge \tau_{i_1}) \wedge \cdots \wedge (\chi_{i_k} \wedge \tau_{i_k})) \Big)$$
$$\to \Big(\wedge_{i < n}((\varphi_i, \psi_i) \geqslant (\chi_i, \tau_i)) \to ((\varphi_n, \psi_n) \geqslant (\chi_n, \tau_n)) \Big) \Big).$$

- *Axiomatic Counterpart of **Finite Cancellation (Multiplication)***:

$$\wedge_{0 \leq i \leq n}\Big(((\varphi_i, \top) \nleftrightarrow (\bot, \top)) \wedge ((\psi_i, \top) \nleftrightarrow (\bot, \top)) \Big)$$
$$\to \Big(\wedge_{0 < k \leq n}((\varphi_k, \wedge_{0 \leq i < k} \varphi_i) \geqslant (\psi_{\beta(k)}, \wedge_{0 \leq i < \beta(k)} \psi_i)) \to ((\wedge_{0 < i \leq n} \varphi_i, \varphi_0) \geqslant (\wedge_{0 < i \leq n} \psi_{\beta(i)}, \psi_0)) \Big),$$

  where β is a permutation on $\{1, \ldots, n\}$.
- *Modus Ponens.*
- *A proof of $\varphi \in \Phi_{\mathscr{L}_{\mathsf{LQCP}}}$ is a finite sequence of $\mathscr{L}_{\mathsf{LQCP}}$-formulae having φ as the last formula such that either each formula is an instance of an axiom or it can be obtained from formulae that appear earlier in the sequence by applying an inference rule.*
- *If there is a proof of φ, we write $\vdash_{\mathsf{LQCP}} \varphi$.*

Remark 3 (Segerberg's Formulation of *Finite Cancellation (Addition)*). *In the proof system of* PK, *Segerberg [30, p. 342] formulates the axiomatic counterpart of the qualitative absolute probabilistic version of **Finite Cancellation (Addition)**. On the basis of his formulation, we can formulate the axiomatic counterpart of the qualitative conditional probabilistic version of this condition. However, the latter formulation is more complicated than our formulation in Definition 4.*

4.5   Metalogic

Soundness. We prove the soundness of LQCP:

Theorem 2 (Soundness of LQCP). *For any $\psi \in \Phi_{\mathscr{L}_{\mathsf{LQCP}}}$, if $\vdash_{\mathsf{LQCP}} \psi$, then $\mathfrak{F} \vDash_{\mathsf{LQCP}} \psi$.*

Proof. It is trivial to verify the conditions except the axiomatic counterparts of **Finite Cancellation (Addition)** and **Finite Cancellation (Multiplication)**. The verification of the former is much the same as Segerberg [30, pp. 344–346]'s. The verification of the latter can be conducted in much the same way as that of the former.　∎

Completeness of LQCP We would like to prove the completeness of LQCP. This proof has the method of *filtration* in common with Segerberg [30]'s, Gärdenfors [8]'s, and Delgrande et al. [3]'s proofs of completeness. In this paper, we prove the completeness of LQCP by using mainly the ideas of Gärdenfors [pp. 177-179] [8]. The model of LQCP has a *qualitative conditional probability space assignment*, whereas Segerberg [30], Gärdenfors [8], and Delgrande et al. [3] each have their own *absolute probability measure space assignment*. Let $\varphi \in \Phi_{\mathscr{L}_{\mathsf{LQCP}}}$ any such formula that $\nvdash_{\mathsf{LQCP}} \varphi$. We would like to construct, from infinite Boolean algebra $\mathscr{B}$ below, a finite model of LQCP $\mathfrak{M}^{\varphi} := (\Omega^{\varphi}, \rho^{\varphi}, V^{\varphi})$ where $\rho^{\varphi} := (\Omega^{\varphi}, \mathscr{B}^{\varphi}, \succeq_{\omega})$ in which $\mathscr{B}^{\varphi}$ is a *finite* Boolean algebra and so an instance of $\mathscr{F}$ of Definition 2 such that $\mathfrak{M}^{\varphi} \nvDash_{\mathsf{LQCP}} \varphi$. First, we define equivalence class $[\varphi]$:

Definition 5 (Equivalence Class). *For any* $\varphi \in \Phi_{\mathscr{L}_{\mathsf{LQCP}}}$, *the* equivalence class $[\varphi] := \{\psi \in \Phi_{\mathscr{L}_{\mathsf{LQCP}}} :\vdash_{\mathsf{LQCP}} \varphi \leftrightarrow \psi\}$.

We define Boolean operators $\overline{}$ and $\cap$ on the set of equivalence classes:

Definition 6 ($\overline{}$ and $\cap$). $\overline{}$ *and* $\cap$ *on the set of equivalence classes are defined in such a way that* $\overline{[\varphi]} := [\neg\varphi]$ *and* $[\varphi] \cap [\psi] := [\varphi \wedge \psi]$ *for any* $\varphi, \psi \in \Phi_{\mathscr{L}_{\mathsf{LQCP}}}$.

Under Definitions 5 and 6, we obtain from $\mathscr{L}_{\mathsf{LQCP}}$ a infinite Boolean algebra $\mathscr{B}$ with $[\top]$ and $[\bot]$ unit and zero elements respectively. The following lemma follows directly from Definitions 5 and 6:

Lemma 1 (Provability and Equivalence Class). $\vdash_{\mathsf{LQCP}} \varphi$ *iff* $[\varphi] = [\top]$.

We would like to obtain from the infinite Boolean algebra $\mathscr{B}$ finite Boolean algebras $\mathscr{B}^{\varphi}$ and $\widetilde{\mathscr{B}^{\varphi}}$. We define *rank* as follows:

Definition 7 (*rank*). *The rank of* $\psi \in \Phi_{\mathscr{L}_{\mathsf{LQCP}}}$ *denoted by* $rank(\psi)$ *is the maximal number of nestings of the* $\succeq$ *in the formulae. If* ψ *does not contain* $\succeq$, *we stipulate that* $rank(\psi) = 0$.

We define $\mathscr{G}^{\varphi}$ and $\widetilde{\mathscr{G}^{\varphi}}$ as follows:

- $\mathscr{G}^{\varphi}$: the set of formulae ψ built up from some subset of sentential symbols in φ where $rank(\varphi) \geq rank(\psi)$,
- $\widetilde{\mathscr{G}^{\varphi}}$: the set of formulae ψ built up from the same sentential symbols where $rank(\varphi) + 1 \geq rank(\psi)$.

We define $\mathscr{B}^{\varphi}$ and $\widetilde{\mathscr{B}^{\varphi}}$ as follows:

- $\mathscr{B}^{\varphi}$: the subalgebra of $\mathscr{B}$ that has as its elements only the equivalence classes of the formulae in $\mathscr{G}^{\varphi}$,
- $\widetilde{\mathscr{B}^{\varphi}}$: the subalgebra of $\mathscr{B}$ that has as its elements only the equivalence classes of the formulae in $\widetilde{\mathscr{G}^{\varphi}}$.

$\mathscr{B}^{\varphi}$ is a subalgebra of $\widetilde{\mathscr{B}^{\varphi}}$. Since φ has a finite *rank* and contains only finitely many sentential symbols, $\widetilde{\mathscr{B}^{\varphi}}$ contains only finitely many elements. We define Ω^{φ} as the set of a *representative element* of an element of $\widetilde{\mathscr{B}^{\varphi}}$.[8] Then we define $([\psi],[\chi]) \succsim_{\omega} ([\tau],[\mu])$ on $\mathscr{B}^{\varphi}$ in terms of $[(\psi,\chi) \geqslant (\tau,\mu)] \in \widetilde{\mathscr{B}^{\varphi}}$:

Definition 8 ($\succsim_{\omega}$). *For any* $\omega \in \Omega^{\varphi}$, *for any* $\psi,\chi,\tau,\mu \in \Phi_{\mathscr{L}_{\mathsf{LQCP}}}$, $([\psi],[\chi]) \succsim_{\omega}$ $([\tau],[\mu])$ *iff* $\omega \in [(\psi,\chi) \geqslant (\tau,\mu)]$.

Then we prove the next lemma:

Lemma 2 (Satisfaction of Conditions for Strong Representation). *For any* $\omega \in \Omega^{\varphi}$, $\succsim_{\omega}$ *on* $\mathscr{B}^{\varphi}$ *defined by Definition 8 satisfies* **Nontriviality, Nonnegativity, Absorption, Connectedness, Finite Cancellation (Addition),** *and* **Finite Cancellation (Multiplication)** *of Hypothesis 1.*

Proof. It is trivial to verify that **Nontriviality, Nonnegativity, Absorption,** and **Connectedness** are satisfied given their axiomatic counterparts. For the verification of **Finite Cancellation (Addition)**, we make the following three assumptions:

1. Assume that $[\varphi_1],\ldots,[\varphi_n],[\chi_1],\ldots,[\chi_n]$ $\in$ $\mathscr{B}^{\varphi}$ and that $[\psi_1],\ldots,[\psi_n],[\tau_1],\ldots,[\tau_n] \in \mathscr{B}_0^{\varphi}$, where $\mathscr{B}_0^{\varphi} := \{A : A \in \mathscr{B}^{\varphi}$ and $(A,[\top]) \nsucc_{\omega} ([\bot],[\top])\}$.
2. Assume on the basis of Theorem 1 that the equation part of the axiomatic counterpart of **Finite Cancellation (Addition)** holds for these elements:

$$\bigcup_{1 \leq i_1 < \cdots < i_k \leq n} (([\varphi_{i_1}] \cap [\psi_{i_1}]) \cap \cdots \cap ([\varphi_{i_k}] \cap [\psi_{i_k}])) = \bigcup_{1 \leq i_1 < \cdots < i_k \leq n} (([\chi_{i_1}] \cap [\tau_{i_1}]) \cap \cdots \cap ([\chi_{i_k}] \cap [\tau_{i_k}]))$$

   for any k with $1 \leq k \leq n$.
3. Assume that $([\varphi_i],[\psi_i]) \succsim_{\omega} ([\chi_i],[\tau_i])$ for any $i < n$.

Then by Assumptions 1 and 2 and Definition 6 and the property of equivalence classes,

$$\left[\bigwedge_{1 \leq i \leq n} \left(((\psi_i, \top) \nsucc (\bot, \top)) \wedge ((\tau_i, \top) \nsucc (\bot, \top)) \right) \right] = [\top]$$

and

$$\left[\bigvee_{1 \leq i_1 < \cdots < i_k \leq n} ((\varphi_{i_1} \wedge \psi_{i_1}) \wedge \cdots \wedge (\varphi_{i_k} \wedge \psi_{i_k})) \leftrightarrow \bigvee_{1 \leq i_1 < \cdots < i_k \leq n} ((\chi_{i_1} \wedge \tau_{i_1}) \wedge \cdots \wedge (\chi_{i_k} \wedge \tau_{i_k})) \right] = [\top],$$

for any k with $1 \leq k \leq n$. So from the axiomatic counterpart of **Finite Cancellation (Addition)** and Lemma 1 and Modus Pones, it follows that

$$\left[\bigwedge_{i < n} ((\varphi_i, \psi_i) \geqslant (\chi_i, \tau_i)) \rightarrow ((\varphi_n, \psi_n) \geqslant (\chi_n, \tau_n)) \right] = [\top].$$

---

[8] Generally speaking, in probability theory, the status of an elementary event ω is not specified. This non-specificity guarantees that we can equate an elementary event ω with a representative element of an element of $\widetilde{\mathscr{B}^{\varphi}}$ that is a syntactic object.

So from Assumption 3 and Definition 8 and the property of equivalence classes and Lemma 1 and Modus Ponens, we have $([\varphi_n], [\psi_n]) \succsim_\omega ([\chi_n], [\tau_n])$. The verification of **Finite Cancellation (Multiplication)** can be conducted in much the same way as that of **Finite Cancellation (Addition)**. ∎

Then we prove Truth Lemma from Lemma 2 as follows:

Lemma 3 (Truth Lemma). *For any $\psi \in \mathscr{G}^\varphi$, $[\psi] = [[\psi]]^{\mathfrak{M}^\varphi}$.*

Proof. Let V^φ be any truth assignment such that, for any sentential variable $s \in \mathscr{G}^\varphi$, $[s] = \{\omega \in \Omega^\varphi : V^\varphi(\omega)(s) = true\} = [[s]]^{\mathfrak{M}^\varphi}$. This equivalence class belongs to $\mathscr{B}^\varphi$. We show by induction on the length of ψ that $[\psi] = [[\psi]]^{\mathfrak{M}^\varphi}$ as follows:

1. When ψ is a sentential variable, it holds by definition.
2. When ψ is a $\top$, trivially $[\top] = [[\top]]^{\mathfrak{M}^\varphi}$ ($[\bot] = [[\bot]]^{\mathfrak{M}^\varphi}$).
3. When ψ is $\neg\chi$, $[\neg\chi] = \overline{[\chi]}$ by Definition 6. $\overline{[\chi]} = \overline{[[\chi]]^{\mathfrak{M}^\varphi}}$ by the induction hypotheses. $\overline{[[\chi]]^{\mathfrak{M}^\varphi}} = [[\neg\chi]]^{\mathfrak{M}^\varphi}$ by Definition 3.
4. When ψ is $\chi\wedge\tau$, $[\chi\wedge\tau] = [\chi]\cap[\tau]$ by Definition 6. $[\chi]\cap[\tau] = [[\chi]]^{\mathfrak{M}^\varphi}\cap[[\tau]]^{\mathfrak{M}^\varphi}$ by the induction hypotheses. $[[\chi]]^{\mathfrak{M}^\varphi}\cap[[\tau]]^{\mathfrak{M}^\varphi} = [[\chi\wedge\tau]]^{\mathfrak{M}^\varphi}$ by Definition 3.
5. When ψ is $(\chi,\tau) \succcurlyeq (\mu,\nu)$, $[(\chi,\tau) \succcurlyeq (\mu,\nu)] = \{\omega \in \Omega^\varphi : \omega \in [(\chi,\tau) \succcurlyeq (\mu,\nu)]\}$ by the property of equivalence classes. $\{\omega \in \Omega^\varphi : \omega \in [(\chi,\tau) \succcurlyeq (\mu,\nu)]\} = \{\omega \in \Omega^\varphi : ([\chi],[\tau]) \succsim_\omega ([\mu],[\nu])\}$ by Definition 8. $\{\omega \in \Omega^\varphi : ([\chi],[\tau]) \succsim_\omega ([\mu],[\nu])\} = \{\omega \in \Omega^\varphi : ([[\chi]]^{\mathfrak{M}^\varphi},[[\tau]]^{\mathfrak{M}^\varphi}) \succsim_\omega ([[\mu]]^{\mathfrak{M}^\varphi},[[\nu]]^{\mathfrak{M}^\varphi})\}$ by the induction hypotheses. By Lemma 2, we can apply the truth clause of $\succcurlyeq$ of Definition 3 to $\succsim_\omega$ on $\mathscr{B}^\varphi$. So $\{\omega \in \Omega^\varphi : ([[\chi]]^{\mathfrak{M}^\varphi},[[\tau]]^{\mathfrak{M}^\varphi}) \succsim_\omega ([[\mu]]^{\mathfrak{M}^\varphi},[[\nu]]^{\mathfrak{M}^\varphi})\} = [[(\chi,\tau) \succcurlyeq (\mu,\nu)]]^{\mathfrak{M}^\varphi}$. ∎

The completeness of LQCP follows from Lemmata 1 and 3:

Theorem 3 (Completeness). *For any $\psi \in \Phi_{\mathscr{L}_{\mathsf{LQCP}}}$, if $\mathfrak{F} \models_{\mathsf{LQCP}} \psi$, then $\vdash_{\mathsf{LQCP}} \psi$.*

Proof. Suppose that $\nvdash_{\mathsf{LQCP}} \psi$, where $\psi \in \mathscr{G}^\varphi$. Then, if $\nvdash_{\mathsf{LQCP}} \psi$, then $[\psi] \neq [\top]$ by Lemma 1. So $[[\psi]]^{\mathfrak{M}^\varphi} \neq [[\top]]^{\mathfrak{M}^\varphi}$ by Lemma 3. So $\mathfrak{M}^\varphi \nvDash_{\mathsf{LQCP}} \psi$. Therefore, there is a model $\mathfrak{M}$ such that $\mathfrak{M} \nvDash_{\mathsf{LQCP}} \psi$. ∎

Remark 4 (Role of Hypothesis 1 in Non-Conditional Proof of Completeness) *The model $\mathfrak{M}$ of LQCP has a qualitative conditional probability space assignment, whereas Segerberg [30], Gärdenfors [8], and Delgrande et al. [3] each have their own (absolute) probability measure space assignment. Since in Lemma 3 (Truth Lemma), no conditional probability measure the existence of which is asserted by Hypothesis 1 is used, the proof of the completeness of LQCP is no conditional proof. Since all the other Technical results in this paper than Corollary 1 do not directly depend on Hypothesis 1, their proofs are not conditional. In LQCP, Hypothesis 1 plays such a role that a qualitative conditional probability relation can hypothetically represent a conditional probability measure. On the other hand, like Segerberg [30], Gärdenfors [8], and Delgrande et al. [3], suppose counterfactually that LQCP had a conditional probability measure space assignment. Then we would replace the clause of "ψ is $(\chi,\tau) \succcurlyeq (\mu,\nu)$" by the following clause:*

1. When ψ is $(\chi,\tau) \geqslant (\mu,\nu)$, $[(\chi,\tau) \geqslant (\mu,\nu)] = \{\omega \in \Omega^\varphi : \omega \in [(\chi,\tau) \geqslant (\mu,\nu)]\}$ by the property of equivalence classes. ... $\{\omega \in \Omega^\varphi : ([[\chi]]^{\mathfrak{M}^\varphi}, [[\tau]]^{\mathfrak{M}^\varphi}) \gtrsim_\omega ([[\mu]]^{\mathfrak{M}^\varphi}, [[\nu]]^{\mathfrak{M}^\varphi})\} = \{\omega \in \Omega^\varphi : P_\omega([[\chi]]^{\mathfrak{M}^\varphi}, [[\tau]]^{\mathfrak{M}^\varphi}) \geq P_\omega([[\mu]]^{\mathfrak{M}^\varphi}, [[\nu]]^{\mathfrak{M}^\varphi})\} = [[(\chi,\tau) \geqslant (\mu,\nu)]]^{\mathfrak{M}^\varphi}$ by Lemma 2 and Corollary 1 following from Hypothesis 1 and Definition 3.

So since in this clause, the conditional probability measure the existence of which is asserted by Hypothesis 1 is used, the proof of the completeness of LQCP *based on this clause would* result in a *conditional proof.*

5 Concluding Remarks

5.1 Summary

Koopman [14–16] introduces such a quaternary relation $\gtrsim$ that $(A, B) \gtrsim (C, D)$ is interpreted to mean that A given B is at least as probable as C given D. Recently Mundici [24] and Ibeling et al. [13] respectively have investigated Koopman's qualitative conditional probability from a logical point of view. However, the complete axiomatization of logic of qualitative conditional probability has been an open problem. In this paper we have proposed a new version of complete logic–Logic of Qualitative Conditional Probability (LQCP) by proving a theorem (Theorem 1) that can bridge the gap between the semantics and the axiomatization of LQCP.

5.2 List of Further Work

1. We would like to carry out the verification of the proof of Domotor's Theorem 10 ([4, p. 85]) in the case of $|\Omega| = n \geq 1$ by appeal to geometry of webs (nets) of Aczél et al. [2] and Aczél [1].
2. Narens [26, pp. 414–415] proves that the same conditions as Scott [29] are *sufficient for the existence of a nonstandard probability measure satisfying weak representation without the limitation of the size of* Ω. When $^*\mathbb{R}$ denotes a *nonstandard field* defined by the usual ultrapower construction on reals[9], with the help of Narens [26, pp. 414–415], we try to confirm our conjecture:

Conjecture 1 (Weak Representation) *There exists a finitely additive* nonstandard-real-valued *conditional probability measure* $P_\omega : \mathscr{F} \times \mathscr{F}_0 \to^* \mathbb{R}$ *relative to* $\omega \in \Omega$ *satisfying* weak representation, *that is, for any* $A, C \in \mathscr{F}$ *and any* $B, D \in \mathscr{F}_0$,

$$\text{if }\ (A, B) \gtrsim_\omega (C, D), \quad \text{then }\ P_\omega(A, B) \geq P_\omega(C, D),$$

if $(\Omega, \mathscr{F}, \gtrsim_\omega)$ *satisfies the conditions of Hypothesis 1.*

---

[9] About the details of ultrapower construction on reals and (qualitative) nonstandard-real-valued probability theory, refer to Narens [27].

On the basis of Conjecture 1, we can provide LQCP with its model without the limitation of the size of Ω.

Acknowledgments. The author would like to thank the three reviewers of LORI-10 for their very helpful comments.

Disclosure of Interests. The author has no competing interests.

References

1. Aczél, J.: Quasigroups, nets, and nomograms. Adv. Math. **1**, 383–450 (1965)
2. Aczél, J., et al.: Nomogramme, Gewebe und Quasigruppen. Mathematica (Cluj) **2**, 5–24 (1960)
3. Delgrande, J.P., et al.: The logic of qualitative probability. Artif. Intell. **275**, 457–486 (2019)
4. Domotor, Z.: Probabilistic relational structures and their applications (1969), technical Report No. 144, Institute for Mathematical Studies in the Social Sciences, Stanford University
5. Domotor, Z.: Qualitative probabilities revisited. In: Humphreys, P. (ed.) Patrick Suppes: Scientific Philosopher, Vol. 1. Probability and Probabilistic Causality, pp. 197–238. Kluwer, Dordrecht (1994)
6. Fine, T.L.: Theories of Probability. An Examination of Foundations. Academic Press, New York (1973)
7. Finetti, B.: La prévision: Ses lois logiques, ses sources subjectives. Annales de l'Institut Henri Poincaré 7, 1–68 (1937), English translation. In: Kyburg, H.E., Jr., Smokler, H.E. (eds.) Studies in Subjective Probability, pp. 94–158. Wiley, New York (1964)
8. Gärdenfors, P.: Qualitative probability as an intensional logic. J. Philos. Log. **4**, 171–185 (1975)
9. Granville, A., Tucker, T.: It's as easy as ABC. Not. AMS **49**, 1224–1231 (2002)
10. Hölder, O.: Die Axiome der Quantität und die Lehre vom Mass. Berichte über die Verhandlungen der Königlich Sächsischen Gesellschaft der Wissenschaften zu Leipzig. Mathematisch-Physikalische Klasse **53**, 1–64 (1901)
11. Holliday, W.H., Icard, III, T.F.: Measure semantics and qualitative semantics for epistemic modals. In: Snider, T. (ed.) Proceedings of SALT 23. pp. 514–534. Linguistic Society of America (2013)
12. Holman, E.W.: Strong and weak extensive measurement. J. Math. Psychol. **6**, 286–293 (1969)
13. Ibeling, D., et al.: Probing the quantitative qualitative divide in probabilistic reasoning. Ann. Pure Appl. Logic **175**, 103339 (2024)
14. Koopman, B.O.: The axioms and algebra of intuitive probability. Ann. Math. **41**, 269–292 (1940)
15. Koopman, B.O.: The bases of probability. Bull. Am. Math. Soc. **46**, 763–774 (1940)
16. Koopman, B.O.: Intuitive probabilities and sequences. Ann. Math. **42**, 169–187 (1941)
17. Krantz, D.H., et al.: Foundations of Measurement, vol. 1. Academic Press, New York (1971)
18. Kratzer, A.: Modality. In: von Stechow, A., Wunderlich, D. (eds.) Semantics: An International Handbook of Contemporary Research, pp. 639–50. de Gruyter, Berlin (1991)

19. Kratzer, A.: Modals and Conditionals. Oxford University Press, Oxford (2012)
20. Lewis, D.: Probabilities of conditionals and conditional probabilities. Philos. Rev. **85**, 297–315 (1976)
21. Lewis, D.: Probabilities of conditionals and conditional probabilities ll. Philos. Rev. **95**, 581–589 (1986)
22. Luce, R.D.: On the numerical representation of qualitative conditional probability. Ann. Math. Stat. **39**, 481–491 (1968)
23. Luce, R.D., et al.: Foundations of Measurement, vol. 3. Academic Press, San Diego (1990)
24. Mundici, D.: Deciding Koopman s qualitative probability. Artif. Intell. **299**, 103524 (2021)
25. Mundy, B.: Faithful representation, physical extensive measurement theory and Archimedean axioms. Synthese **70**, 373–400 (1987)
26. Narens, L.: Minimal conditions for additive conjoint measurement and qualitative probability. J. Math. Psychol. **11**, 404–430 (1974)
27. Narens, L.: Theories of Probability. World Scientific, Singapore (2007)
28. Roberts, F.S.: Measurement Theory. Addison-Wesley, Reading (1979)
29. Scott, D.: Measurement structures and linear inequalities. J. Math. Psychol. **1**, 233–247 (1964)
30. Segerberg, K.: Qualitative probability in a modal setting. In: Fenstad, J.E. (ed.) Proceedings of the Second Scandinavian Logic Symposium, pp. 341–352. North-Holland, Amsterdam (1971)
31. Stalnaker, R.C.: Probability and conditionals. Philos. Sci. **37**, 64–80 (1970)
32. Suppes, P., Zanotti, M.: Necessary and sufficient qualitative axioms for conditional probability. Zeitschrift für Wahrscheinlichkeitstheorie und Verwandte Gebiete **60**(2), 163–169 (1982). https://doi.org/10.1007/BF00531820
33. Suppes, P., et al.: Foundations of Measurement, vol. 2. Academic Press, San Diego (1989)
34. Suzuki, S.: Measurement-theoretic foundations of logic of inexact knowledge. In: Aiswarya, C., et al. (eds.) Logic and Its Applications, LNCS, vol. 15402, pp. 218–232. Springer-Verlag, Heidelberg (2025)
35. Yalcin, S.: Probability operators. Philos Compass **5**, 916–937 (2010)

Craig Interpolation Property in ∃□-Bundled Fragment of First-Order Modal Logic

Xun Wang[(✉)]

School of Philosophy and Religious Studies, Minzu University of China,
Beijing, China
`wangxun@muc.edu.cn`

Abstract. By extending the quantifier-free predicate logic (without equality, constant and function symbols) with a bundled modality $\exists x\square$, which packs the quantifier $\exists x$ and the modality $\square$ together, we obtain a fragment of first-order modal logic, namely, $\exists\square$-bundled fragment. In this paper, we prove that the Craig interpolation theorem holds for systems of the $\exists\square$-bundled fragment based on K/D/T/4/S4, while it fails for the system based on S5.

Keywords: Bundled fragment · Craig interpolation · First-order modal logic

1 Introduction

We say that a logic L has the *Craig Interpolation Property* (CIP), if for any sentences φ and ψ, the condition $\vDash_L (\varphi \to \psi)$ implies that there is a sentence θ such that $\vDash_L \varphi \to \theta$, $\vDash_L \theta \to \psi$, and θ contains no nonlogical symbols except those which are both in φ and in ψ. In addition, we call such a θ an *interpolant* between φ and ψ. CIP is a seminal theoretical problem combining model and proof theory. Meanwhile, interpolation problems have found numerous unexpected practical applications in computer science, such as hardware and software specification, model checking and automated reasoning. At present there are a significant number of works that establish CIP or disprove it in various types of logic. For example, CIP holds for a number of well-known first-order modal logics (FOMLs), such as first-order modal K, T, D, 4, S4 [4], while it fails for their counterparts with constant domain as well as first-order modal S5 [3].

This paper focuses on CIP in a particular fragment of FOML, namely $\exists\square$-bundled fragment, or simply $\exists\square$-fragment, EB-fragment (to mean ExistsBox). The $\exists\square$-fragment originated from the study of decidable fragments of FOML, and is one of the few decidable fragments that are known to be decidable [11]. This fragment is obtained by extending the quantifier-free predicate logic (without equality, constant and function symbols) with a bundled operator $\exists x\square$, that is, the quantifier $\exists x$ and the modality $\square$ are always bundled together to appear as

V. Goranko et al. (Eds.): LORI 2025, LNCS 16010, pp. 151–164, 2026.
https://doi.org/10.1007/978-981-95-2481-5_11

a single quantifier-modality pair ($\exists x \Box$). The motivation of the $\exists x \Box$-combination came from a formal treatment of knowledge. Wang used the $\exists x \Box$ operator to capture the logical structure of various know-wh[1] expressions (cf. [12]), e.g., knowing how to achieve φ is rendered as there exists a method x such that the agent knows that x can guarantee φ [13]. Subsequently, more types of such quantifier-modality combinations have been proposed, and (un)decidability, axiomatization as well as model-theoretical studies about these bundled fragments are explored (see [5,6,8,9,14]).

This paper is organized as follows. In Sect. 2 we provide essential definitions/facts about the $\exists\Box$-fragment. In Sect. 3 we demonstrate that CIP holds for systems of the $\exists\Box$-fragment based on K/D/T/4/S4, while it fails for the system based on S5. Finally Sect. 4 is devoted to discuss some open questions for further research.

2 $\exists\Box$-Fragment of First Order Modal Logic

In this section, we review the basic concepts and facts about the $\exists\Box$-fragment. For a thorough treatment the reader is referred to [8,11].

2.1   Syntax and Semantics

We assume a countably infinite set X of first-order variables. We do not consider function symbols or constants, so all terms are variables; moreover, we do not consider equality, and leave them to a future occasion. And, we assume a countably infinite set of predicate symbols of various arities. As usual, 0-place predicate symbols correspond to propositional symbols. The $\exists\Box$-*bundled fragment* of FOML denoted by $\mathscr{L}_{\exists\Box}$ is defined as follows:

$$\varphi ::= P(x_1, \cdots, x_n) \mid \neg\varphi \mid (\varphi \vee \varphi) \mid \exists x \Box \varphi$$

where P is an n-place predicate symbol, $x_1, \cdots, x_n$ and x are variables. $\top, \bot, \wedge, \rightarrow, \leftrightarrow$ are defined as usual. $\forall x \Diamond = \neg\exists x \Box\neg$ is the dual of $\exists x \Box$. The notions of *free* and *bound* occurrences of variables are defined in the obvious way. A formula is called a *sentence* if no free variable appears in it. We denote $Var(\varphi)$ as the set of variables of φ, and $FV(\varphi)$ as the set of free variables of φ. Given a set Γ of formulas, we denote $Var(\Gamma)$ as the union of all $Var(\varphi)$ for $\varphi \in \Gamma$ and $FV(\Gamma)$ as the union of all $FV(\varphi)$ for $\varphi \in \Gamma$. We use $\bar{x}$ to denote a finite sequence of variables. We write $\varphi(\bar{x})$ if all the free variables in φ are included in $\bar{x}$. If $\bar{x}$ is a list of distinct variables and $\bar{y}$ is a list of variables of the same length as $\bar{x}$, then $\varphi[\bar{y}/\bar{x}]$ is the formula φ where the variables $\bar{y}$ have been simultaneously substituted for all free occurrences of the variables $\bar{x}$. In particular, $\varphi[y/x]$ is the formula obtained by replacing every free occurrence of x by y in φ. We say $\varphi[\bar{y}/\bar{x}]$ is *admissible* if no variable x_i in $\bar{x}$ occurs free in φ within the scope of $\exists y_i \Box$.

---

[1] Know-wh stands for verb know followed by an interrogative.

We interpret $\mathscr{L}_{\exists\Box}$ in *first-order Kripke models* with *increasing domains* of the form $\mathcal{M} = (W, R, D, \delta, \rho)$, where W is a non-empty set of *worlds*, $R \subseteq W \times W$ is an *accessibility relation* on W, D is a non-empty *domain* of the model, $\delta : W \to 2^D$ assigns to each $w \in W$ a non-empty *local domain* such that $\delta(w) \subseteq \delta(v)$ whenever wRv, and we also write D_w for $\delta(w)$, ρ is an *interpretation* of the model, such that on each world $w \in W$, ρ assigns to each n-ary predicate an n-ary relation on D. The tuple (W, R) is called a *frame* and the model is said to be based on this frame. A *constant domain model* is a model such that $D_w = D$ for any $w \in W$. For brevity, we usually denote a constant domain model by a 4-tuple (W, R, D, ρ). Given a model $\mathcal{M}$, we denote its components as $W^{\mathcal{M}}, D^{\mathcal{M}}, \delta^{\mathcal{M}}, R^{\mathcal{M}}$ and $\rho^{\mathcal{M}}$.

To interpret free variables, we need a variable assignment $\sigma : X \to D$. Call σ *relevant* at $w \in W$ if $\sigma(x) \in \delta(w)$ for all $x \in X$. The increasing domain condition ensures that whenever σ is relevant at w and we have wRv, then σ is relevant at v as well. Note that in a constant domain model, every assignment σ is relevant at all worlds. Given a model $\mathcal{M} = (W, R, D, \delta, \rho)$, $w \in W$, and an assignment $\sigma : X \to D$, the *truth relation* $\mathcal{M}, w, \sigma \vDash \varphi$ is defined inductively by taking

- $\mathcal{M}, w, \sigma \vDash P(x_1, \cdots, x_n) \Leftrightarrow (\sigma(x_1), \cdots, \sigma(x_n)) \in \rho(P, w)$
- $\mathcal{M}, w, \sigma \vDash \exists x \Box \varphi \Leftrightarrow$ there exists $a \in \delta(w)$ such that $\mathcal{M}, v, \sigma[x \mapsto a] \vDash \varphi$ for each v such that wRv, where $\sigma[x \to a]$ denotes an assignment just like σ except mapping x to a

and the standard clauses for $\neg$ and $\vee$.[2] Notice that the standard semantics for $\Box$ is defined as:

- $\mathcal{M}, w, \sigma \vDash \Box \varphi \Leftrightarrow \mathcal{M}, v, \sigma \vDash \varphi$ for each v such that wRv

Clearly, $\Box \varphi$ is equivalent to $\exists x \Box \varphi$ if $x \notin FV(\varphi)$. In this sense, we regard the standard modal formulas $\Box \varphi$ and $\Diamond \varphi$ as abbreviations of $\exists x \Box \varphi$ and $\forall x \Diamond \varphi$ where $x \notin FV(\varphi)$, respectively.

We say φ is *valid*, if φ is true on any $\mathcal{M}, w$ with respect to any relevant σ at w. We say φ is *satisfiable* if $\neg \varphi$ is not valid. It is easy to see that if $\sigma(x) = \sigma'(x)$ for all the free x in φ, $\mathcal{M}, w, \sigma \vDash \varphi \Leftrightarrow \mathcal{M}, w, \sigma' \vDash \varphi$. In this light, we write $\mathcal{M}, w, \sigma \vDash \varphi(x_1, \cdots, x_n)$ as $\mathcal{M}, w \vDash \varphi[a_1, \cdots, a_n]$ given σ assigns a_i to x_i for each i; or simply we write $\mathcal{M}, w, \sigma \vDash \varphi(\bar{x})$ as $\mathcal{M}, w \vDash \varphi[\bar{a}]$ given σ assigns free variables $\bar{x}$ in φ the corresponding objects in $\bar{a}$ for $|\bar{x}| = |\bar{a}|$, where $|\cdot|$ denotes the length.

2.2   Axiom Systems

In FOML, the distinction between increasing domain models and constant domain models is captured by the Barcan formula $\forall x \Box \varphi \to \Box \forall x \varphi$. However,

---

[2] The variables in our logic designate *rigidly*, that is, a variable designates the same object in all worlds. On the other hand, predicates and quantifier domains are both relativized to worlds, that is, a predicate might have different extensions in different worlds, and quantified domains might be different in different worlds.

the $\exists\Box$-fragment generally cannot distinguish between increasing domain and constant domain, in the sense that most systems for the $\exists\Box$-fragment of increasing domains are the same as those of constant domains.[3]

The sound and strongly complete system $\mathbf{K}_{\exists\Box}$ w.r.t. the class of all increasing (or constant) domain models is defined as follows:

Axioms:

TAUT	all axioms of propositional logic
DISTK	$\Box(\varphi \to \psi) \to (\Box\varphi \to \Box\psi)$
$\Box$to$\exists\Box$	$\Box\varphi[y/x] \to \exists x \Box\varphi$ (if $\varphi[y/x]$ is admissible)

Rules:

$$\text{MP } \frac{\varphi, \varphi \to \psi}{\psi} \qquad \text{NEC } \frac{\varphi}{\Box\varphi} \qquad \text{R}\Box\text{to}\exists\Box \; \frac{\Box\varphi \to \psi}{\exists x \Box\varphi \to \psi}(x \notin FV(\psi))$$

Moreover, the corresponding systems w.r.t. frame classes D/T/4/S4/S5 are defined as follows:

$$\mathbf{D}_{\exists\Box} := \mathbf{K}_{\exists\Box} \oplus \Box\varphi \to \Diamond\varphi$$
$$\mathbf{T}_{\exists\Box} := \mathbf{K}_{\exists\Box} \oplus \Box\varphi \to \varphi$$
$$\mathbf{4}_{\exists\Box} := \mathbf{K}_{\exists\Box} \oplus \Box\varphi \to \Box\Box\varphi$$
$$\mathbf{S4}_{\exists\Box} := \mathbf{T}_{\exists\Box} \oplus \Box\varphi \to \Box\Box\varphi$$
$$\mathbf{S5}_{\exists\Box} := \mathbf{T}_{\exists\Box} \oplus \Diamond\varphi \to \Box\Diamond\varphi$$

2.3 Bisimulation

Given a model $\mathcal{M}$, let $D^*_{\mathcal{M}}$ be the set of (possibly empty) finite sequences of objects in $D^{\mathcal{M}}$. For $w \in W^{\mathcal{M}}$ and $a_1 \cdots a_n \in D^*_{\mathcal{M}}$, we abbreviate $(w, a_1 \cdots a_n)$ by $w\bar{a}$. Given two models $\mathcal{M}$ and $\mathcal{N}$, the relation $Z \subseteq (W^{\mathcal{M}} \times D^*_{\mathcal{M}}) \times (W^{\mathcal{N}} \times D^*_{\mathcal{N}})$ is called an $\exists\Box$-*bisimulation*, if for every $(w\bar{a}, v\bar{b}) \in Z$, $|\bar{a}| = |\bar{b}|$ and the following holds:

PISO $\mathcal{M}, w \vDash P[\bar{a}]$ iff $\mathcal{N}, v \vDash P[\bar{b}]$, for any atomic formula $P\bar{x}$ such that $|\bar{x}| = |\bar{a}| = |\bar{b}|$;

$\exists\Box$Zig For any $c \in D^{\mathcal{M}}_w$, there is a $d \in D^{\mathcal{N}}_v$ such that for any $v' \in W^{\mathcal{N}}$ if $vR^{\mathcal{N}}v'$ then there exists $w' \in W^{\mathcal{M}}$ such that $wR^{\mathcal{M}}w'$ and $w'\bar{a}cZv'\bar{b}d$;

$\exists\Box$Zag For any $d \in D^{\mathcal{N}}_v$, there is a $c \in D^{\mathcal{M}}_w$ such that for any $w' \in W^{\mathcal{M}}$ if $wR^{\mathcal{M}}w'$ then there exists $v' \in W^{\mathcal{N}}$ such that $vR^{\mathcal{N}}v'$ and $w'\bar{a}cZv'\bar{b}d$.

We say $\mathcal{M}, w\bar{a}$ and $\mathcal{N}, v\bar{b}$ are $\exists\Box$-*bisimilar* (notation: $\mathcal{M}, w\bar{a} \leftrightarrow_{\exists\Box} \mathcal{N}, v\bar{b}$), if there is an $\exists\Box$-bisimulation Z between $\mathcal{M}$ and $\mathcal{N}$ such that $w\bar{a}Zv\bar{b}$. Moreover, we say $\mathcal{M}, w$ and $\mathcal{N}, v$ are $\exists\Box$-*bisimilar* if $\mathcal{M}, w \leftrightarrow_{\exists\Box} \mathcal{N}, v$, i.e., when $|\bar{a}| = |\bar{b}| = 0$.

Wang [11] has shown that $\mathscr{L}_{\exists\Box}$ is invariant under $\exists\Box$-bisimilarity, by induction on the structure of formulas.

---

[3] Except frame classes 5, D5, D45, KD45 and S4.2, the axiom systems for the common frame classes from K to S5 with increasing domains are the same as those with constant domains [10,14].

Theorem 1. *If* $\mathcal{M}, w\bar{a} \; \underline{\leftrightarrow}_{\exists\Box} \; \mathcal{N}, v\bar{b}$*, then* $\mathcal{M}, w \vDash \varphi[\bar{a}] \Leftrightarrow \mathcal{N}, v \vDash \varphi[\bar{b}]$ *for each* $\varphi(\bar{x})$ *such that* $|\bar{x}| = |\bar{a}| = |\bar{b}|$*. In particular, if* $\mathcal{M}, w \; \underline{\leftrightarrow}_{\exists\Box} \; \mathcal{N}, v$*, then* $\mathcal{M}, w \vDash \varphi \Leftrightarrow \mathcal{N}, v \vDash \varphi$ *for any sentence* φ*.*

3   Craig Interpolation Property

We include $\top, \bot$ in our language as logical constants, since it may happen that we need to allow $\top, \bot$ to serve as an interpolant. For example, $\exists x\Box Qx$ implies $\exists x\Box(Px \vee \neg Px)$, but there are no sentences at all that contain only predicates that occur both in φ and in ψ, since there are no such predicates. Clearly, $\top$ could do for an interpolant in any case where the consequent is valid, and $\bot$ in any case where the antecedent is unsatisfiable. In the following, we restrict our attention only to cases where the antecedent is satisfiable and the consequent is not valid.

3.1   Positive Results

This subsection is influenced by the work in Chaps. 13 and 17 of [1]. As we shall see below, we adapt the definitions of satisfaction properties and closure properties for first-order logic in [1] to our $\exists\Box$-fragment.

Our goal is to establish the Craig Interpolation Theorem for the systems $\mathbf{K}_{\exists\Box}$, $\mathbf{D}_{\exists\Box}$, $\mathbf{T}_{\exists\Box}$, $\mathbf{4}_{\exists\Box}$ and $\mathbf{S4}_{\exists\Box}$. The hypothesis of the theorem, it will be recalled, is that φ implies ψ, i.e., $\{\varphi, \neg\psi\}$ is unsatisfiable, and the conclusion we want to prove is that there is an interpolant θ between φ and ψ. It suffices to show that if there is no such interpolant θ, then $\{\varphi, \neg\psi\}$ is after all satisfiable, or, in other words, $\{\varphi, \neg\psi\}$ belongs to the set S of all satisfiable sets of formulas. To construct this target set S, first we give some properties enjoyed by S. We denote $\Box^-(\Gamma)$ as the set $\{\varphi \mid \Box\varphi \in \Gamma\}$.

Lemma 1 (Satisfaction properties lemma). *Let* S *be the set of all satisfiable sets* Γ *of formulas of a given* $\exists\Box$*-bundled language. Then* S *has the following properties:*

(S1) *If* $\Gamma \in S$*, then there is no formula* α *such that* α *and* $\neg\alpha$ *are both in* Γ*.*

(S2) *If* $\Gamma \in S$ *and* $\neg\neg\alpha \in \Gamma$*, then* $\Gamma \cup \{\alpha\} \in S$*.*

(S3) *If* $\Gamma \in S$ *and* $(\alpha_1 \vee \alpha_2) \in \Gamma$*, then either* $\Gamma \cup \{\alpha_1\} \in S$ *or* $\Gamma \cup \{\alpha_2\} \in S$*.*

(S4) *If* $\Gamma \in S$ *and* $\neg(\alpha_1 \vee \alpha_2) \in \Gamma$*, then* $\Gamma \cup \{\neg\alpha_1, \neg\alpha_2\} \in S$*.*

(S5) *If* $\Gamma \in S$ *and* $\exists x\Box\alpha(x) \in \Gamma$*, and the variable* y *does not occur in* Γ*, then* $\Gamma \cup \{\Box\alpha[y/x]\} \in S$*.*

(S6) *If* $\Gamma \in S$ *and* $\neg\exists x\Box\alpha(x) \in \Gamma$*, then:*
 - *for each* y *such that* $\alpha[y/x]$ *is admissible,* $\Box^-(\Gamma) \cup \{\neg\alpha[y/x]\} \in S$;[4]
 - *for each* y *such that* $\alpha[y/x]$ *is not admissible, then there exists a formula* α' *obtained by relettering the quantifiers of* y *in* α *with some fresh variable such that:* $\vDash \alpha \leftrightarrow \alpha'$*,* $\alpha'[y/x]$ *is admissible, and* $\Box^-(\Gamma) \cup \{\neg\alpha'[y/x]\} \in S$*.*

---

[4] By (S6), if $\Gamma \in S$ and $\neg\Box\alpha \in \Gamma$, then $\Box^-(\Gamma) \cup \{\neg\alpha\} \in S$.

(S7) *If $\Gamma \in S$ and $\Gamma_0 \subseteq \Gamma$, then $\Gamma_0 \in S$.*
(S8) *If every finite subset Γ_0 of Γ is in S, then Γ is in S.*

Furthermore, if S is the set of all sets Γ such that Γ is T/D/4/S4-satisfiable, then S also has the corresponding satisfaction property (D)/(T)/(4)/(T4)[5]:

(D) *If $\Gamma \in S$ and $\Box\alpha \in \Gamma$, then $\Gamma \cup \{\neg\Box\neg\alpha\} \in S$.*
(T) *If $\Gamma \in S$ and $\Box\alpha \in \Gamma$, then $\Gamma \cup \{\alpha\} \in S$;*
(4) *If $\Gamma \in S$ and $\Box\alpha \in \Gamma$, then $\Gamma \cup \{\Box\Box\alpha\} \in S$.*
(T4) *If $\Gamma \in S$ and $\Box\alpha \in \Gamma$, then $\Gamma \cup \{\alpha, \Box\Box\alpha\} \in S$.*

Proof. Trivial.

We call (S1)–(S8) the *satisfaction properties*, and call (S1)–(S8) with the property (T)/(D)/(4)/(T4) the T/D/4/S4-*satisfaction properties*.

Remark 1. Assume that S has the satisfaction properties, $\Gamma_0, \Gamma_1, \Gamma_2, \cdots \in S$ and $\Gamma_0 \subseteq \Gamma_1 \subseteq \Gamma_2 \subseteq \cdots$. Then by satisfaction properties (S7) and (S8), $\bigcup_{k\in\mathbb{N}} \Gamma_k \in S$.

Next we reduce proving the satisfiability of the target set S mentioned above to proving the following Lemma 2, which is a kind of converse to Lemma 1. Before stating it, we introduce a concept to be used later. We say a language $\mathscr{L}$ is an *infinitely proper sublanguage* of another language $\mathscr{L}^+$ (notation: $\mathscr{L} \prec \mathscr{L}^+$), if $\mathscr{L}^+$ is obtained by adding infinitely many new variables to $\mathscr{L}$.

Lemma 2 (Model existence lemma). *Let $\mathscr{L}_{\exists\Box} \prec \mathscr{L}^+$. Let S be the set of sets Γ of $\mathscr{L}^+$-formulas with the satisfaction properties. Then:*

(i) for each set Γ of $\mathscr{L}_{\exists\Box}$-formulas in S, Γ is satisfiable on a model with the identity assignment σ, i.e., σ interprets each variable to itself.
(ii) if S has the T/D/4/S4-satisfaction properties, then for each set Γ of $\mathscr{L}_{\exists\Box}$-formulas in S, Γ is satisfiable on a T/D/4/S4-model with the identity assignment σ.

Before proving Lemma 2, we introduce some important concepts. The conclusion of Lemma 2 asserts the existence of some models. Every world w in these models corresponds to a set Γ of formulas that are true at w with σ. Now we set down some properties which we require the set of all such sets Γ to satisfy.

Definition 1 (Closure properties). *Let S^* be a set of sets Γ of formulas of a given $\exists\Box$-bundled language[6]. We say S^* has the closure properties[7], if it satisfies the followings:*

---

[5] Notice that the fourth condition above is denoted as (S4), so the condition about the system S4 is named (T4).

[6] We mean an $\exists\Box$-bundled language based on a particular variable set and/or a particular predicate set.

[7] A set with the closure properties is a Hintikka set w.r.t. the $\exists\Box$-bundled logic. For more on Hintikka sets, see [7].

(C1) *If $\Gamma \in S^*$, then there is no formula α such that α and $\neg\alpha$ are both in Γ.*
(C2) *If $\Gamma \in S^*$ and $\neg\neg\alpha \in \Gamma$, then $\alpha \in \Gamma$.*
(C3) *If $\Gamma \in S^*$ and $(\alpha_1 \vee \alpha_2) \in \Gamma$, then either $\alpha_1 \in \Gamma$ or $\alpha_2 \in \Gamma$.*
(C4) *If $\Gamma \in S^*$ and $\neg(\alpha_1 \vee \alpha_2)$ is in Γ, then $\neg\alpha_1, \neg\alpha_2 \in \Gamma$.*
(C5) *If $\Gamma \in S^*$ and $\exists x\Box\alpha(x) \in \Gamma$, then there exists a variable y such that $\Box\alpha[y/x]$ is admissible and $\Box\alpha[y/x] \in \Gamma$.*
(C6) *If $\Gamma \in S^*$ and $\neg\exists x\Box\alpha(x) \in \Gamma$, then*
  - *for each variable y such that $\alpha[y/x]$ is admissible, there exists a $\Delta_y \in S^*$ such that $\Box^-(\Gamma) \cup \{\neg\alpha[y/x]\} \subseteq \Delta_y$;*
  - *for each variable y such that $\alpha[y/x]$ is not admissible, there exist a $\Delta_y \in S^*$ and a formula α' obtained by relettering the quantifiers of y in α with some fresh variable such that: $\vDash \alpha \leftrightarrow \alpha'$, $\alpha'[y/x]$ is admissible, and $\Box^-(\Gamma) \cup \{\neg\alpha'[y/x]\} \subseteq \Delta_y$.*

Furthermore, we say S^* has the $D/T/4/S4$-closure properties, if it also satisfies the corresponding closure property $(D)/(T)/(4)/(T4)$:

(D) *If $\Gamma \in S^*$ and $\Box\alpha \in \Gamma$, then $\neg\Box\neg\alpha \in \Gamma$.*
(T) *If $\Gamma \in S^*$ and $\Box\alpha \in \Gamma$, then $\alpha \in \Gamma$.*
(4) *If $\Gamma \in S^*$ and $\Box\alpha \in \Gamma$, then $\Box\Box\alpha \in \Gamma$.*
(T4) *If $\Gamma \in S^*$ and $\Box\alpha \in \Gamma$, then $\alpha, \Box\Box\alpha \in \Gamma$.*

Proposition 1. *Let S^* be a set of sets Γ of formulas of a given $\exists\Box$-bundled language $\mathscr{L}$. Then:*

(i) if S^* has the closure properties, then every set Γ in S^* is satisfiable on a model with the identity assignment.

(ii) if S^* has the $D/T/4/S4$-closure properties, then every set Γ in S^* is satisfiable on a $D/T/4/S4$-model with the identity assignment.[8]

Proof. (i) Assume that S^* has the closure properties. Define a constant domain model $\mathcal{M} = (W, X, R, \rho)$ as:

  - $W = S^*$
  - $X = Var(\mathscr{L})$
  - $\Gamma R\Delta$ iff $\Box^-(\Gamma) \subseteq \Delta$
  - $\bar{x} \in \rho(P, \Gamma)$ iff $P\bar{x} \in \Gamma$

ρ is well-defined by the closure property (C1). Let $\sigma : X \to X$ be the identity function. Next we show that for any formula $\varphi \in \mathscr{L}$, any $\Gamma \in S^*$:

$$\varphi \in \Gamma \Rightarrow \mathcal{M}, \Gamma, \sigma \vDash \varphi$$

---

[8] This method does not apply to S5. Assume that we define the closure property (T5) as "if $\Gamma \in S^*$ and $\Box\alpha, \Diamond\beta \in \Gamma$, then $\alpha, \Box\Diamond\beta \in \Gamma$". (T5) cannot guarantee that $\mathcal{M}$ constructed in the proof is an S5-model. More specifically, if $\Gamma R\Delta_1$, $\Gamma R\Delta_2$ and $\beta(y) \in \Delta_2$, note that y may not occur in Γ, so we cannot obtain $\Diamond\beta(y) \in \Gamma$, thus we cannot obtain $\Diamond\beta(y) \in \Delta_1$ based on (T5).

Assume that $\varphi \in \Gamma$. Next we show $\mathcal{M}, \Gamma, \sigma \vDash \varphi$ by induction on φ. The case of $P\bar{x}$ follows from the definition of ρ and σ.

For the case of $\psi_1 \vee \psi_2$, it follows by (C3) that $\psi_1 \in \Gamma$ or $\psi_2 \in \Gamma$. By IH, $\mathcal{M}, \Gamma, \sigma \vDash \psi_1$ or $\mathcal{M}, \Gamma, \sigma \vDash \psi_2$, it follows that $\mathcal{M}, \Gamma, \sigma \vDash \psi_1 \vee \psi_2$.

For the case of $\exists x \Box \psi$, it follows by (C5) that there exists y such that $\Box \psi[y/x]$ is admissible and $\Box \psi[y/x] \in \Gamma$. For any Δ such that $\Gamma R \Delta$, we have that $\psi[y/x] \in \Delta$, then by IH, $\mathcal{M}, \Delta, \sigma \vDash \psi[y/x]$. Thus $\mathcal{M}, \Gamma, \sigma \vDash \Box \psi[y/x]$. Due to the definition of σ and the fact that $\Box \psi[y/x]$ is admissible, $\mathcal{M}, \Gamma, \sigma[x \mapsto y] \vDash \Box \psi$, it follows that $\mathcal{M}, \Gamma, \sigma \vDash \exists x \Box \psi$.

For the case where φ is of the form $\neg \psi$, there are four cases to consider:

Case 1. $\varphi = \neg P\bar{x}$. It is immediate from (C1) and the definition of ρ and σ.

Case 2. $\varphi = \neg \neg \psi$. By (C2), $\psi \in \Gamma$. By IH, $\mathcal{M}, \Gamma, \sigma \vDash \psi$, so $\mathcal{M}, \Gamma, \sigma \vDash \neg \neg \psi$.

Case 3. $\varphi = \neg(\psi_1 \vee \psi_2)$. By (C4), $\neg \psi_1 \in \Gamma$ and $\neg \psi_2 \in \Gamma$. By IH, $\mathcal{M}, \Gamma, \sigma \vDash \neg \psi_1$ and $\mathcal{M}, \Gamma, \sigma \vDash \neg \psi_2$, so $\mathcal{M}, \Gamma, \sigma \vDash \neg(\psi_1 \vee \psi_2)$.

Case 4. $\varphi = \neg \exists x \Box \psi$. First consider any variable y such that $\psi[y/x]$ is admissible. By (C6), there exists a $\Delta_y \in S^*$ such that $\Box^{\neg}(\Gamma) \cup \{\neg \psi[y/x]\} \subseteq \Delta_y$. By IH, $\mathcal{M}, \Delta_y, \sigma \vDash \neg \psi[y/x]$, then $\mathcal{M}, \Delta_y, \sigma[x \mapsto y] \vDash \neg \psi$. It follows by definition of R that $\Gamma R \Delta_y$. Thus $\mathcal{M}, \Gamma, \sigma[x \mapsto y] \vDash \neg \Box \psi$.

Now consider any variable y' such that $\psi[y'/x]$ is not admissible. We have that there exist a $\Delta_{y'} \in S^*$ and a formula ψ' obtained by relettering the quantifiers of y' in ψ with some fresh variable such that: $\vDash \psi \leftrightarrow \psi'$, $\psi'[y'/x]$ is admissible, and $\Box^{\neg}(\Gamma) \cup \{\neg \psi'[y'/x]\} \subseteq \Delta_{y'}$. We can then repeat the reasoning above to obtain $\mathcal{M}, \Gamma, \sigma[x \mapsto y'] \vDash \neg \Box \psi'$, it follows that $\mathcal{M}, \Gamma, \sigma[x \mapsto y'] \vDash \neg \Box \psi$. Therefore, for each variable y, no matter whether $\psi[y/x]$ is admissible, we have $\mathcal{M}, \Gamma, \sigma[x \mapsto y] \vDash \neg \Box \psi$. Therefore, $\mathcal{M}, \Gamma, \sigma \vDash \neg \exists x \Box \psi$.

(ii) Similar to (i). The only point that requires comment is that $\mathcal{M}$ is a D/T/4/S4-model, since S^* has the closure property (D)/(T)/(4)/(S4).

With the aid of Proposition 1, now we give the proof of Lemma 2:

Proof (of Lemma 2). (i) As $\mathscr{L}_{\exists \Box} \prec \mathscr{L}^+$, there exists a sequence of languages $\mathscr{L}_0 = \mathscr{L}_{\exists \Box} \prec \mathscr{L}_1 \prec \mathscr{L}_2 \prec \cdots$ such that $\mathscr{L}^+ = \bigcup_{k \in \mathbb{N}} \mathscr{L}_k$. Enumerate all the variables in $\mathscr{L}^+$. Enumerate all the formulas in $\mathscr{L}^+$.

Take a set Γ of the $\mathscr{L}_0$-formulas in S. Let ϵ be the empty sequence. Enumerate the $\exists x \Box$-formulas in $\mathscr{L}_1$ as $\exists x_0 \Box \alpha_0, \exists x_1 \Box \alpha_1, \cdots$. For each $k \geq 0$, define $\Gamma^{0,k}_{(\epsilon, \epsilon)}$ as follows:

$$\Gamma^{0,0}_{(\epsilon, \epsilon)} = \Gamma$$

$$\Gamma^{0,k+1}_{(\epsilon, \epsilon)} = \begin{cases} \Gamma^{0,k}_{(\epsilon, \epsilon)} \cup \{\exists x_k \Box \alpha_k, \Box \alpha_k[x'/x_k]\} & \text{if } \Gamma^{0,k}_{(\epsilon, \epsilon)} \cup \{\exists x_k \Box \alpha_k\} \in S; \\ \Gamma^{0,k}_{(\epsilon, \epsilon)} & \text{otherwise} \end{cases}$$

where x' is the first variable in the above fixed enumeration such that x' is in $\mathscr{L}_1$ but does not occur in $\Gamma^{0,k}_{(\epsilon, \epsilon)} \cup \{\exists x_k \Box \alpha_k\}$. Such x' always exists, since $\mathscr{L}_1$ contains infinitely more new variables than $\Gamma^{0,0}_{(\epsilon, \epsilon)}$ and each $\Gamma^{0,k+1}_{(\epsilon, \epsilon)}$ contains only at most finitely more new variables than $\Gamma^{0,k}_{(\epsilon,\epsilon)}$. By the satisfaction property (S5), $\Gamma^{0,k}_{\epsilon,\epsilon} \in S$ for each k. Now, let

$$\Gamma^0_{(\epsilon,\epsilon)} = \bigcup_{k\in\mathbb{N}} \Gamma^{0,k}_{(\epsilon,\epsilon)}$$

Then by Remark 1, $\Gamma^0_{(\epsilon,\epsilon)} \in S$.

Enumerate the formulas not of the form $\exists x\Box\alpha$ in $\mathscr{L}_1$ as $\varphi_0, \varphi_1, \cdots$. For each $k \geq 0$, define $\Gamma^{k+1}_{(\epsilon,\epsilon)}$ as follows:

$$\Gamma^{k+1}_{(\epsilon,\epsilon)} = \begin{cases} \Gamma^k_{(\epsilon,\epsilon)} \cup \{\varphi_k\} & \text{if } \Gamma^k_{(\epsilon,\epsilon)} \cup \{\varphi_k\} \in S; \\ \Gamma^k_{(\epsilon,\epsilon)} & \text{otherwise} \end{cases}$$

Define

$$\Gamma_{(\epsilon,\epsilon)} = \bigcup_{k\in\mathbb{N}} \Gamma^k_{(\epsilon,\epsilon)}$$

Three remarks are in order here. First, by construction, for each $\exists x\Box\alpha \in \Gamma_{(\epsilon,\epsilon)}$, there exists $\Box\alpha[x'/x]$ such that $\Box\alpha[x'/x]$ is admissible and $\Box\alpha[x'/x] \in \Gamma^0_{(\epsilon,\epsilon)}$, and it follows that $\Box\alpha[x'/x] \in \Gamma_{(\epsilon,\epsilon)}$. Second, by construction $\Gamma^k_{(\epsilon,\epsilon)} \in S$ for each k, then as we noted in Remark 1, we have that $\Gamma_{(\epsilon,\epsilon)} \in S$. Finally, note that $\Gamma_{(\epsilon,\epsilon)}$ is a set of $\mathscr{L}_1$-formulas.

Suppose $\Gamma_{(s_1,s_2)} \in S$ has been defined, and $\Gamma_{(s_1,s_2)}$ is a set of $\mathscr{L}_i$-formulas ($i \in \mathbb{N}$). Take $\neg\exists x\Box\alpha \in \Gamma_{(s_1,s_2)}$ and $y \in \mathscr{L}^+$. Now we define $\Gamma_{(s_1+\neg\exists x\Box\alpha,s_2+y)}$. For brevity, we abbreviate $s_1 + \neg\exists x\Box\alpha$ by s_1', $s_2 + y$ by s_2'. If $\alpha[y/x]$ is not admissible, then by the satisfaction property (S6), there exists a formula α' obtained by relettering the quantifiers of y in α with some fresh variable such that: $\vDash \alpha \leftrightarrow \alpha'$, $\alpha'[y/x]$ is admissible, and $\Box^-(\Gamma) \cup \{\neg\alpha'[y/x]\} \in S$. If $\alpha[y/x]$ is admissible, let $\alpha' = \alpha$. We have that $\Box^-(\Gamma)\cup\{\neg\alpha'[y/x]\} \in S$. Let j be the least number such that $\Box^-(\Gamma) \cup \{\neg\alpha'[y/x]\} \subseteq \mathscr{L}_j$. Enumerate the $\exists y\Box$-formulas in $\mathscr{L}_{j+1}$ as $\exists y_0\Box\beta_0, \exists y_1\Box\beta_1, \cdots$. For each $k \geq 0$, define $\Gamma^{0,k}_{(s_1',s_2')}$ as follows:

$$\Gamma^{0,0}_{(s_1',s_2')} = \Box^-(\Gamma) \cup \{\neg\alpha'[y/x]\}$$

$$\Gamma^{0,k+1}_{(s_1',s_2')} = \begin{cases} \Gamma^{0,k}_{(s_1',s_2')} \cup \{\exists y_k\Box\beta_k, \Box\beta_k[y'/y_k]\} & \text{if } \Gamma^{0,k}_{(s_1',s_2')} \cup \{\exists y_k\Box\beta_k\} \in S; \\ \Gamma^{0,k}_{(s_1',s_2')} & \text{otherwise} \end{cases}$$

where y' is the first variable in the ordering such that y' is in $\mathscr{L}_{j+1}$ but does not occur in $\Gamma^{0,k}_{(s_1',s_2')} \cup \{\exists y_k\Box\beta_k\}$. Similarly to $\Gamma^{0,k+1}_{(\epsilon,\epsilon)}$, such y' always exists.

Enumerate the formulas not of the form $\exists y\Box\beta$ in $\mathscr{L}_{j+1}$ as $\psi_0, \psi_1, \cdots$. For each $k \geq 0$, define $\Gamma^k_{(s_1',s_2')}$ as follows:

$$\Gamma^0_{(s_1',s_2')} = \bigcup_{k\in\mathbb{N}} \Gamma^{0,k}_{(s_1',s_2')}$$

$$\Gamma^{k+1}_{(s'_1,s'_2)} = \begin{cases} \Gamma^k_{(s'_1,s'_2)} \cup \{\psi_k\} & \text{if } \Gamma^k_{(s'_1,s'_2)} \cup \{\psi_k\} \in S; \\ \Gamma^k_{(s'_1,s'_2)} & \text{otherwise} \end{cases}$$

Define

$$\Gamma_{(s'_1,s'_2)} = \bigcup_{k\in\mathbb{N}} \Gamma^k_{(s'_1,s'_2)}$$

Similarly, for each $\exists y\Box\beta \in \Gamma_{(s'_1,s'_2)}$, there exists $\Box\beta[y'/y]$ such that $\Box\beta[y'/y]$ is admissible and $\Box\beta[y'/y] \in \Gamma_{(s'_1,s'_2)}$. Moreover, $\Gamma_{(s'_1,s'_2)} \in S$, and $\Gamma_{(s'_1,s'_2)}$ is a set of $\mathscr{L}_{j+1}$-formulas.

Let S^* be the set of all such $\Gamma_{(s_1,s_2)}$, more precisely, $S^* = \{\Gamma_{(s_1,s_2)} \mid n \geq 0; s_2 = z_1 \cdots z_n$ where each z_i is a variable in $\mathscr{L}^+; s_1 = \gamma_1 \cdots \gamma_n$, where for each $1 \leq i \leq n : \gamma_i$ is a formula of the form $\neg\exists z\Box\gamma$, and $\gamma_i \in \Gamma_{(\gamma_1\cdots\gamma_{i-1},z_1\cdots z_{i-1})}\}$.

Next we show that S^* has the closure properties. Take $\Gamma_{(s_1,s_2)} \in S^*$. We have that $\Gamma_{(s_1,s_2)} \in S$.

(C1) It follows by the satisfaction property (S1).

(C2) Suppose $\neg\neg\alpha \in \Gamma_{(s_1,s_2)}$. If $\alpha \notin \Gamma_{(s_1,s_2)}$, then by construction $\Gamma_{(s_1,s_2)} \cup \{\alpha\} \notin S$, which contradicts the satisfaction property (S2).

(C3), (C4) Similar to (C2).

(C5) Suppose $\exists x\Box\alpha \in \Gamma_{(s_1,s_2)}$. By the construction of $\Gamma^0_{(s_1,s_2)}$, there exists a y such that $\Box\alpha[y/x]$ is admissible and $\Box\alpha[y/x] \in \Gamma^0_{(s_1,s_2)} \subseteq \Gamma_{(s_1,s_2)}$.

(C6) Suppose $\neg\exists x\Box\alpha \in \Gamma_{(s_1,s_2)}$. Take a variable $y \in \mathscr{L}^+$. We have that $\Gamma_{(s_1+\neg\exists x\Box\alpha,s_2+y)} \in S^*$. By the construction of $\Gamma_{(s_1+\neg\exists x\Box\alpha,s_2+y)}$, we have that: if $\Box\alpha[y/x]$ is admissible, then $\Box^-(\Gamma_{(s_1,s_2)}) \cup \{\neg\alpha[y/x]\} \subseteq \Gamma_{(s_1+\neg\exists x\Box\alpha,s_2+y)} \in S^*$; if $\Box\alpha[y/x]$ is not admissible, then there exists α' obtained by relettering the quantifiers of y in α with some fresh variable such that: $\vDash \alpha \leftrightarrow \alpha'$, $\alpha'[y/x]$ is admissible, and $\Box^-(\Gamma) \cup \{\neg\alpha'[y/x]\} \subseteq \Gamma_{(s_1+\neg\exists x\Box\alpha',s_2+y)} \in S^*$. This completes the proof of (C6), and establishes the required result for S^*.

For item (ii), it remains to check S^* has the closure property (D)/(T)/(4)/(T4). Obviously, this is guaranteed by the satisfaction property (D)/(T)/(4)/(T4) of S.

By these preliminary manoeuvres, we are ready for the final step – to prove CIP for $\mathbf{K}_{\exists\Box}, \mathbf{D}_{\exists\Box}, \mathbf{T}_{\exists\Box}, \mathbf{4}_{\exists\Box}$ and $\mathbf{S4}_{\exists\Box}$.

Theorem 2 (Craig interpolation theorem). *Let φ,ψ be two sentences of $\mathscr{L}_{\exists\Box}$. For any $L \in \{\mathbf{K}_{\exists\Box}, \mathbf{D}_{\exists\Box}, \mathbf{T}_{\exists\Box}, \mathbf{4}_{\exists\Box}, \mathbf{S4}_{\exists\Box}\}$, if $\vDash_L \varphi \to \psi$ then there exists an interpolant θ between φ and ψ.*

Proof. Suppose that there is no such interpolant θ. Then we show that $\{\varphi, \neg\psi\}$ is satisfiable – a contradiction. By Lemma 2, it suffices to show that if $\mathscr{L}$ is the $\exists\Box$-bundled language containing all the predicate symbols of φ and of ψ, and if $\mathscr{L} \prec \mathscr{L}^+$, then $\{\varphi, \neg\psi\}$ will be satisfiable provided that it belongs to some set S of sets of $\mathscr{L}^+$-formulas having the satisfaction properties.

Now we define a set S, use the assumption that there is no interpolant θ of φ and ψ to show $\{\varphi, \neg\psi\}$ is in S, and establish the desired satisfaction properties for S. Here we need to introduce some terminology. Call a formula α of $\mathscr{L}^+$ a *left* formula (respectively, a *right* formula) if every predicate in α is in φ (respectively, is in ψ). If Γ_l is a satisfiable set of left formulas and Γ_r a satisfiable set of right formulas, we say that β *bars* the pair Γ_l, Γ_r if β satisfies the following conditions:

- β is both a left and a right formula;[9]
- every free variable in β is free both in Γ_l and in Γ_r;
- Γ_l implies β while Γ_r implies $\neg\beta$.

Our assumption that there is no interpolant, restated in this terminology, is the assumption that no formula β in $\mathscr{L}^+$ bars $\{\varphi\}, \{\neg\psi\}$. Let S be the set of all Γ that admit an *unbarred division* in the sense that we can write Γ as a union $\Gamma_l \cup \Gamma_r$ of two sets of formulas where Γ_l consists of left formulas and Γ_r of right formulas, each of Γ_l and Γ_r is satisfiable, and no formula in $\mathscr{L}^+$ bars the pair Γ_l, Γ_r. Clearly, $\{\varphi, \neg\psi\} \in S$; all that remains to be checked is that S has the satisfaction properties.

For (S1), suppose $\Gamma = \Gamma_l \cup \Gamma_r$ is an unbarred division. Then Γ_l is satisfiable, so there is no formula α with both α and $\neg\alpha$ in Γ_l. Similarly for Γ_r. Nor can there be a α with α in Γ_l and $\neg\alpha$ in Γ_r, for in that case α would bar Γ_l, Γ_r. Similarly, the reverse case with $\neg\alpha$ in Γ_l and α in Γ_r is impossible.

For (S2), suppose $\Gamma = \Gamma_l \cup \Gamma_r$ is an unbarred division and $\neg\neg\alpha$ is in Γ. Without loss of generality, we may assume $\neg\neg\alpha$ is in Γ_l. Obviously, α is a left formula, $\Gamma_l \cup \{\alpha\}$ is satisfiable and the set of all syntax consequences of $\Gamma_l \cup \{\alpha\}$ is the same as that of Γ_l. It follows that $\Gamma \cup \{\alpha\} = (\Gamma_l \cup \{\alpha\}) \cup \Gamma_r$ is an unbarred division, so $\Gamma \cup \{\alpha\}$ is in S.

For (S3), suppose $\Gamma = \Gamma_l \cup \Gamma_r$ is an unbarred division and $(\alpha_1 \vee \alpha_2)$ is in Γ. Again we assume that $(\alpha_1 \vee \alpha_2)$ is in Γ_l. Obviously, α_1 and α_2 are left formulas. We claim that $\Gamma \cup \{\alpha_i\} = (\Gamma_l \cup \{\alpha_i\}) \cup \Gamma_r$ is an unbarred division for at least one i $(i = 1, 2)$. If $\Gamma_l \cup \{\alpha_1\}$ is unsatisfiable, it follows from $(\alpha_1 \vee \alpha_2) \in \Gamma_l$ that Γ_l implies α_2, note that we now encounter a situation similar to (S2), then clearly $(\Gamma_l \cup \{\alpha_2\}) \cup \Gamma_r$ is an unbarred division. Similarly, if $\Gamma_l \cup \{\alpha_2\}$ is unsatisfiable, then $(\Gamma_l \cup \{\alpha_1\}) \cup \Gamma_r$ is an unbarred division. Now we are left to treat the case where $\Gamma_l \cup \{\alpha_i\}$ is satisfiable for both i. In this case, assume for the sake of a contradiction that neither $(\Gamma_l \cup \{\alpha_1\}) \cup \Gamma_r$ nor $(\Gamma_l \cup \{\alpha_2\}) \cup \Gamma_r$ is an unbarred division. Then there exist formulas β_1 and β_2 such that β_i bars $\Gamma_l \cup \{\alpha_i\}, \Gamma_r$ for $i = 1, 2$. So β_i is implied by $\Gamma_l \cup \{\alpha_i\}$ for each i. Note that $(\alpha_1 \vee \alpha_2) \in \Gamma_l$, whence $\beta_1 \vee \beta_2$ is implied by Γ_l. And, since each $\neg\beta_i$ is implied by Γ_r, so is $\neg(\beta_1 \vee \beta_2)$. It is easy to see that $\beta_1 \vee \beta_2$ is both a left and a right formula and every free variable in $\beta_1 \vee \beta_2$ is free both in Γ_l and in Γ_r. Thus $\beta_1 \vee \beta_2$ bars Γ_l, Γ_r, contradiction.

For (S4), similar to (S2).

For (S5), suppose $\Gamma = \Gamma_l \cup \Gamma_r$ is an unbarred division and $\exists x \Box \varphi \in \Gamma$. Assume that $\exists x \Box \varphi \in \Gamma_l$. Take an arbitrary y in $\mathscr{L}^+$ but not in $Var(\Gamma)$.[10] Next we show

---

[9] That is, every predicate in α is both in φ and in ψ.

[10] Maybe such y does not exist, then (S5) trivially holds.

that $(\Gamma_l \cup \{\Box\varphi[y/x]\}) \cup \Gamma_r$ is an unbarred division. For suppose not. It is not hard to see that $\Gamma_l \cup \{\Box\varphi[y/x]\}$ is satisfiable by R$\Box$to$\exists\Box$. So there is a formula β_3 such that β_3 bars the pair $\Gamma_l \cup \{\Box\varphi[y/x]\}, \Gamma_r$. That is, there exist $\theta_{l,1}, \cdots, \theta_{l,m} \in \Gamma_l$ and $\theta_{r,1}, \cdots, \theta_{r,n} \in \Gamma_r$ such that $\Box\varphi[y/x] \wedge \theta_{l,1} \wedge \cdots \wedge \theta_{l,m} \to \beta_3$ and $\theta_{r,1} \wedge \cdots \wedge \theta_{r,n} \to \neg\beta_3$. It follows that $\Box\varphi[y/x] \to (\theta_{l,1} \wedge \cdots \wedge \theta_{l,m} \to \beta_3)$. Clearly $y \notin FV(\beta_3)$, for otherwise $y \in FV(\Gamma_l) \subseteq Var(\Gamma)$. And we have that y does not occur in $\theta_{l,1}, \cdots, \theta_{l,m}$. Then by R$\Box$to$\exists\Box$, $\exists y\Box\varphi[y/x] \to (\theta_{l,1} \wedge \cdots \wedge \theta_{l,m} \to \beta_3)$, so $\exists x\Box\varphi \wedge \theta_{l,1} \wedge \cdots \wedge \theta_{l,m} \to \beta_3$. Note that $y \notin FV(\beta_3)$, so every free variable in β_3 is free both in Γ_l and in Γ_r. Thus β_3 bars Γ_l, Γ_r, contradiction.

For (S6), suppose $\Gamma = \Gamma_l \cup \Gamma_r$ is an unbarred division and $\neg\exists x\Box\varphi \in \Gamma$. Assume that $\neg\exists x\Box\varphi \in \Gamma_l$. First consider any variable y in $\mathscr{L}^+$ such that $\varphi[y/x]$ is admissible. Next we show that $(\Box^-(\Gamma_l) \cup \{\neg\varphi[y/x]\}) \cup \Box^-(\Gamma_r)$ is an unbarred division. For suppose not. It is routine to show that $(\Box^-(\Gamma_l) \cup \{\neg\varphi[y/x]\})$ is satisfiable based on NEC, DISTK and $\Box$to$\exists\Box$. So there is a formula β_4 such that β_4 bars the pair $\Box^-(\Gamma_l) \cup \{\neg\varphi[y/x]\}, \Box^-(\Gamma_r)$. That is, there exist $\gamma_{l,1}, \cdots, \gamma_{l,m'} \in \Box^-(\Gamma_l)$ and $\gamma_{r,1}, \cdots, \gamma_{r,n'} \in \Box^-(\Gamma_r)$ such that $\neg\varphi[y/x] \wedge \gamma_{l,1} \wedge \cdots \wedge \gamma_{l,m'} \to \beta_4$ and $\gamma_{r,1} \wedge \cdots \wedge \gamma_{r,n'} \to \neg\beta_4$. It follows that $\Diamond(\neg\varphi[y/x] \wedge \gamma_{l,1} \wedge \cdots \wedge \gamma_{l,m'}) \to \Diamond\beta_4$ and $\Box\gamma_{r,1} \wedge \cdots \wedge \Box\gamma_{r,n'} \to \Box\neg\beta_4$. Note that $\neg\exists x\Box\varphi \wedge \Box\gamma_{l,1} \wedge \cdots \wedge \Box\gamma_{l,m'} \to \Diamond(\neg\varphi[y/x] \wedge \gamma_{l,1} \wedge \cdots \wedge \gamma_{l,m'})$, so $\neg\exists x\Box\varphi \wedge \Box\gamma_{l,1} \wedge \cdots \wedge \Box\gamma_{l,m'} \to \Diamond\beta_4$. (i) If y is free in Γ_l, then $\Diamond\beta_4$ bars Γ_l, Γ_r, contradiction; (ii) If y is not free in Γ_l, then by R$\Box$to$\exists\Box$, $\neg\exists x\Box\varphi \wedge \Box\gamma_{l,1} \wedge \cdots \wedge \Box\gamma_{l,m'} \to \forall y\Diamond\beta_4$. And we have that $\Box\gamma_{r,1} \wedge \cdots \wedge \Box\gamma_{r,n'} \to \exists y\Box\neg\beta_4$. Thus, $\forall y\Diamond\beta_4$ bars Γ_l, Γ_r, contradiction.

Now consider any variable $y' \in \mathscr{L}^+$ such that $\varphi[y'/x]$ is not admissible. We can reletter the quantifiers of y' in φ with some fresh variable to obtain φ' such that $\vdash \varphi \leftrightarrow \varphi'$ and $\varphi'[y'/x]$ is admissible. We can then repeat the reasoning above to obtain that $(\Box^-(\Gamma_l) \cup \{\neg\varphi'[y'/x]\}) \cup \Box^-(\Gamma_r)$ is an unbarred division.

(S7) and (S8) are obvious.

This completes the proof when L is $\mathbf{K}_{\exists\Box}$. For the cases of $\mathbf{D}_{\exists\Box}/\mathbf{T}_{\exists\Box}/\mathbf{4}_{\exists\Box}/\mathbf{S4}_{\exists\Box}$, the proof of the satisfiability property (D)/(T)/(4)/(T4) is the same as in (S2). And this completes the whole proof.

3.2 Negative Result

We shall now show that CIP fails for $\mathbf{S5}_{\exists\Box}$. It is not hard to check that[11]

$$\vDash_{\mathbf{S5}_{\exists\Box}} p \wedge \forall x\Diamond\Box(p \to Fx) \to (q \to \forall x\Diamond(q \wedge Fx))$$

But we have that

Proposition 2. *In* $\mathbf{S5}_{\exists\Box}$, $p \wedge \forall x\Diamond\Box(p \to Fx) \to (q \to \forall x\Diamond(q \wedge Fx))$ *has no interpolation.*

---

[11] Recall that $\forall x\Diamond$ and $\Box$ are abbreviations of $\neg\exists x\Box\neg$ and $\exists y\Box$ whenever y is not free in the scope of $\exists y\Box$, respectively. So $\forall x\Diamond\Box(p \to Fx)$ and $\forall x\Diamond(q \wedge Fx)$ are abbreviations of $\neg\exists x\Box\neg\exists y\Box(p \to Fx)$ and $\neg\exists x\Box\neg(q \wedge Fx)$, respectively.

Proof. Suppose not. Then the interpolant θ between $p \wedge \forall x \Diamond \Box (p \to Fx)$ and $q \to \forall x \Diamond (q \wedge Fx)$ only contains the predicate symbol F. Consider the $\exists\Box$-bundled language $\mathscr{L}_F$ such that $\mathscr{L}_F$ contains the only predicate symbol F. Consider the following pointed S5-models $\mathcal{M}, w$ and $\mathcal{N}, s$:

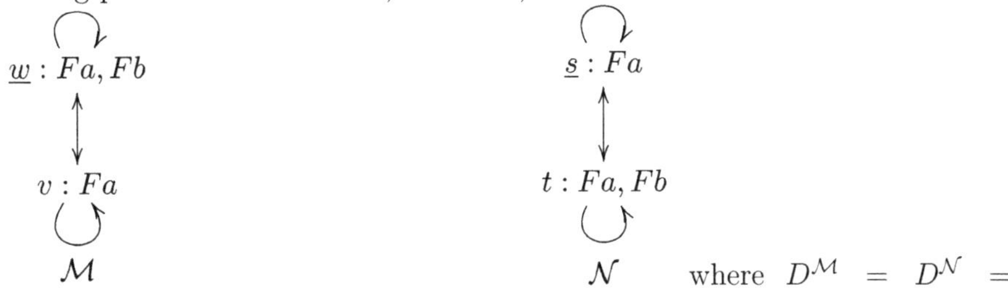

where $D^{\mathcal{M}} = D^{\mathcal{N}} = \{a, b\}$; $\rho^{\mathcal{M}}(F, w) = \rho^{\mathcal{N}}(F, t) = \{a, b\}$, $\rho^{\mathcal{M}}(F, v) = \rho^{\mathcal{N}}(F, s) = \{a\}$. It is easy to verify that $Z = \{(w, s), (w\bar{c}, t\bar{c}), (v\bar{c}, s\bar{c}) \mid \bar{c} \in D_{\mathcal{M}}^*\}$ is a bisimulation between $\mathcal{M}$ and $\mathcal{N}$ w.r.t. the language $\mathscr{L}_F$. We have that $\mathcal{M}, w \leftrightarrow_{\exists\Box} \mathcal{N}, s$ w.r.t. $\mathscr{L}_F$. As $\theta \in \mathscr{L}_F$,

$$\mathcal{M}, w \vDash \theta \iff \mathcal{N}, s \vDash \theta \tag{1}$$

Now we extend $\mathscr{L}_F$ with the nullary predicate symbols p, q, and call the new language $\mathscr{L}_F^+$. Moreover, we expand the models $\mathcal{M}$ and $\mathcal{N}$ with the interpretation about p, q, respectively. And, we call the new models $\mathcal{M}'$ and $\mathcal{N}'$, respectively. Define $\rho(p, \mathcal{M}') = \{w\}$, $\rho(q, \mathcal{N}') = \{s\}$, $\rho(q, \mathcal{M}') = \rho(p, \mathcal{N}') = \emptyset$:

We have $\mathcal{M}', w \vDash p \wedge \forall x \Diamond \Box (p \to Fx)$, so $\mathcal{M}', w \vDash \theta$. Note that $\mathcal{M}', w$ and $\mathcal{M}, w$ agree on all $\mathscr{L}_F$-sentences, so $\mathcal{M}, w \vDash \theta$. By [1], $\mathcal{N}, s \vDash \theta$. Likewise, $\mathcal{N}', s$ and $\mathcal{N}, s$ agree on all $\mathscr{L}_F$-sentences, so $\mathcal{N}', s \vDash \theta$. Thus, $\mathcal{N}', s \vDash q \to \forall x \Diamond (q \wedge Fx)$, but that is not the case. Contradiction.

4   Future Work

In this paper, we study the Craig Interpolation Property in some basic $\exists\Box$-bundled logics, including $\mathbf{K}_{\exists\Box}, \mathbf{D}_{\exists\Box}, \mathbf{T}_{\exists\Box}, \mathbf{4}_{\exists\Box}, \mathbf{S4}_{\exists\Box}$ and $\mathbf{S5}_{\exists\Box}$. For future research, we can explore the interpolation property, including the uniform interpolation property, in more $\exists\Box$-bundled logics. We notice that the Craig Interpolation Theorem fails for the S5-systems of both the $\exists\Box$-fragment and FOML. It is worth investigating the general reason for this failure and repairing the interpolation theorem in $\mathbf{S5}_{\exists\Box}$. In addition, as one referee suggested, it would be interesting to interpret an interpolant as something similar to a Dedekind cut.

Acknowledgments. The author thanks the anonymous reviewers for their valuable comments, in particular for pointing out some related work and future work. The author also thanks Wei Zhao for helpful feedback on the paper. This study was funded by "Fundamental Research Funds for the Central Universities" (No. 2024KYQD26).

Disclosure of Interests. The author has no competing interests to declare that are relevant to the content of this article.

References

1. Boolos, G., Burgess, J., Jeffrey, R.: Computability and Logic, 5th edn. Cambridge University Press, New York (2007)
2. Braüner, T., Ghilardi, S.: First-order modal logic. In: Handbook of Modal Logic, pp. 549–620. Elsevier (2007). https://doi.org/10.1016/S1570-2464(07)80012-7
3. Fine, K.: Failures of the interpolation lemma in quantified modal logic. J. Symb. Log. **44**(2), 201–206 (1979). https://doi.org/10.2307/2273727
4. Gabbay, D.M.: Craig's interpolation theorem for modal logics. In: Hodges, W. (ed.) Conference in Mathematical Logic — London '70. LNM, vol. 255, pp. 111–127. Springer, Heidelberg (1972). https://doi.org/10.1007/BFb0059541
5. Liu, M., Padmanabha, A., Ramanujam, R., Wang, Y.: Are bundles good deals for first-order modal logic? Inf. Comput. **293** (2023). https://doi.org/10.1016/j.ic.2023.105062
6. Padmanabha, A., Ramanujam, R., Wang, Y.: Fragments of first-order modal logic: (un)decidability. In: Proceedings of FSTTCS 2018 (2018). https://doi.org/10.48550/arXiv.1803.10508
7. Smullyan, R.M.: First-Order Logic, 2nd printing edn. Springer, New York (1971)
8. Wang, X.: Completeness theorems for $\exists\Box$-bundled fragment of first-order modal logic. Synthese **201**(123) (2023). https://doi.org/10.1007/s11229-023-04104-7
9. Wang, X.: Lindström theorems for $\exists\Box$-bundled fragment of first-order modal logic. Stud. Logic **17**(3), 1–23 (2024)
10. Wang, X.: Axiomatization of $\forall\Box$ and $\exists\Box$ bundles, Working manuscript
11. Wang Y.: A new modal framework for epistemic logic. In: Lang, J. (ed.) Proceedings TARK 2017, pp. 515–534 (2017). https://doi.org/10.4204/EPTCS.251.38
12. Wang, Y.: Beyond knowing that: a new generation of epistemic logics. In: van Ditmarsch, H., Sandu, G. (eds.) Jaakko Hintikka on Knowledge and Game-Theoretical Semantics. OCL, vol. 12, pp. 499–533. Springer, Cham (2018). https://doi.org/10.1007/978-3-319-62864-6_21
13. Wang, Y.: A logic of goal-directed knowing how. Synthese **195**(10), 4419–4439 (2018)
14. Yang, Y.: Knowledge-WH and false belief sensitivity: a logical study (an extended abstract). In: Verbrugge, R. (ed.) Proceedings TARK 2023, pp. 527—544 (2023). https://doi.org/10.4204/EPTCS.379.40

Finite Model Property in Normal Extensions of Euclidean Quasi-Boolean Modal Logics

Yiheng Wang$^{(\boxtimes)}$ and Yu Peng$^{(\boxtimes)}$

Department of Philosophy, Institute of Logic and Cognition, Sun Yat-sen University,
Guangzhou 510275, China
ianwang747@gmail.com, amberlogos@163.com

Abstract. In this paper, we show that every normal extension of Euclidean quasi-Boolean modal logic has the finite model property. Invariance results regarding disjoint union, generated submodel, and p-morphism are established. Our results generalize and essentially echo the well-known similar result of normal extensions of classical Euclidean modal logic K5 (cf. [17]).

Keywords: quasi-Boolean modal logic · Euclidean modal logic · finite model property · invariance result

1 Introduction

The algebraic counterpart of classical modal logic is Boolean algebras with modal operators (cf. e.g. [4]). Similarly, quasi-Boolean modal logic is based on quasi-Boolean algebra with modal operators. A quasi-Boolean algebra[1] ([3,21], qBa for short) is a structure $(A, \wedge, \vee, \sim, 0, 1)$ where $(A, \wedge, \vee, 0, 1)$ is a bounded distributive lattice and $\sim$ is a De Morgan negation on A i.e., it satisfies $\sim 1 = 0$, $\sim(a \wedge b) = \sim a \vee \sim b$, and $\sim\sim a = a$. The well-known BelnapDunn four-valued logic is based on the qBa (cf. [2,10,12]). By enriching qBa with modal operators $\Box, \Diamond$ satisfying $\Box 1 = 1$, $\Box(a \wedge b) = \Box a \wedge \Box b$, and $\Box a \wedge \Diamond b \leq \Diamond(a \wedge b)$[2], one obtains its modal extension. Such a structure and its logic form the foundation for certain algebraic and logical frameworks that emerge when rough set theory is applied to knowledge representation (cf. [22,25]). Topics regarding Kripke-completeness, duality between algebraic and relational structures, correspondence theory, etc. related to quasi-Boolean modal logic can be found in the rich literature [5,8,13,16,20].

It is well-known that under classical modal logic, modal axiom 5: $\Diamond p \rightarrow \Box \Diamond p$ corresponds to a frame condition titled *Euclidean property*: $\forall w, u, v \in W(wRu \wedge$

---

[1] Also known as De Morgan algebra.
[2] This axiom is called interaction axiom in the study of positive modal logic (cf. [9]).

Y. Wang and Y. Peng—Both authors contributed equally to this work and should be considered co-first authors.

V. Goranko et al. (Eds.): LORI 2025, LNCS 16010, pp. 165–178, 2026.
https://doi.org/10.1007/978-981-95-2481-5_12

$wRv \to uRv$). Classical Euclidean modal logic has been well-investigated in the literature, especially by Nagle [17–19]. In [17,18], Nagle showed that each normal extension of K5 enjoys finite model property (FMP for short), finitely axiomatizable property, and is thus decidable. Nagle [19] further considered the lattice of extensions of K5, giving results regarding tabularity, post-completeness, and explicit finite axiomatizations. Various other studies about Euclidean modal logic can also be found in the literature [1,24]. On the other hand, modal axiom 5, known as the *negative introspective axiom* in both epistemology and doxastic/epistemic logic [14], is widely discussed in various philosophical discussions related to agents' belief and knowledge (cf. [15,23]).

We continue both research lines, and this paper may be viewed as a direct follow-up of [16,17]. We adopt quasi-Boolean modal logic QK defined in [16] whose relational semantics and Kripke-completeness were established by the canonical model method. The difference between the quasi-Boolean frame and the classical relational frame is that there is an extra involution g (cf. Definition 3) in the former[3]. Invariance results regarding disjoint union, generated submodel, and p-morphism are established. We consider normal extensions of Q5 = QK$\oplus$5. Inspired by methods from [17], we will show that all these extensions enjoy FMP by presenting features of its generated submodels. Lemma 1 showing that the universe of generated submodel can be equivalently obtained by first taking transitive closure and then taking its g-image (cf. Remark 1) is crucial in our FMP proof. Our results essentially generalize and echo what have been established in classical Euclidean modal logic.

This paper is structured as follows. Section 2 provides necessary backgrounds on quasi-Boolean modal logic and its relational semantics. Section 3 presents the invariance results regarding disjoint union, generated submodel and p-morphism. Section 4 presents the main results on the finite model property, detailing the proof techniques. Finally, we give some concluding remarks in Sect. 5.

2   Normal Quasi-Boolean Modal Logics

In this section, we introduce syntactical and semantical preliminaries about normal quasi-Boolean modal logic. We start with the definition of language.

Definition 1. *Let* **Var** $= \{p_i : i < \omega\}$ *be a denumerable set of propositional variables. The* set *of all formulas* $\mathcal{F}$ *is defined inductively as follows:*

$$\mathcal{F} \ni \alpha ::= p \mid \bot \mid \alpha_1 \wedge \alpha_2 \mid \sim\alpha \mid \Box\alpha, \ \text{where} \ p \in \textbf{Var}.$$

We use abbreviations $\top := \sim\bot$, $\alpha_1 \vee \alpha_2 := \sim(\sim\alpha_1 \wedge \sim\alpha_2)$ and $\Diamond\alpha := \sim\Box\sim\alpha$. Formulas in **Var** $\cup \{\bot\}$ are called *atomic*. The *complexity* of any formula α is defined in a usual way. Let $Sub(\alpha)$ be the set of all subformulas of α. For a

---

[3] There are other kinds of definitions of relational frames about quasi-Boolean algebra with operators in the study of duality [6,11]; here we adopt the definition of frame in [16] where QK and Q5 were shown to be Kripke-complete.

set of formulas Γ, let $Sub(\Gamma) = \bigcup_{\alpha \in \Gamma} Sub(\alpha)$, Γ is closed under subformulas if $Sub(\Sigma) = \Sigma$. A *substitution* is an endomorphism $\sigma : \mathcal{F} \to \mathcal{F}$.

A *basic sequent* is an expression of the form $\alpha \Rightarrow \beta$ where $\alpha, \beta \in \mathcal{F}$ and α, β are called the antecedent and the succedent, respectively. Let s, t, etc., denote any basic sequents and $\mathcal{BS}$ the set of all basic sequents. A *basic sequent rule* is an expression of the form as follows:

$$\frac{s_1 \dots s_n}{s_0}(\mathrm{R})$$

where $s_1, \dots, s_n$ are *premisses* and s_0 is the *conclusion* of (R). The sequential quasi-Boolean modal logic is defined as follows:

Definition 2. *A quasi-Boolean modal logic (qBml for short)* L *is a set of basic sequents such that the following conditions hold:*

(1) L *contains all instances of the following axiom schemata:*

$$(\mathrm{ID})\ \alpha \Rightarrow \alpha \quad (\top)\ \alpha \Rightarrow \top \quad (\bot)\ \bot \Rightarrow \alpha \quad (\mathrm{DN_1})\ \alpha \Rightarrow \mathord{\sim}\mathord{\sim}\alpha$$

$$(\mathrm{DN_2})\ \mathord{\sim}\mathord{\sim}\alpha \Rightarrow \alpha \quad (\mathrm{Dis})\ \alpha \wedge (\beta \vee \gamma) \Rightarrow (\alpha \wedge \beta) \vee (\alpha \wedge \gamma)$$

$$(\Box\top)\ \top \Rightarrow \Box\top \quad (\mathrm{K_\Box})\ \Box\alpha \wedge \Box\beta \Rightarrow \Box(\alpha \wedge \beta) \quad (\mathrm{IA})\ \Box\alpha \wedge \Diamond\beta \Rightarrow \Diamond(\alpha \wedge \beta)$$

(2) L *is closed under the following rules: for* $i \in \{1, 2\}$,

$$\frac{\alpha_i \Rightarrow \beta}{\alpha_1 \wedge \alpha_2 \Rightarrow \beta}(\wedge\Rightarrow) \qquad \frac{\alpha \Rightarrow \beta_1 \quad \alpha \Rightarrow \beta_2}{\alpha \Rightarrow \beta_1 \wedge \beta_2}(\Rightarrow\wedge)$$

$$\frac{\mathord{\sim}\alpha_1 \Rightarrow \beta \quad \mathord{\sim}\alpha_2 \Rightarrow \beta}{\mathord{\sim}(\alpha_1 \wedge \alpha_2) \Rightarrow \beta}(\mathord{\sim}\wedge\Rightarrow) \qquad \frac{\alpha \Rightarrow \mathord{\sim}\beta_i}{\alpha \Rightarrow \mathord{\sim}(\beta_1 \wedge \beta_2)}(\Rightarrow\mathord{\sim}\wedge)$$

$$\frac{\alpha \Rightarrow \beta}{\mathord{\sim}\beta \Rightarrow \mathord{\sim}\alpha}(\mathrm{CP}) \qquad \frac{\alpha \Rightarrow \beta}{\Box\alpha \Rightarrow \Box\beta}(\mathrm{M_\Box}) \qquad \frac{\alpha \Rightarrow \chi \quad \chi \Rightarrow \beta}{\alpha \Rightarrow \beta}(\mathrm{Cut})$$

(3) L *is closed under uniform substitution: if* $\alpha \Rightarrow \beta \in \mathsf{L}$*, then* $\sigma(\alpha) \Rightarrow \sigma(\beta) \in \mathsf{L}$ *for any substitution* σ*.*

A basic sequent s is a *theorem* of L or *provable* from L (notation: $\mathsf{L} \vdash s$) if $s \in \mathsf{L}$. We write $\mathsf{L} \vdash \alpha \Leftrightarrow \beta$ if $\mathsf{L} \vdash \alpha \Rightarrow \beta$ and $\mathsf{L} \vdash \beta \Rightarrow \alpha$. Let $\{\mathsf{L}_i : i \in I\}$ be a family of qBmls. Then $\bigcap_{i \in I} \mathsf{L}_i$ is a qBml. Let the minimal qBml be denoted as $\mathsf{QK} = \bigcap\{\mathsf{L} : \mathsf{L} \text{ is a qBml}\}$. For every set of basic sequents $\mathcal{S}$ and qBml L, let $\mathsf{L} \oplus \mathcal{S} = \bigcap\{\mathsf{J} : \mathsf{J} \text{ is a qBml and } \mathsf{L} \cup \mathcal{S} \subseteq \mathsf{J}\}$, i.e., the smallest qBml containing $\mathsf{L} \cup \mathcal{S}$. If $\mathcal{S} = \{s_1, \dots, s_n\}$, we write $\mathsf{L} \oplus s_1 \oplus \dots \oplus s_n$ for $\mathsf{L} \oplus \mathcal{S}$. Let $\mathsf{NExt}(\mathsf{L})$ be the set of all qBmls containing L.

Fact 1 ([16], **Prop. 2.8**). *The following basic sequents are provable in any qBml:*

$$\mathord{\sim}(\alpha \wedge \beta) \Leftrightarrow \mathord{\sim}\alpha \vee \mathord{\sim}\beta \quad \mathord{\sim}(\alpha \vee \beta) \Leftrightarrow \mathord{\sim}\alpha \wedge \mathord{\sim}\beta \quad (\alpha \wedge \beta) \vee (\alpha \wedge \gamma) \Rightarrow \alpha \wedge (\beta \vee \gamma)$$

$$\mathord{\sim}\Box\alpha \Leftrightarrow \Diamond\mathord{\sim}\alpha \quad \mathord{\sim}\Diamond\alpha \Leftrightarrow \Box\mathord{\sim}\alpha \quad \Box(\alpha \wedge \beta) \Rightarrow \Box\alpha \wedge \Box\beta \quad \Box(\alpha \vee \beta) \Rightarrow \Box\alpha \vee \Diamond\beta$$

Next, we introduce the $\underline{\text{frame semantics}}$ for qBmls. We will use common set-theoretic operations: $\cap, \cup, \overline{(\cdot)}$ (complement) and $\mathcal{P}(\cdot)$ (powerset). For $X \subseteq W$, let $w \in W \backslash X$ iff $w \in W$ and $w \notin X$.

Definition 3. *A qBm-frame is a tuple $\mathfrak{F} = (W, g, R)$ where $g : W \to W$ is an involution, i.e., $g(g(w)) = w$ for all $w \in W$ and R is a binary relation on W such that for all $w, u \in W$,*

$$(\dagger) \quad \text{if } wRu, \text{ then } g(w)Rg(u).$$

For every $w \in W$ and $X \subseteq W$, let $R(w) = \{u \in W : wRu\}$, $R(X) = \bigcup_{w \in X} R(w)$, and $g(X) = \{g(w) : w \in X\}$. The operations $\sim_g$ and $\Box_R$ on $\mathcal{P}(W)$ are defined as $\sim_g X = \overline{g(X)}$ and $\Box_R X = \{w \in W : R(w) \subseteq X\}$. Then $\mathfrak{F}^+ = (\mathcal{P}(W), \cap, \cup, \sim_g, \Box_R, \varnothing, W)$ is a quasi-Boolean algebra with modal operator.

Fact 2. *Let $g : W \to W$ be an involution on W. Clearly, g is actually a bijection. For any subsets $X, Y \subseteq W$, the following properties hold: $g(W) = W$, $g(\varnothing) = \varnothing$, $g(X \cap Y) = g(X) \cap g(Y)$, $g(X \cup Y) = g(X) \cup g(Y)$, and $x \in X$ iff $g(x) \in g(X)$.*

Definition 4. *A qBm-model is a tuple $\mathfrak{M} = (\mathfrak{F}, V)$ where $\mathfrak{F} = (W, g, R)$ is a qBm-frame and $V : \mathbf{Var} \to \mathcal{P}(W)$ is a valuation in $\mathfrak{F}$. The satisfaction relation of a formula α at w in $\mathfrak{M}$ (notation: $\mathfrak{M}, w \models \alpha$) is defined as follows:*

$$\mathfrak{M}, w \not\models \bot,$$
$$\mathfrak{M}, w \models p \text{ iff } w \in V(p) \text{ where } p \in \mathbf{Var},$$
$$\mathfrak{M}, w \models \alpha \wedge \beta \text{ iff } \mathfrak{M}, w \models \alpha \text{ and } \mathfrak{M}, w \models \beta,$$
$$\mathfrak{M}, w \models \sim\alpha \text{ iff } \mathfrak{M}, g(w) \not\models \alpha,$$
$$\mathfrak{M}, w \models \Box\alpha \text{ iff for all } u \in R(w), \mathfrak{M}, u \models \alpha.$$

Accordingly, the *truth set* $V(\alpha)$ of α is defined as follows:

$$V(\bot) = \varnothing, \quad V(\sim\alpha) = \sim_g V(\alpha), \quad V(\alpha \wedge \beta) = V(\alpha) \cap V(\beta), \quad V(\Box\alpha) = \Box_R V(\alpha)$$

Clearly $\mathfrak{M}, w \models \alpha$ iff $w \in V(\alpha)$, and $\mathfrak{M} \models \alpha$ iff $V(\alpha) = W$. For any basic sequent $\alpha \Rightarrow \beta$, we write $\mathfrak{M}, w \models \alpha \Rightarrow \beta$ if $w \notin V(\alpha)$ or $w \in V(\beta)$. Particularly, $\mathfrak{M} \models \top \Rightarrow \alpha$ is often abbreviated as $\mathfrak{M} \models \alpha$. Let Σ be a set of formulas. Then $\mathfrak{M} \models \Sigma$ iff $\mathfrak{M} \models \alpha$ for any $\alpha \in \Sigma$. We write $\mathfrak{M} \models s$ if $\mathfrak{M}, w \models s$ for all $w \in W$, and $\mathfrak{M} \models \mathcal{S}$ if $\mathfrak{M} \models s$ for all $s \in \mathcal{S}$.

A basic sequent s is *valid* in a qBm-frame (notation: $\mathfrak{F} \models s$) if $\mathfrak{F}, V \models s$ for every valuation V in $\mathfrak{F}$. Let $\mathcal{K}$ be a class of qBm-frames. A basic sequent s is valid in $\mathcal{K}$ (notation: $\mathcal{K} \models s$) if $\mathfrak{F} \models s$ for all $\mathfrak{F} \in \mathcal{K}$. The sequential theory of $\mathcal{K}$ is defined as the set $Th(\mathcal{K}) = \{s \in \mathcal{BS} : \mathcal{K} \models s\}$. For a set of basic sequents $\mathcal{S}$, we write $\mathfrak{F} \models \mathcal{S}$ if $\mathfrak{F} \models s$ for all $s \in \mathcal{S}$. Let $Fr(\mathcal{S}) = \{\mathfrak{F} : \mathfrak{F} \models \mathcal{S}\}$. A qBml L is *Kripke-complete* if $\mathsf{L} = Th(Fr(\mathsf{L}))$.

The Kripke-completeness of qBmls QK and some of its common extensions are shown in [16] by canonical model method. Here we list results from [16], details are omitted. Let $\mathfrak{F}^\mathsf{L} = (W^\mathsf{L}, g^\mathsf{L}, R^\mathsf{L})$ be the canonical frame for L and

$\mathfrak{M}^{\mathsf{L}} = (\mathfrak{F}^{\mathsf{L}}, V^{\mathsf{L}})$ be the canonical model. A qBml L is *canonical* if $\mathfrak{F}^{\mathsf{L}} \models \mathsf{L}$. A set of basic sequents $\mathcal{S}$ is *canonical* if for every qBml L, if $\mathcal{S} \subseteq \mathsf{L}$, then L is canonical. We have the following Canonical Model Theorem:

Theorem 1 (cf. [16], Coro. 3.10). *For any qBml L and any basic sequent s,* $\mathsf{L} \vdash s$ *iff* $\mathfrak{M}^{\mathsf{L}} \models s$.

Corollary 1 ([16], Coro. 3.12). QK *is canonical and hence Kripke-complete.*

Let $\mathfrak{F} = (W, g, R)$ be a qBm-frame. We say $\mathfrak{F}$ is *Euclidean* if for all $w, u, v \in W$, wRu and wRv implies uRv or vRu. It has been shown in [16, Lemma 3.13 and Proposition 3.14] that the modal axiom $5 : \Diamond\alpha \Rightarrow \Box\Diamond\alpha$ is canonical and $\mathfrak{F} \models \Diamond\alpha \Rightarrow \Box\Diamond\alpha$ if and only if R is Euclidean. A qBml L is *Euclidean* if $\mathfrak{F}^{\mathsf{L}} \models 5$. Let $\mathsf{Q5} = \mathsf{QK} \oplus 5$, it is Kripke-complete (cf. [16, Corollary 3.15]).

Fact 3. *Let $\mathfrak{F} = (W, g, R)$ be a qBm-frame. If R is Euclidean, then the following hold: for any $w, u, v, z \in W$,*

    (1) *if wRu, then uRu (shift reflexivity).*
    (2) *if wRu, then uRv implies vRu (shift symmetry).*
    (3) *if wRu, then uRv and vRz imply uRz (shift transitivity).*

3   Invariance Results

In this section, similar to constructing new models from old ones that do not affect modal satisfaction for classical modal logics, we show some common invariance results in terms of disjoint union, generated submodel, and p-morphism for quasi-Boolean modal logics.

Definition 5. (Disjoint Union). *Let $\{\mathfrak{M}_i = (W_i, g_i, R_i, V_i)\}_{i \in I}$ be a set of disjoint qBm-models, i.e., for any $i, j \in I$, if $i \neq j$, then $W_i \cap W_j = \varnothing$. Let $\mathfrak{F}_i = (W_i, g_i, R_i)$ where $i \in I$. $\biguplus_{i \in I} \mathfrak{F}_i := (W, g, R)$ is the* disjoint union *of frames $\{\mathfrak{F}_i\}_{i \in I}$ where $W := \bigcup_{i \in I} W_i$, $g(w) := g_i(w)$ for $w \in W_i$ and $R := \bigcup_{i \in I} R_i$. $\biguplus_{i \in I} \mathfrak{M}_i := (W, g, R, V)$ is the* disjoint union *of models $\{\mathfrak{M}_i\}_{i \in I}$ where (W, g, R) is the disjoint union of frames $\{\mathfrak{F}_i\}_{i \in I}$ and $V(p) := \bigcup_{i \in I} V(p)$ for all $p \in \mathbf{Var}$.*

Proposition 1. *For every formula α, for each $i \in I$, and each $w \in W_i$, $\mathfrak{M}_i, w \models \alpha \Rightarrow \beta$ iff $\biguplus_{i \in I} \mathfrak{M}_i, w \models \alpha \Rightarrow \beta$.*

Proof. It suffices to show that $\mathfrak{M}_i, w \models \beta$ iff $\biguplus_{i \in I} \mathfrak{M}_i, w \models \beta$. The proof proceeds by induction on the complexity of β. We only show the $\sim$ case, others can be checked easily. If $\beta = \sim\phi$. Suppose that $\mathfrak{M}_i, w \models \sim\phi$, i.e., $\mathfrak{M}_i, g(w) \not\models \phi$ by Definition 4. Then by the induction hypothesis, $\biguplus_{i \in I} \mathfrak{M}_i, g(w) \not\models \phi$. Hence $\biguplus_{i \in I} \mathfrak{M}_i, w \models \sim\phi$. Another direction is similar.

The definition of the generated submodel for quasi-Boolean modal logics shall take the involution g of a qBm-model into consideration since the conditions relative to g may affect the satisfiability when taking the generated submodel. We introduce the following definitions.

Definition 6 (Generated Submodel). *Let* $\mathfrak{M} = (W, g, R, V)$ *be a qBm-model. Let* $\mathfrak{F} = (W, g, R)$ *and* $X \subseteq W$. *Then the* submodel $\mathfrak{M}_X = (W_X, g_X, R_X, V_X)$ *of* $\mathfrak{M}$ *generated by* X *is defined as follows:*

(G1) W_X *is constructed inductively: Let* $W_0 = X$; *For any* $n \geq 0$, *let* $W_{n+1} = W_n \cup g(W_n) \cup R(W_n)$. *Then* $W_X = \bigcup_{n \geq 0} W_n$,
(G2) $g_X : W_X \to W_X$ *where* $g_X(w) = g(w)$ *for any* $w \in W_X$,
(G3) $R_X = R \cap (W_X \times W_X)$,
(G4) $V_X(p) = V(p) \cap W_X$ *where* $p \in \mathbf{Var}$.

Let $\mathfrak{F}_X = (W_X, g_X, R_X)$ be the *generated subframe of* $\mathfrak{F}$ *by* X. Particularly, if X is a singleton set $\{w\}$, we write $\mathfrak{M}_w$ $(\mathfrak{F}_w)$ and call $\mathfrak{M}_w$ $(\mathfrak{F}_w)$ the *point-generated submodel (subframe) from* $\mathfrak{M}$ $(\mathfrak{F})$ *by* $\{w\}$.

Let R^+ be the transitive closure of R (also called the *ancestral* of R). Now we show that any qBm-submodel $\mathfrak{M}_X$ generated by $X \subseteq W$ can be equivalently obtained by firstly taking X and its R^+-successors, then incorporating what we call a g-image of it. Lemma 1 will be essential in the FMP proof of Sect. 4.

Lemma 1. *Let* $\mathfrak{M}_X = (W_X, g_X, R_X, V_X)$ *be the generated submodel from* $\mathfrak{M} = (W, g, R, V)$ *by* $X \subseteq W$. *Let* $W' = (R^+(X) \cup X) \cup g(R^+(X) \cup X)$, *then* $W_X = W'$.

Proof. By Definition 6, $X \subseteq W_X$ and $R^+(X) \subseteq W_X$. By Fact 2, $g(R^+(X) \cup X) = g(R^+(X)) \cup g(X)$. Clearly $g(X) \subseteq W_X$. Since $R^+(X) \subseteq W_X$, $g(R^+(X)) \subseteq W_X$. Thus $W' \subseteq W_X$.

Conversely, observe that W_X is actually the smallest set of points that contains X and is closed under g and R i.e. if $w \in W_X$, then $g(w) \in W_X$ and $R(w) \subseteq W_X$. Thus, it suffices to show that W' contains X and is closed under g and R. Clearly $X \subseteq W'$. Suppose that any $w \in W'$. If $w \in R^+(X) \cup X$, then by Fact 2, $g(w) \in g(R^+(X) \cup X) \subseteq W'$ and $R(w) \subseteq R^+(X) \cup X$ since R^+ is the transitive closure of R. If $w \in g(R^+(X) \cup X)$, then $g(w) \in R^+(X) \cup X \subseteq W'$. To show that $R(w)$ is contained in W', we consider the following two cases:

Case (i): Let $w \in g(R^+(X))$. We need to show $R(w) \subseteq W'$. Since $w \in g(R^+(X))$, $R(w) \subseteq R(g(R^+(X)))$. Then it suffices to show that $R(g(R^+(X))) \subseteq g(R^+(X))$. Suppose that any $u \in R(g(R^+(X)))$, then there is a $v \in g(R^+(X))$ such that vRu. Then $g(v) \in R^+(X)$ and by Definition 3 (†), $g(v)Rg(u)$. Thus there is a point $x \in X$ such that $xR^+g(v)$. Then $xR^+g(u)$ and thus $g(u) \in R^+(x) \subseteq R^+(X)$. Then $u \in g(R^+(X)) \subseteq W'$. Therefore, one has $R(w) \subseteq R(g(R^+(X))) \subseteq g(R^+(X)) \subseteq W'$.

Case (ii): Otherwise let $w \in g(X)$, then $R(w) \subseteq R(g(X))$. To prove $R(w) \subseteq W'$, it suffices to show that $R(g(X)) \subseteq g(R^+(X))$. Suppose that any $u \in R(g(X))$. Then there is a point $v \in g(X)$ such that vRu. Then $g(v) \in X$ and by Definition 3 (†), $g(v)Rg(u)$. Thus $g(u) \in R(X)$, and clearly $u \in g(R^+(X))$. Therefore, one has $R(w) \subseteq R(g(X)) \subseteq g(R^+(X)) \subseteq W'$.

Remark 1. Intuitively, Lemma 1 tells us that W_X, the smallest set of points that contains X and is closed under g and R, can be equivalently obtained by firstly taking X with its transitive closure $R^+(X) \cup X$, then taking its g-*image*: $g(R^+(X) \cup X)$. The g-image can be viewed as a kind of copy.

Theorem 2 (Generation Theorem). *For every formula α and each $w \in W_w$, $\mathfrak{M}, w \models \alpha \Rightarrow \beta$ iff $\mathfrak{M}_w, w \models \alpha \Rightarrow \beta$.*

Proof. It suffices to show that $\mathfrak{M}, w \models \beta$ iff $\mathfrak{M}_w, w \models \beta$. The proof proceeds by induction on the complexity of β. We only show the proof of the $\sim$ case. Suppose that $\mathfrak{M}, w \models \sim\phi$, then $\mathfrak{M}, g(w) \not\models \phi$. By Definition 6, $g(w) \in W_w$. Then by the induction hypothesis, $\mathfrak{M}_w, g(w) \not\models \phi$. Hence $\mathfrak{M}_w, w \models \sim\phi$. The inverse direction is quite similar.

Definition 7. *Let $\mathfrak{M}$ and $\mathfrak{M}' = (W', g', R', V')$ be two qBm-models where $\mathfrak{F} = (W, g, R)$ and $\mathfrak{F}' = (W', g', R')$. A mapping $f : W \to W'$ is a p-morphism from $\mathfrak{F}$ to $\mathfrak{F}'$ (notation: $\mathfrak{F} \twoheadrightarrow \mathfrak{F}'$) if it satisfies the following conditions: for any $w, u \in W$,*

 (P1) *f is surjective,*
 (P2) *if wRu, then $f(w)R'f(u)$,*
 (P3) *$f(g(w)) = g'(f(w))$,*
 (P4) *if $f(w)R'u'$, then there is some $u \in W$ such that $f(u) = u'$ and wRu.*

f is a p-morphism from $\mathfrak{M}$ to $\mathfrak{M}'$ (notation: $\mathfrak{M} \twoheadrightarrow \mathfrak{M}'$) if a condition is further satisfied: (P5) *$w \in V(p)$ iff $f(w) \in V'(p)$.*

Theorem 3 (P-morphism Theorem). *Let $\mathfrak{M}, \mathfrak{M}'$ be two qBm-models and $f : \mathfrak{M} \twoheadrightarrow \mathfrak{M}'$. For any $w \in W$, $\mathfrak{M}, w \models \alpha \Rightarrow \beta$ iff $\mathfrak{M}', f(w) \models \alpha \Rightarrow \beta$.*

Proof. It suffices to show that $\mathfrak{M}, w \models \beta$ iff $\mathfrak{M}', f(w) \models \beta$. The proof proceeds by induction on the complexity of β. We only provide the case for $\sim$. Let $\beta = \sim\phi$. Suppose that $\mathfrak{M}, w \models \sim\phi$, then $\mathfrak{M}, g(w) \not\models \phi$. Then by the induction hypothesis, $\mathfrak{M}', f(g(w)) \not\models \phi$. By Definition 7 (P3), $f(g(w)) = g'(f(w))$. Then $\mathfrak{M}', g'(f(w)) \not\models \phi$. Therefore, $\mathfrak{M}', f(w) \models \sim\phi$. Another direction is quite similar.

Corollary 2. *Let $\mathfrak{M}$ and $\mathfrak{M}'$ be two qBm-models and $f : \mathfrak{M} \twoheadrightarrow \mathfrak{M}'$. $\mathfrak{M} \models \Sigma$ iff $\mathfrak{M}' \models \Sigma$ where Σ is a set of basic sequents.*

Corollary 3. *Let $\mathfrak{M}$ be a qBm-model and $\{\mathfrak{M}_w\}_{w \in W}$ a family of point-generated models from $\mathfrak{M}$ by any $w \in W$, then there is a $f : \biguplus_{w \in W} \mathfrak{M}_w \twoheadrightarrow \mathfrak{M}$.*

4   Finite Model Property

In this section, we show that every normal extension of $\mathsf{Q5}$ has finite model property. To obtain FMP of any $\mathsf{L} \in \mathsf{NExt}(\mathsf{Q5})$, it suffices to show that for any non-theorem $\alpha \Rightarrow \beta$ of L, there exists a finite model $\mathfrak{M}^F$ of L that refutes $\alpha \Rightarrow \beta$ i.e., $\mathfrak{M}^F \models \alpha$ and $\mathfrak{M}^F \not\models \beta$. Suppose that $\mathsf{L} \nvdash \alpha \Rightarrow \beta$, by Theorem 1, $\mathfrak{M}^\mathsf{L} \not\models \alpha \Rightarrow \beta$. Then $\mathfrak{M}^\mathsf{L} \models \alpha$ and $\mathfrak{M}^\mathsf{L} \not\models \beta$. For $\mathfrak{M}^\mathsf{L} \not\models \beta$, there is a $w_0 \in W^\mathsf{L}$ such that $\mathfrak{M}^\mathsf{L}, w_0 \not\models \beta$. By Theorem 2, one has $\mathfrak{M}^\mathsf{L}_{w_0}, w_0 \not\models \beta$ where $\mathfrak{M}^\mathsf{L}_{w_0} = (W^\mathsf{L}_{w_0}, g^\mathsf{L}_{w_0}, R^\mathsf{L}_{w_0}, V^\mathsf{L}_{w_0})$ is a submodel of $\mathfrak{M}^\mathsf{L}$ generated by w_0. Then according to the features of $\mathfrak{M}^\mathsf{L}_{w_0}$, we shall obtain its corresponding finite model under different cases. Notice that R^L and $R^\mathsf{L}_{w_0}$ are both Euclidean, since L is a normal extension of $\mathsf{Q5}$ and axiom 5 is canonical.

Point-generated submodels of any normal extensions of Euclidean classical modal logic $\mathsf{K5}$ can be divided into three cases according to the shape of its generated subframe (cf. [17, Section 2]). Here, although we are dealing with quasi-Boolean modal logic rather than classical modal logic, situations are the same when we omit the additional involution g in any qBm-frames. By Lemma 1, we have shown that to obtain a generated submodel, it suffices to generate its R-transitive closure submodel first and then incorporate a so-called g-image to it. Following this observation, when considering $\mathfrak{M}^{\mathsf{L}}_{w_0}$ with the help of Fact 3, one can divide different situations into several cases listed as follows:

Case (A): Suppose that $R^{\mathsf{L}+}(w_0) = \varnothing$. That is, there is no $R^{\mathsf{L}}_{w_0}$-successor of w_0. Then clearly $\{w_0\} \cup R^{\mathsf{L}+}(w_0) = \{w_0\}$ and the $g^{\mathsf{L}}_{w_0}$-image only depends on whether $g^{\mathsf{L}}_{w_0}(w_0)$ is w_0. Hence, one has two subcases and the figures of their frames are depicted in Fig. 1[4]:

$$(A1) : g^{\mathsf{L}}_{w_0}(w_0) = w_0 \quad (A2) : g^{\mathsf{L}}_{w_0}(w_0) \neq w_0$$

(a) Subcase (A1).   (b) Subcase (A2).

Fig. 1. Frames of Case (A).

Now we suppose that $R^{\mathsf{L}+}(w_0) \neq \varnothing$. We shall consider the following two cases about whether $w_0 \in R^{\mathsf{L}}_{w_0}(w_0)$.

Case (B): Suppose that $w_0 \in R^{\mathsf{L}}_{w_0}(w_0)$ i.e. w_0 is reflexive. Then $\{w_0\} \cup R^{\mathsf{L}+}(w_0) = R^{\mathsf{L}+}(w_0)$. Then the relation $R^{\mathsf{L}}_{w_0}$ restricted to $R^{\mathsf{L}+}(w_0)$ i.e. $R^{\mathsf{L}}_{w_0} {\restriction} R^{\mathsf{L}+}(w_0)$ is a total relation by Fact 3. Then by Definition 3 (†), the $g^{\mathsf{L}}_{w_0}$-image also has a total relation. Now further suppose that there is a $w \in R^{\mathsf{L}+}(w_0)$ such that $g^{\mathsf{L}}_{w_0}(w) = w$, then w is both in $R^{\mathsf{L}+}(w_0)$ and its $g^{\mathsf{L}}_{w_0}$-image. Since $w R^{\mathsf{L}}_{w_0} w$, it is not hard to see that the entire $R^{\mathsf{L}}_{w_0}$ will be a total relation. Otherwise, the subframe $\mathfrak{F}^{\mathsf{L}}_{w_0}$ would consist of two $R^{\mathsf{L}}_{w_0}$-disjoint parts i.e. $R^{\mathsf{L}+}(w_0)$ and its $g^{\mathsf{L}}_{w_0}$-image, with each of their relation restricted to their parts is a total relation, but no $R^{\mathsf{L}}_{w_0}$-relation from any points of each other. We denote these two

---

[4] We use a rectangular box in following figures to indicate what is within it is a frame. $R^{\mathsf{L}}_{w_0}$ and $g^{\mathsf{L}}_{w_0}$ are represented as solid line and dashed line, respectively. ○ or ● are used to denote reflexive or irreflexive points, respectively.

different subcases as (B1) and (B2), respectively. Figure 2 depicts their frames as follows[5]:

$$(B1) : \exists w \in R^{\mathsf{L}+}(w_0) \text{ s.t. } g^{\mathsf{L}}_{w_0}(w) = w \qquad (B2) : \forall w \in R^{\mathsf{L}+}(w_0) \text{ s.t. } g^{\mathsf{L}}_{w_0}(w) \neq w$$

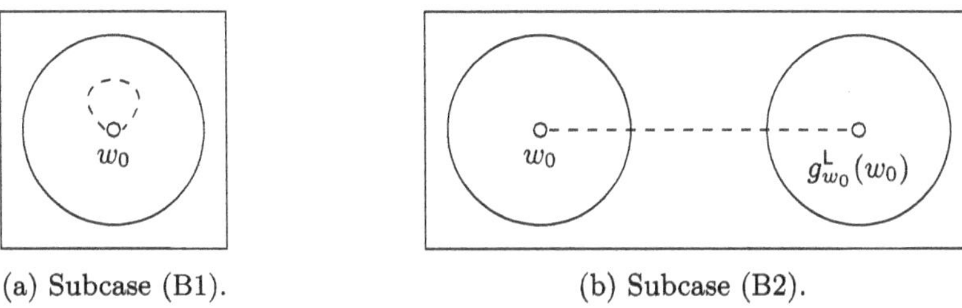

(a) Subcase (B1).　　　　　　　(b) Subcase (B2).

Fig. 2. Frames of Case (B).

Case (C): Suppose that $w_0 \notin R^{\mathsf{L}}_{w_0}(w_0)$ i.e. w_0 is irreflexive. It is clear that for $\{w_0\} \cup R^{\mathsf{L}+}(w_0)$ and its $g^{\mathsf{L}}_{w_0}$-image, the relation restricted to them is a total relation. Now, depending on whether $g^{\mathsf{L}}_{w_0}(w_0)$ is w_0 and whether $\exists w \in R^{\mathsf{L}+}(w_0)$ such that $g^{\mathsf{L}}_{w_0}(w) = w$, Case (C) can be divided into three subcases. First, let $g^{\mathsf{L}}_{w_0}(w_0) = w_0$. Then for any $w_1 \in R^{\mathsf{L}}_{w_0}(w_0)$ and $w_2 \in R^{\mathsf{L}}_{w_0}(g^{\mathsf{L}}_{w_0}(w_0))$, one has $w_1 R^{\mathsf{L}}_{w_0} w_2$ and $w_2 R^{\mathsf{L}}_{w_0} w_1$ by the Euclidean property. Thus $R^{\mathsf{L}}_{w_0}(w_0)$ restricted to $W^{\mathsf{L}}_{w_0}(w_0) \backslash \{w_0\}$ is a total relation. Now let $g^{\mathsf{L}}_{w_0}(w_0) \neq w_0$. Second, suppose that there is a $w \in R^{\mathsf{L}+}(w_0)$ such that $g^{\mathsf{L}}_{w_0}(w) = w$. By an argument similar to subcase (B1), one has $R^{\mathsf{L}}_{w_0}(w_0)$ restricted to $W^{\mathsf{L}}_{w_0}(w_0) \backslash \{w_0, g^{\mathsf{L}}_{w_0}(w_0)\}$ is a total relation. Third, further suppose that for any $w \in R^{\mathsf{L}+}(w_0)$, $g^{\mathsf{L}}_{w_0}(w) \neq w$. Then, by an argument similar to subcase (B2), one has $R^{\mathsf{L}+}(w_0)$ and its $g^{\mathsf{L}}_{w_0}$-image, with each of their relation restricted to themselves is a total relation, but no $R^{\mathsf{L}}_{w_0}$-relation from any points of each other. We denote these three different subcases as (C1), (C2), and (C3), respectively. Figure 3 depicts their frames as follows[6]:

$$(C1) : g^{\mathsf{L}}_{w_0}(w_0) = w_0 \qquad (C2) : g^{\mathsf{L}}_{w_0}(w_0) \neq w_0 \ \& \ \exists w \in R^{\mathsf{L}+}(w_0) \text{ s.t. } g^{\mathsf{L}}_{w_0}(w) = w$$

$$(C3) : g^{\mathsf{L}}_{w_0}(w_0) \neq w_0 \ \& \ \forall w \in R^{\mathsf{L}+}(w_0) \text{ s.t. } g^{\mathsf{L}}_{w_0}(w) \neq w$$

Remark 2. It should be noted that Cases (A), (B), and (C) are similar to Cases (1), (2a), and (2b) of [17, Section 2].

Before we enter into the FMP proofs of any normal extensions of Q5, let us define some notions that will be used throughout the proof. Let Ψ be a finite set

---

[5] We use big circles to denote that it is a frame with a total relation.

[6] Note that w_0 may have many $R^{\mathsf{L}}_{w_0}$-successor in Case (C), but here we use only one solid line to indicate that fact for the sake of neatness.

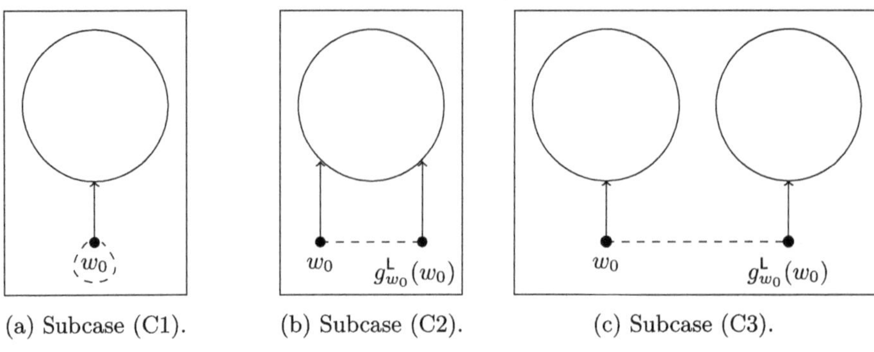

(a) Subcase (C1).   (b) Subcase (C2).   (c) Subcase (C3).

Fig. 3. Frames of Case (C).

of formulas closed under subformulas. A binary relation $\approx_\Psi$ on W induced by Ψ can be defined as follows:

$$w \approx_\Psi u \text{ if and only if } \forall \alpha \in \Psi, \ \mathfrak{M}, w \models \alpha \text{ if and only if } \mathfrak{M}, u \models \alpha$$

Clearly, $\approx_\Psi$ is an equivalence relation. The equivalence class of any $w \in W$ is denoted by $[w]_\Psi$, and $[W]_\Psi = \{[w]_\Psi : w \in W\}$[7].

Lemma 2. *Let Ψ be a finite set of formulas closed under subformulas. For any* $\mathsf{L} \in \mathsf{NExt}(\mathsf{Q5})$, *let* $\mathfrak{M}^\mathsf{L}_{w_0}$ *be the submodel generated from the canonical model of* L *by w_0. Then, there is a finite model* $\mathfrak{M}^F = (W^F, g^F, R^F, V^F)$ *of* L *such that* $\mathfrak{M}^F$ *and* $\mathfrak{M}^\mathsf{L}_{w_0}$ *are equivalent modulo Ψ.*

Proof. According to previous discussions, we prove this claim holds case by case.

Case (A): $R^{\mathsf{L}+}(w_0) = R^{\mathsf{L}+}_{w_0}(w_0) = \varnothing$. Clearly, $\mathfrak{M}^\mathsf{L}_{w_0} = \{w_0\}$ or $\{w_0, g^\mathsf{L}_{w_0}(w_0)\}$ is the required $\mathfrak{M}^F$.

Case (B): $w_0 \in R^\mathsf{L}_{w_0}(w_0)$. We shall consider subcases (B1) and (B2) separately.

Subcase (B1): We define the corresponding $\mathfrak{M}^F$ as follows:

$$W^F = [W^\mathsf{L}_{w_0}] \quad R^F = \{([w], [u]) : w, u \in W^\mathsf{L}_{w_0}\} \quad g^F([w]) = [g^\mathsf{L}_{w_0}(w)]$$

$$V^F(p) = \begin{cases} \{[w] : w \in W^\mathsf{L}_{w_0}\}, & \text{if } p \in \Psi, \\ \varnothing, & \text{otherwise.} \end{cases}$$

We further claim that the function $f : W^\mathsf{L}_{w_0} \to W^F$ defined as $f(w) = [w]$ is a p-morphism from $\mathfrak{M}^\mathsf{L}_{w_0}$ to $\mathfrak{M}^F$.

For (P1), clearly f is surjective. For (P2), it holds because R^F is a total relation and the fact that $R^\mathsf{L}_{w_0}$ is also a total relation in subcase (B1). For (P3), $f(g^\mathsf{L}_{w_0}(w)) = [g^\mathsf{L}_{w_0}(w)] = g^F([w]) = g^F(f(w))$. For (P4), suppose $[w] R^F u'$. Then

---

[7] The subscripts $_\Psi$ in $\approx_\Psi$, $[w]_\Psi$, and $[W]_\Psi$ are omitted if it is clear from the context.

$u' = [u]$ for some $u \in W^{\mathsf{L}}_{w_0}$. Clearly $f(u) = [u] = u'$ and $wR^{\mathsf{L}}_{w_0}u$ since $R^{\mathsf{L}}_{w_0}$ is a total relation. For (P5), it holds by the definition of $V^F(p)$. Thus by Definition 7, f is a p-morphism from $\mathfrak{M}^{\mathsf{L}}_{w_0}$ to $\mathfrak{M}^F$. Further by Theorem 3, for any $w \in W^{\mathsf{L}}_{w_0}$ and $\alpha \in \Psi$, $\mathfrak{M}^{\mathsf{L}}_{w_0}, w \models \alpha$ iff $\mathfrak{M}^F, f(w) \models \alpha$. Thus, under subcase (B1), $\mathfrak{M}^F$ and $\mathfrak{M}^{\mathsf{L}}_{w_0}$ are equivalent modulo Ψ.

Subcase (B2): The corresponding $\mathfrak{M}^F$ is defined the same as in subcase (B1) except for $R^F = \{([w], [u]) : \exists w' \in [w], \exists u' \in [u] \text{ s.t. } w'R^{\mathsf{L}}_{w_0}u'\}$. For $f : W^{\mathsf{L}}_{w_0} \to W^F$ defined as $f(w) = [w]$, (P1), (P3), and (P5) still hold. For (P2), it holds by our definition of R^F. For (P4), suppose that $[w]R^F u'$, then without loss of generality, let $u' = [u]$ for some point $u \in W^{\mathsf{L}}_{w_0}$ and thus $[w]R^F[u]$. Then by the definition of R^F, there are $x \in [w]$ and $y \in [u]$ such that $xR^{\mathsf{L}}_{w_0}y$. Since $f(x) = [w]$ and $f(y) = [u]$, (P4) holds. Then function $f : W^{\mathsf{L}}_{w_0} \to W^F$ defined as $f(w) = [w]$ is a p-morphism from $\mathfrak{M}^{\mathsf{L}}_{w_0}$ to $\mathfrak{M}^F$. By Theorem 3, under subcase (B2), $\mathfrak{M}^F$ and $\mathfrak{M}^{\mathsf{L}}_{w_0}$ are equivalent modulo Ψ.

Case (C): $w_0 \notin R^{\mathsf{L}}_{w_0}(w_0)$. We shall consider subcases (C1), (C2), and (C3) separately.

Subcase (C1): $g^{\mathsf{L}}_{w_0}(w_0) = w_0$. We define the corresponding $\mathfrak{M}^F$ as follows:

$$W^F = \{w_0\} \cup [W^{\mathsf{L}}_{w_0}\setminus\{w_0\}] \quad g^F([w]) = [g^{\mathsf{L}}_{w_0}(w)] \quad R^F = R^F_1 \cup R^F_2$$

$$R^F_1 = \{([w], [u]) : w, u \in W^{\mathsf{L}}_{w_0}\setminus\{w_0\}\} \quad R^F_2 = \{(w_0, [w]) : w_0 R^{\mathsf{L}}_{w_0}w' \text{ for some } w' \in [w]\}$$

$$V^F(p) = \begin{cases} \{[w] : w \in W^{\mathsf{L}}_{w_0}\setminus\{w_0\}\} \cup \{w_0\}, & \text{if } w_0 \in V^{\mathsf{L}}_{w_0}(p) \ \& \ p \in \Psi, \\ \{[w] : w \in W^{\mathsf{L}}_{w_0}\setminus\{w_0\}\}, & \text{if } w_0 \notin V^{\mathsf{L}}_{w_0}(p) \ \& \ p \in \Psi, \\ \varnothing, & \text{otherwise.} \end{cases}$$

A function $f : W^{\mathsf{L}}_{w_0} \to W^F$ is defined as follows:

$$f(w) = \begin{cases} [w], & \text{if } w \in W^{\mathsf{L}}_{w_0}\setminus\{w_0\}, \\ w_0, & \text{otherwise.} \end{cases}$$

We claim that f is a p-morphism from $\mathfrak{M}^{\mathsf{L}}_{w_0}$ to $\mathfrak{M}^F$. It is not hard to see (P1), (P2), (P3), and (P5) hold. For (P4), suppose that $f(w)R^F u'$. Let $f(w) = [w]$, then $u' = [u]$ for some $u \in W^{\mathsf{L}}_{w_0}\setminus\{w_0\}$ since otherwise one has $w_0 R^{\mathsf{L}}_{w_0}w_0$ by the definition of R^F_2, which contradicts to $w_0 \notin R^{\mathsf{L}}_{w_0}(w_0)$. Thus, similar to the proof of subcase (B1), (P4) holds. Let $f(w) \neq [w]$ i.e., $f(w) = w_0 = w$. Then $u' = [u]$ for some $u \in W^{\mathsf{L}}_{w_0}\setminus\{w_0\}$ and $w_0 R^F[u]$. By the definition of R^F_2, there is a $v \in [u]$ such that $w_0 R^{\mathsf{L}}_{w_0}v$ and $f(v) = [u] = u'$. Then (P4) holds. By Theorem 3, under subcase (C1), $\mathfrak{M}^F$ and $\mathfrak{M}^{\mathsf{L}}_{w_0}$ are equivalent modulo Ψ.

Subcase (C2): $g^{\mathsf{L}}_{w_0}(w_0) \neq w_0$ and $\exists w \in R^{\mathsf{L}+}(w_0)$ s.t. $g^{\mathsf{L}}_{w_0}(w) = w$. The required $\mathfrak{M}^F$ can be obtained by replacing the corresponding definition under subcase (C1) with the followings:

$$W^F = \{w_0, g^{\mathsf{L}}_{w_0}(w_0)\} \cup [W^{\mathsf{L}}_{w_0}\setminus\{w_0\}] \quad R^F = R^F_1 \cup R^F_2 \cup R^F_3$$

$$R^F_3 = \{(g^{\mathsf{L}}_{w_0}(w_0), [w]) : g^{\mathsf{L}}_{w_0}(w_0)R^{\mathsf{L}}_{w_0}w' \text{ for some } w' \in [w]\}$$

Let $V' = \{[w] : w \in W^{\mathsf{L}}_{w_0} \setminus \{w_0, g^{\mathsf{L}}_{w_0}(w_0)\}\}$, $V^F(p)$ is defined as follows:

$$V^F(p) = \begin{cases} V' \cup \{w_0, g^{\mathsf{L}}_{w_0}(w_0)\}, & \text{if } w_0, \in V^{\mathsf{L}}_{w_0}(p) \ \& \ g^{\mathsf{L}}_{w_0}(w_0) \in V^{\mathsf{L}}_{w_0}(p) \ \& \ p \in \Psi, \\ V' \cup \{w_0\}, & \text{if } w_0, \in V^{\mathsf{L}}_{w_0}(p) \ \& \ g^{\mathsf{L}}_{w_0}(w_0) \notin V^{\mathsf{L}}_{w_0}(p) \ \& \ p \in \Psi, \\ V' \cup \{g^{\mathsf{L}}_{w_0}(w_0)\}, & \text{if } w_0, \notin V^{\mathsf{L}}_{w_0}(p) \ \& \ g^{\mathsf{L}}_{w_0}(w_0) \in V^{\mathsf{L}}_{w_0}(p) \ \& \ p \in \Psi, \\ V', & \text{if } w_0 \notin V^{\mathsf{L}}_{w_0}(p) \ \& \ g^{\mathsf{L}}_{w_0}(w_0) \notin V^{\mathsf{L}}_{w_0}(p) \ \& \ p \in \Psi, \\ \varnothing, & \text{otherwise.} \end{cases}$$

A function $f : W^{\mathsf{L}}_{w_0} \to W^F$ is defined as follows:

$$f(w) = \begin{cases} [w], & \text{if } w \in W^{\mathsf{L}}_{w_0} \setminus \{w_0, g^{\mathsf{L}}_{w_0}(w_0)\}, \\ w_0, & \text{if } w = w_0, \\ g^{\mathsf{L}}_{w_0}(w_0), & \text{if } w = g^{\mathsf{L}}_{w_0}(w_0). \end{cases}$$

The proof of f is a p-morphism from $\mathfrak{M}^{\mathsf{L}}_{w_0}$ to $\mathfrak{M}^F$ is similar to the proof in subcase (C1). By Theorem 3, under subcase (C2), $\mathfrak{M}^F$ and $\mathfrak{M}^{\mathsf{L}}_{w_0}$ are equivalent modulo Ψ.

Subcase (C3): $g^{\mathsf{L}}_{w_0}(w_0) \neq w_0$ and $\forall w \in R^{\mathsf{L}+}(w_0)$ s.t. $g^{\mathsf{L}}_{w_0}(w) \neq w$. The corresponding $\mathfrak{M}^F$ is defined as follows:

$$W^F = [R^{\mathsf{L}+}_{w_0}(w_0)] \cup [R^{\mathsf{L}+}_{w_0}(g^{\mathsf{L}}_{w_0}(w_0))] \cup \{w_0, g^{\mathsf{L}}_{w_0}(w_0)\} \quad R^F = R^F_1 \cup R^F_2 \cup R^F_3$$

$$R^F_1 = \{([w], [u]) : \exists w' \in [w], \exists u' \in [u] \text{ s.t. } w' R^{\mathsf{L}}_{w_0} u'\}$$

$$R^F_2 = \{(w_0, [w]) : w_0 R^{\mathsf{L}}_{w_0} w' \text{ for some } w' \in [w]\}$$

$$R^F_3 = \{(g^{\mathsf{L}}_{w_0}(w_0), [w]) : g^{\mathsf{L}}_{w_0}(w_0) R^{\mathsf{L}}_{w_0} w' \text{ for some } w' \in [w]\}$$

The $g^F([w])$, V^F, and function $f : W^{\mathsf{L}}_{w_0} \to W^F$ are defined the same as in subcase (C2). The proof of f is a p-morphism from $\mathfrak{M}^{\mathsf{L}}_{w_0}$ to $\mathfrak{M}^F$ is similar to proofs in subcases (B1) and (C1). By Theorem 3, under subcase (C3), $\mathfrak{M}^F$ and $\mathfrak{M}^{\mathsf{L}}_{w_0}$ are equivalent modulo Ψ.

$\mathfrak{M}^F$ are clearly finite in case (A). In cases (B) and (C), observe that since Ψ is a finite set of formulas closed under subformulas, all $\mathfrak{M}^F$ are finite models. Therefore, in all cases, $\mathfrak{M}^F$ and $\mathfrak{M}^{\mathsf{L}}_{w_0}$ are equivalent modulo Ψ.

Theorem 4 (FMP). *For any normal extension* $\mathsf{L} \in \mathsf{NExt}(\mathbf{Q5})$, *if* $\mathsf{L} \nvdash \alpha \Rightarrow \beta$, *then there is a finite model* $\mathfrak{M}^F$ *of* L *such that* $\mathfrak{M}^F \nvDash \alpha \Rightarrow \beta$.

Proof. It suffices to show that there is a finite model of L such that $\mathfrak{M}^F \vDash \alpha$ and $\mathfrak{M}^F \nvDash \beta$. Suppose that $\mathsf{L} \nvdash \alpha \Rightarrow \beta$. By Theorem 1, $\mathfrak{M}^{\mathsf{L}} \nvDash \alpha \Rightarrow \beta$. Then $\mathfrak{M}^{\mathsf{L}} \vDash \alpha$ and $\mathfrak{M}^{\mathsf{L}} \nvDash \beta$, and there is a $w_0 \in W^{\mathsf{L}}$ such that $\mathfrak{M}^{\mathsf{L}}, w_0 \vDash \alpha$ and $\mathfrak{M}^{\mathsf{L}}, w_0 \nvDash \beta$. By Theorem 2, $\mathfrak{M}^{\mathsf{L}}_{w_0}, w_0 \vDash \alpha$ and $\mathfrak{M}^{\mathsf{L}}_{w_0}, w_0 \nvDash \beta$. Let Ψ be the set of all subformulas of α and β. By Lemma 2, there is a finite model $\mathfrak{M}^F$ such that $\mathfrak{M}^F$ and $\mathfrak{M}^{\mathsf{L}}_{w_0}$ are equivalent modulo Ψ. Thus $\mathfrak{M}^F \vDash \alpha$, $\mathfrak{M}^F \nvDash \beta$ and $\mathfrak{M}^F \nvDash \alpha \Rightarrow \beta$. By Corollary 2, $\mathfrak{M}^F \vDash \mathsf{L}$[8]. This completes the proof.

---

[8] Note that p-morphism is used under all subcases in the proof of Lemma 2 except for (A1) and (A2). Nevertheless, clearly $\mathfrak{M}^F \vDash \mathsf{L}$ under subcases (A1) and (A2).

5   Concluding Remarks

In this paper, we show the FMP for any normal extensions of Euclidean normal quasi-Boolean modal logics, which echoes the well-known similar result under classical modal logic in [17]. Invariance results of disjoint union, generated submodel, and p-morphism are also established. One immediate future work is the finitely axiomatizable property (FAP for short) of all these extensions, as what has been done in [17] under classical modal logic. However, it should be noted that following the method of [17], the proof for FAP relies on the equivalence between finite frame property and finite model property. Although this is a well-known result in classical modal logic (cf. e.g. [7, pp. 252–255]), things are a bit different under the quasi-Boolean setting. In the literature (cf. [16, Theorem 4.8]), it has only been shown that qBmls that are involutive-closed (cf. [16, Definition 4.1]) enjoy such equivalence. Thus one cannot directly apply the method in [17] to obtain FAP for each normal extension and thus decidability. Nevertheless, this challenge is left to be addressed in the future.

Acknowledgments. Thanks are given to anonymous reviewers for their helpful comments on this paper. This work of both authors was supported by the Chinese National Funding of Social Sciences (Grant no. 18ZDA033).

Disclosure of Interests. The authors have no competing interests to declare that are relevant to the content of this article.

References

1. Alizadeh, M., Ardeshir, M., Balbiani, P., Mojtahedi, M.: Unification types in Euclidean modal logics. Logic J. IGPL **31**(3), 422–440 (2023)
2. Belnap, N.: A useful four-valued logic. In: Dunn, M., Epstein, G. (eds.) Modern Uses of Multiple-valued Logic, pp. 5–37. Springer (1977)
3. Białynicki-Birula, A., Rasiowa, H.: On the representation of quasi-Boolean algebras. Bull. Polish Acad. Sci. **Cl. III 5**(3), 259–261 (1957)
4. Blackburn, P., Rijke, M., Venema, Y.: Modal Logic, vol. 53. Cambridge University Press, Cambridge, UK (2001)
5. Celani, S.A.: Classical modal De Morgan algebras. Stud. Logica. **98**(1), 251–266 (2011)
6. Celani, S.A.: Lógicas modales distributivas y de De Morgan. Ph.D. thesis, Universidad de Barcelona (1995)
7. Chagrov, A., Zakharyaschev, M.: Modal Logic. Clarendon Press (1997)
8. Dunn, J.M.: A relational representation of quasi-Boolean algebras. Notre Dame J. Formal Logic **23**(4), 353–357 (1982)
9. Dunn, J.M.: Positive modal logic. Stud. Logica. **55**(2), 301–317 (1995)
10. Dunn, J.M.: The algebra of intensional logics. Ph.D. thesis, University of Pittsburgh (1966)
11. Dzik, W., Orlowska, E., Alten, C.: Relational representation theorems for general lattices with negations. In: Schmidt, R.A. (ed.) RelMiCS 2006. LNCS, vol. 4136, pp. 162–176. Springer, Heidelberg (2006). https://doi.org/10.1007/11828563_11

12. Font, J.M.: Belnap's four-valued logic and De Morgan lattices. Log. J. IGPL **5**(3), 1–29 (1997)
13. Gregori, V.: Discrete duality for De Morgan algebras with operators. Asian-Eur. J. Math. **12**(01), 1950010 (2019)
14. Hintikka, J.: Knowledge and Belief: An Introduction to the Logic of the Two Notions. Cornell University Press (1962)
15. Lenzen, W.: Recent work in epistemic logic. Acta Philosophica Fennica **30** (1978)
16. Ma, M., Guo, J.: Kripke-completeness and sequent calculus for quasi-Boolean modal logic. Studia Logica 1–30 (2024)
17. Nagle, M.C.: The decidability of normal K5 logics. J. Symbolic Logic **46**(2), 319–328 (1981)
18. Nagle, M.C.: Normal K5 Logics. Ph.D. thesis, University of Calgary (1981)
19. Nagle, M.C., Thomason, S.K.: The extensions of the modal logic K5. J. Symbolic Logic **50**(1), 102–109 (1985)
20. Odintsov, S.P., Wansing, H.: Modal logics with Belnapian truth values. J. Appl. Non-Classical Logics **20**(3), 279–301 (2010)
21. Rasiowa, H.: An Algebraic Approach to Non-classical Logics, Studies in logic and the foundations of mathematics, vol. 78. PWN-Polish Scientific Publishers, Warszawa (1974)
22. Saha, A., Sen, J., Chakraborty, M.K.: Algebraic structures in the vicinity of pre-rough algebra and their logics I. Inf. Sci. **282**, 296–320 (2014)
23. San, W.K.: KK, knowledge, knowability. Mind **132**(527), 605–630 (2023)
24. Shehtman, V.: On Kripke completeness of modal predicate logics around quantified K5. Ann. Pure Appl. Logic **174**(2), 103202 (2023)
25. Wang, Y., Lin, Z., Ma, M.: Decidability of topological quasi-Boolean algebras. J. Appl. Non-Classical Logics **34**(2–3), 269–293 (2024)

Hypothesis-Driven Disjunctive Reasoning in Logical Argumentation

Zheng Zhou[1]([✉]), Christian Straßer[2]([✉]), and Kees van Berkel[3]([✉])

[1] School of Philosophy, Beijing Normal University, Beijing, China
zhouzhenglogic@mail.bnu.edu.cn
[2] Institute for Philosophy II, Ruhr University Bochum, Bochum, Germany
christian.strasser@rub.de
[3] Institute of Logic and Computation, TU Wien, Vienna, Austria
kees.van.berkel@tuwien.ac.at

Abstract. Disjunctive reasoning, or reasoning by cases, plays a central role in defeasible frameworks and normative systems. This paper adopts a cautious approach, suggesting that rebutting disjunctive paths may render the inference defeasible, particularly in the context of default-based reasoning. To formalize this, we introduce explicit hypotheses to track reasoning paths. Building on prior work, we refine a class of Argument Calculi into a robust method for disjunctive reasoning in nonmonotonic logics, for both normative and doxastic interpretations of our system. We propose meta-theoretical properties tailored to disjunctive reasoning: disjunctive consistency and hypothetical consistency, which constrain argument behavior involving disjunction. By systematically comparing Argument Calculi against several properties, we analyze their implications for consistency and nonmonotonic inference in default-based frameworks.

Keywords: logical argumentation · sequent calculi · normative systems · hypothetical reasoning · disjunctive reasoning · meta-theoretical properties

1 Introduction

A central challenge in knowledge representation lies in managing incomplete and inconsistent information. This has resulted in a wide variety of defeasible reasoning systems with nonmonotonic (nm) inference relations. Formal argumentation, pioneered by Dung [14], offers a conflict-explicit framework that unifies diverse nm approaches (see [2] for an overview) and enhances explainability in reasoning [12]. In particular, [7] establishes proof-theoretic bridges between nm Input/Output (I/O) logic [20] – a formalism widely used for deontic reasoning and normative systems – and formal argumentation. More recently, this framework has been extended [8] to argumentatively characterize Default Logic [22], highlighting some problematic disjunctive reasoning scenarios for these formalisms.

Disjunction exhibits distinct properties across various logical systems, often deviating from classical logic due to divergent foundational assumptions (see [1]

V. Goranko et al. (Eds.): LORI 2025, LNCS 16010, pp. 179–194, 2026.
https://doi.org/10.1007/978-981-95-2481-5_13

for a comprehensive overview). Disjunctive reasoning has been thoroughly examined for some nm logics. For instance, [18] explores an "Or" rule and Disjunctive Rationality in preferential models. A similar "Or" rule also appears in Input/Output Logic [20] and [15] proposes Disjunctive Default Logic (DDL). Still, these systems remain problematically susceptible to challenging reasoning scenarios [5,8]. In this paper, we will also identify several additional challenging scenarios. To accommodate these challenges, we develop a hypothesis-driven approach to disjunctive reasoning, introducing the concept of hypothesis within the framework of logical argumentation.

Our contributions are conceptual and technical. Conceptually, we propose a novel approach to disjunctive reasoning in nm logics, featuring (i) a hypothesis-based formalism extending the class of Argument Calculi (AC) from [7,8], and (ii) a refined understanding of disjunction's role in defeasible inference. Our framework explicitly tracks disjunctive reasoning paths, offering insights into how disjunctive reasoning interacts with default rules. Technically, we introduce a specialized disjunction rule, a hypothesis-driven attack relation, and novel meta-theoretical properties for disjunctive reasoning: i.e., disjunctive consistency and hypothetical consistency, which constrain argument behavior involving disjunction. This modular approach enriches AC and systematizes disjunctive reasoning in defeasible systems and logical argumentation.

Outline: Sect. 2 presents examples, highlighting disjunctive limitations of I/O logic. Section 3 reviews AC and introduces AC-induced argumentation frameworks. Section 4 proposes a hypothesis-driven framework extending AC with explicit hypotheses and a refined disjunctive rule. In Sect. 5, meta-theoretical properties are used to evaluate consistency and nm entailment. Section 6 discusses related and future work. Proofs are found in the Appendix.

2   Illustrative Running Examples

We introduce two running examples, using the Input/Output (I/O) framework [20]. We take $\mathscr{L}$ to be a full propositional language containing falsum $\bot$. Propositional atoms are denoted by $p, q, \ldots$, formulas by $\varphi, \psi, \ldots$, and $\Gamma, \Delta, \ldots$ represent finite sets of formulas. Let $\mathscr{L}^d = \{(\varphi, \psi) \mid \varphi, \psi \in \mathscr{L}\}$ be a *language of defaults* which can be given a doxastic or a normative reading. Doxastically, we read elements (φ, ψ) from $\mathscr{L}^d$ as "given φ, ψ is plausible/likely/etc." Normatively, $\mathscr{L}^d$ takes (φ, ψ) as a norm expressing "given φ, it is obligatory that ψ". Defaults are denoted by δ and σ (possibly indexed). Last, let $\mathbb{K} = \langle \mathcal{F}, \mathcal{D} \rangle$ be a knowledge base with factual input $\mathcal{F} \subseteq \mathscr{L}$ and a default set $\mathcal{D} \subseteq \mathscr{L}^d$.

Example 1 (Lunch Options, Fig. 1). Consider the following doxastic scenario. For lunch, we usually have pizza or burgers ($\delta_1 : (l, p \vee b)$). If we decide to have pizza, it is likely that we go to Luigi's ($\delta_2 : (p, lu)$). However, if we choose burgers, we usually go to McBurger ($\delta_3 : (b, mc)$). Both places normally only offer cash payment $\delta_4 : (lu, c)$ and $\delta_5 : (mc, c)$. Today is Monday, so McBurger is likely to be closed ($\delta_6 : (m, \neg mc)$). Now, suppose it is Monday m and we go for lunch l. The resulting knowledge base is $\mathbb{K}_1 = \langle \mathcal{F} : \{m, l\}, \mathcal{D} : \{\delta_i \mid 1 \leq i \leq 6\} \rangle$.

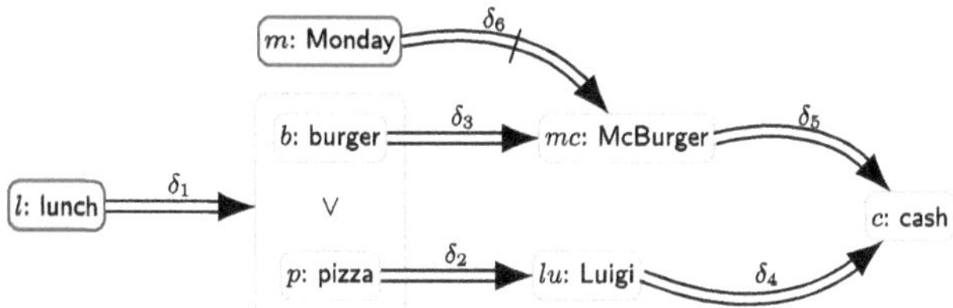

Fig. 1. The lunch options example $\mathbb{K}_1$ (Example 1). Arrows represent defaults, and the arrow with a bar represents $\delta_6 = (m, \neg mc)$. The modules l and m represent input.

Example 2 (Chemical Spill, Fig. 2). Consider a normative scenario: If a chemical spill occurs, you should either call the safety manager or initiate an automatic containment procedure ($\sigma_1 : (c, m \vee p)$). If you initiate the automatic containment procedure, you must seal off the contaminated area ($\sigma_2 : (p, s)$). If you call the safety manager, you must provide a full incident report ($\sigma_3 : (m, r)$). Additionally, if you call the safety manager, you must keep the area open for inspection ($\sigma_5 : (m, o)$). Providing a full incident report requires sealing off the contaminated area ($\sigma_4 : (r, s)$). However, keeping the area open for inspection necessitates not sealing off the contaminated area ($\sigma_6 : (o, \neg s)$). Suppose a chemical spill occurs (c). The knowledge base is $\mathbb{K}_2 = \langle \mathcal{F} : \{c\}, \mathcal{D} : \{\sigma_i \mid 1 \leq i \leq 6\}\rangle$.

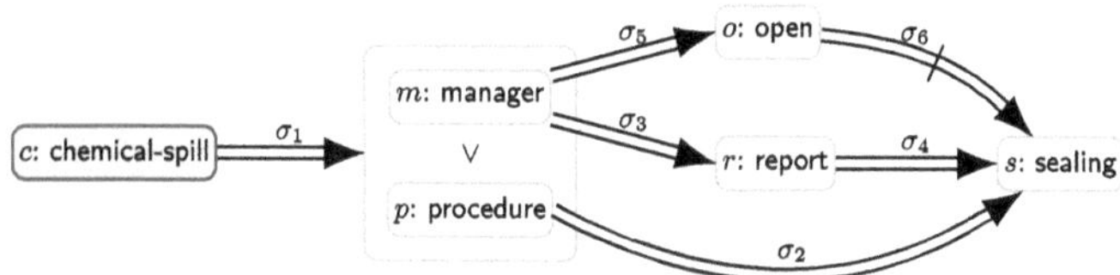

Fig. 2. The chemical spill example $\mathbb{K}_2$ (Example 2). Arrows are read as in Fig. 1.

Let us discuss both examples from an I/O logic perspective. Suppose system deriv_4, a proof theoretic characterizations of I/O logic with disjunctive reasoning "by cases" and successive detachment [20]. Successive detachment allows for the construction of "default chains" (see also Sect. 3). We assume classical consistency as a reasoning constraint [20]. Intuitively, deriv_4 uses rules to derive new defaults from existing ones.[1] We provide an example below but first consider the depiction of Example 1 in Fig. 1: we can use reasoning by cases to construct two reasoning paths leading to conclusion c: the paths $\delta_1-\delta_2-\delta_4$ and $\delta_1-\delta_3-\delta_5$. Through δ_1 these two paths represent reasoning by cases both leading to the conclusion c. The next deriv_4-proof expresses this reasoning (let $\vdash$ be classical inference):

---

[1] In particular, it contains the rules SI $((\varphi, \psi)$, if (φ', ψ) and $\varphi \vdash \varphi')$, WO $((\varphi, \psi)$, if (φ, ψ') and $\psi \vdash \varphi')$, AND $((\varphi, \psi \wedge \psi')$, if (φ, ψ) and $(\varphi, \psi'))$, CT $((\varphi, \psi)$, if (φ, ψ') and $(\psi' \wedge \varphi, \psi))$, and OR $((\varphi \vee \varphi', \psi)$, if (φ, ψ) and $(\varphi', \psi))$. See [20] for an introduction.

$$
\cfrac{
  \delta_1 : (l, p \vee b) \qquad
  \cfrac{
    \cfrac{
      \delta_2 : (p, lu) \quad
      \cfrac{\delta_4 : (lu, c) \quad p \wedge lu \vdash lu}{(p \wedge lu, c)}\ \text{SI}
    }{(p, c)}\ \text{CT}
    \quad
    \cfrac{
      \delta_3 : (b, mc) \quad
      \cfrac{\delta_5 : (mc, c) \quad b \wedge mc \vdash mc}{(b \wedge mc, c)}\ \text{SI}
    }{(b, c)}\ \text{CT}
  }{(p \vee b, c)}\ \textbf{OR}
  \qquad
  \cfrac{l \wedge (p \vee b) \vdash p \vee b}{(l \wedge (p \vee b), c)}\ \text{SI}
}{(l, c)}\ \text{CT}
$$

The branches of the proof starting with δ_3 and δ_5 denote one reasoning path and those starting with δ_2 and δ_4 the other. These branches are utilized with the **OR** rule, through which a new default (l, c) is derived. Then, due to factual input l, we conclude c. In fact, deriv_4 together with logical consistency as a constraint (formally, identifying maximal consistent families of defaults), concludes c skeptically. This is a surprisingly unintuitive outcome since, given the factual input m and the default $(m, \neg mc)$ detaching $\neg mc$, the second reasoning path $\delta_1 - \delta_3 - \delta_5$ using the intermediate conclusion mc, is attacked. If one of the paths in reasoning by cases is compromised, we expect to suspend its conclusion. I/O logics fail to capture this suspension and conclude c instead. To accommodate such cases, we must adopt a cautious perspective on disjunctive default reasoning: an inference from disjunctive reasoning is legitimate only if none of its disjunctive paths is refuted.[2]

Now consider Example 2. Using similar reasoning, deriv_4 derives the norm (c, s) from $\mathbb{K}_2$ and, subsequently, detaches the obligation s since c is factual input. We consider the conclusion overly strong and unintuitive since two distinct disjunctive reasoning paths ($\sigma_1 - \sigma_3 - \sigma_4$ and $\sigma_1 - \sigma_5 - \sigma_6$ in Fig. 6) that are both rooted in the same disjunct m, yield contradictory conclusions: s and $\neg s$. Hence, the disjunctive reasoning is not decisive for s. Ideally, the conclusion s should be cautiously suspended in skeptical reasoning but I/O logic fails to do so. A more sophisticated mechanism is needed for the complexities of disjunctive reasoning.

3   Argument Calculi and Argumentation

Deontic Argument Calculi (DAC) [7] are a family of sequent-style systems adequate for a class of I/O logics. We focus on DAC_4 and DAC_4^+, corresponding to deriv_4 and deriv_4^+ in I/O logic [20].[3] This family was extended to include a Default Calculus (DC) adequate for normal Default Logic [22] and a Disjunctive

---

[2] On the basis of Gricean conversational implicatures [17], one may argue that the decision-problem in Example 1 contextually implies only two options: going to McBurger or Luigi. This reading, by a contrapositive inference of $\neg b$ from $\neg mc$, followed by a disjunctive syllogism, would conclude c. However, the use of contraposition on defeasible rules is highly controversial [10] and such reasoning need not adopt a closed-world assumption: if we extend Example 1 with "If we eat burgers without going to McBurger, we go to KingBurger" $(\neg mc \wedge b, k)$ and "if we go to KingBurger, we do not pay with cash" $(k, \neg c)$, the unintuitive conclusion c remains deriv_4-derivable.

[3] We use these systems since they include disjunctive reasoning and successive detachment. The '+' in deriv_4^+ ensures that all input is also output (**TP** in AC, Definition 1).

Default Calculus (DDC) [8]. Their difference lies in the defeat, disjunction, and detachment rules. We jointly refer to these systems as Argument Calculi (AC).

To provide transparent arguments, AC operates with labeled propositional formulas as (factual) inputs $\mathscr{L}^i$ and outputs $\mathscr{L}^o$.[4] Defaults are also embedded in the object-level language. We adopt negation over defaults through $\overline{\mathscr{L}^d} = \{\neg(\varphi, \psi) \mid (\varphi, \psi) \in \mathscr{L}^d\}$, forming the language $\mathscr{L}^{\mathsf{AC}} = \mathscr{L}^i \cup \mathscr{L}^o \cup \mathscr{L}^d \cup \overline{\mathscr{L}^d}$. The expression $\neg(\varphi, \psi)$ states that "(φ, ψ) is inapplicable." For a default (φ, ψ), we say φ constitutes its body and ψ its head. We write $\Delta^x \subseteq \mathscr{L}^{\mathsf{AC}}$, where $x \in \{i, o, d\}$. We denote classical entailment by $\vdash$ and define $\mathbf{Cn}(\Gamma) = \{\varphi \in \mathscr{L} \mid \Gamma \vdash \varphi\}$ as the deductive closure of Γ. We assume that $\vdash$ has an adequate *sequent calculus* LC, such that $\Delta \vdash \varphi$ holds if and only if $\Delta \Rightarrow \varphi$ is LC-derivable, i.e., $\vdash_{\mathrm{LC}} \Delta \Rightarrow \varphi$. Last, we say $\mathbb{K} = \langle \mathcal{F}, \mathcal{D} \rangle$ is a *labeled* $\mathbb{K}$ whenever $\mathcal{F} \subseteq \mathscr{L}^i$ and we assume that $\mathcal{F}$ is classically consistent, i.e., $\mathcal{F} \not\vdash \bot$.

$$\mathbf{Ax} \ \frac{}{\Gamma^x \Rightarrow \Delta^x} \ \text{where } \vdash_{\mathrm{LC}} \Gamma \Rightarrow \Delta, \text{ and } x \in \{i, o\} \qquad \mathbf{TP} \ \frac{}{\varphi^i \Rightarrow \varphi^o} \qquad \mathbf{Det} \ \frac{}{\varphi^i, (\varphi, \psi) \Rightarrow \psi^o}$$

$$\mathbf{SDet}^{①} \ \frac{\varphi^i, \Gamma \Rightarrow \Xi}{\varphi^o, \Gamma \Rightarrow \Xi} \qquad \mathbf{Cut}^{②} \ \frac{\Gamma_1 \Rightarrow \varphi \qquad \varphi, \Gamma_2 \Rightarrow \Delta}{\Gamma_1, \Gamma_2 \Rightarrow \Delta}$$

$- - - - - - - - - - - - - - - - - -$ **Rules for** $\mathsf{DAC}_4/\mathsf{DAC}_4^+$ $- - - - - - - - - - - - - - - - - -$

$$\mathbf{Def} \ \frac{\Delta, (\varphi, \psi) \Rightarrow}{\Delta \Rightarrow \neg(\varphi, \psi)} \qquad \mathbf{OR}^{③} \ \frac{\Delta, \varphi^i \Rightarrow \Xi \qquad \Delta', \psi^i \Rightarrow \Xi}{\Delta, \Delta', (\varphi \vee \psi)^i \Rightarrow \Xi}$$

$- - - - - - - - - - - - - - - - - -$ **Rules for** DC/DDC $- - - - - - - - - - - - - - - - - -$

$$\mathbf{DDef} \ \frac{}{\varphi^i, \neg\psi^o \Rightarrow \neg(\varphi, \psi)} \qquad \mathbf{DisDet} \ \frac{}{(\bigvee_{i=1}^n \varphi_i)^i, \langle(\varphi_i, \psi_i)\rangle_{i=1}^n \Rightarrow (\bigvee_{i=1}^n \psi_i)^o}$$

$$\mathbf{DisDef} \ \frac{}{(\bigvee_{i=1}^n \varphi_i)^i, \langle(\varphi_i, \psi_i)\rangle_{i=2}^n, \neg\psi_1^o \Rightarrow \neg(\varphi_1, \psi_1)}$$

Fig. 3. Rules of AC (Definition 1). The rules **Ax**, **Det**, **DDef**, **DisDet**, **DisDef**, and **TP** introduce initial sequents. Side-condition ① denotes $\Gamma \cap \mathscr{L}^d \neq \emptyset$; ② denotes $\varphi \in \mathscr{L}^{\mathsf{AC}}$; and ③ denotes if **TP** $\notin$ S, then $\Delta \cap \mathscr{L}^d \neq \emptyset$ and $\Delta' \cap \mathscr{L}^d \neq \emptyset$.

Definition 1 (Argument Calculi (AC)). *The class of argument calculi, denoted* AC, *consists of* AC_0 *and its extensions: as shown in Fig. 3, the base system is* $\mathsf{AC}_0 = \{\boldsymbol{Ax}, \boldsymbol{Det}, \boldsymbol{Cut}\}$, *with extensions:* $\mathsf{DAC}_4 = \mathsf{AC}_0 \cup \{\boldsymbol{OR}, \boldsymbol{Def}, \boldsymbol{SDet}\}$, $\mathsf{DAC}_4^+ = \mathsf{DAC}_4 \cup \{\boldsymbol{TP}\}$, $\mathsf{DC} = \mathsf{AC}_0 \cup \{\boldsymbol{DDef}, \boldsymbol{SDet}, \boldsymbol{TP}\}$ *and* $\mathsf{DDC} = \mathsf{AC}_0 \cup \{\boldsymbol{DisDet}, \boldsymbol{DisDef}, \boldsymbol{SDet}, \boldsymbol{TP}\}$. *Let* S $\in$ AC. *An* S-*argument is an expression of the form* $\Gamma \Rightarrow \Delta$,[5] *where* $\Gamma \subseteq \mathscr{L}^{\mathsf{AC}}$ *is a finite regular set, and* Δ *contains at most one element of* $\mathscr{L}^{\mathsf{AC}}$. *An* S-*derivation of* $\Gamma \Rightarrow \Delta$ *is a tree structure with initial sequents as leaves, root* $\Gamma \Rightarrow \Delta$, *and rule applications as instances of* S *rules. If* $\Gamma \Rightarrow \Delta$ *is derivable in* S, *we write* $\vdash_S \Gamma \Rightarrow \Delta$.

---

[4] DAC also adopts constraint labels in characterizing constrained I/O logic. For readability, we omit such labels, using logical consistency as our only constraint.

[5] For readability, we sometimes enclose arguments within square brackets: $[\Gamma \Rightarrow \Delta]$.

We briefly discuss the AC rules but refer to [7,8] for an extensive discussion. **Ax** initializes a derivation with i- or o-labeled LC-derivable sequents. **TP** ensures that all input is also considered output. **Det** expresses detachment, that is, from a fact φ^i and norm (φ, ψ), we conclude ψ^o. **SDet** enables successive detachment of formulas by allowing output to serve as input as well. The rule **Cut** is standard [21]. **Def** stipulates that if a sequent contains inconsistent premises (that is, it has an empty right-hand side), then at least one of its defaults is not applicably. The rule **OR** reflects reasoning by cases over the input. The rule **DDef** expresses a more restrictive version of **Def**: if the output is considered to be $\neg\psi^o$, then in light of φ^i the default (φ, ψ) is not applicable. Last, **DisDet** and **DisDef** are disjunctive generalizations of **Det** and **Def**, respectively.

AC generates two types of arguments: those detaching conclusions from triggered defaults and those showing why certain defaults are inapplicable due to logical consistency. The latter type captures the defeasible nature of I/O and Default Logic, imposing consistency constraints on detachment. This defeasible interaction is exploited in AC-induced argumentation frameworks ($\mathscr{AF}$s) (Definition 2): for $\mathsf{S} \in \mathsf{AC}$, we collect all S-derivable arguments using only premises from a given knowledge base $\mathbb{K}$ and let an argument $\Delta \Rightarrow \neg(\varphi, \psi)$ attack any argument $\Gamma, (\varphi, \psi) \Rightarrow \Sigma$ using the inapplicable (φ, ψ) as a premise.

Definition 2. *Let $\mathsf{S} \in \mathsf{AC}$ and $\mathbb{K} = \langle \mathcal{F}, \mathcal{D} \rangle$ be a labeled knowledge base. An S-induced argumentation framework $\mathscr{AF}_{\mathsf{S}}(\mathbb{K}) = \langle \mathsf{Args}_{\mathsf{S}}, \mathsf{Atts}_{\mathsf{S}} \rangle$ is defined as:*

- $[\Gamma \Rightarrow \Delta] \in \mathsf{Args}_{\mathsf{S}}$ *iff* $[\Gamma \Rightarrow \Delta]$ *is S-derivable and $\Gamma \subseteq \mathcal{F} \cup \mathcal{D}$;*
- $(a, b) \in \mathsf{Atts}_{\mathsf{S}}$ *iff* $a = [\Gamma \Rightarrow \neg(\varphi, \psi)] \in \mathsf{Args}_{\mathsf{S}}$ *and* $b = [(\varphi, \psi), \Sigma \Rightarrow \Theta] \in \mathsf{Args}_{\mathsf{S}}$.

Semantic extensions are used to identify jointly justifiable sets of arguments [4]. The notion of stable semantics is often sufficient to characterize nm formalisms [2].

Definition 3. *Let $\mathsf{S} \in \mathsf{AC}$, $\mathbb{K} = \langle \mathcal{F}, \mathcal{D} \rangle$ be a labeled knowledge base, and let $\mathscr{AF}_{\mathsf{S}}(\mathbb{K}) = \langle \mathsf{Args}_{\mathsf{S}}, \mathsf{Atts}_{\mathsf{S}} \rangle$ be S-induced. For $\mathcal{A} \subseteq \mathsf{Args}_{\mathsf{S}}$, the following holds:*

- *$\mathcal{A}$ is conflict-free if for all $(a, b) \in \mathsf{Atts}_{\mathsf{S}}$, whenever $a \in \mathcal{A}$, then $b \notin \mathcal{A}$;*
- *$\mathcal{A}$ is stable if $\mathcal{A}$ is conflict-free, and for every $b \in \mathsf{Args}_{\mathsf{S}} \setminus \mathcal{A}$, there exists an $a \in \mathcal{A}$ such that $(a, b) \in \mathsf{Atts}_{\mathsf{S}}$.*

We define credulous (c) and skeptic (s) nonmonotonic inference as follows:

- *$\mathbb{K} \hspace{1pt}\vdash\hspace{-9pt}\sim^{c}_{\mathsf{S},\mathsf{sta}} \varphi$ if there exists a stable set $\mathcal{A} \subseteq \mathsf{Args}_{\mathsf{S}}$ such that $\Delta \Rightarrow \varphi \in \mathcal{A}$;*
- *$\mathbb{K} \hspace{1pt}\vdash\hspace{-9pt}\sim^{s}_{\mathsf{S},\mathsf{sta}} \varphi$ if for every stable set $\mathcal{A} \subseteq \mathsf{Args}_{\mathsf{S}}$, there exists $\Delta \Rightarrow \varphi \in \mathcal{A}$.*

A correspondence between maximal consistent families of constrained I/O logics and stable sets in DAC-instantiated $\mathscr{AF}$s is proven in [7]. An extended result, establishing the correspondence between (disjunctive) default extensions and stable sets in DC-instantiated $\mathscr{AF}$s is given in [8]. These systems, thus, inherit the disjunctive reasoning problems identified for I/O Logics in Sect. 2, a limitation passed on to AC-instantiated $\mathscr{AF}$s. While DDC can deal with Example 1 in which an arm of a disjunctive argument is attacked by an argument in $\mathsf{Arg}(\mathbb{K}_1)$, it also cannot adequately model Example 2. A new argumentation theory is needed to address these problems arising from disjunctions.

4 Hypothesis-Driven Disjunctive Reasoning

The main reason why I/O logics fail to properly handle Example 1 and 2 is the inability to effectively handle conflicts *within* disjunctive reasoning. Recall that a characteristic feature of formal argumentation is its capacity to explicitly model conflicts. In this respect, it is particularly promising to have a method that explicitly represents the reasoning, and corresponding conflicts, triggered by different disjuncts. Inspired by natural deduction, we propose a solution that introduces hypotheses to track reasoning paths explicitly.[6] We introduce variants of $\mathsf{DAC}_4 \backslash \mathsf{DAC}_4^+$, named $\mathsf{DAC}_4^{\mathbb{H}} \backslash \mathsf{DAC}_4^{\mathbb{H}+}$. It is clear that the same methods can be extended to other AC systems, such as DDC (see Sect. 3 and [8]).

We define $\mathscr{L}^h$ with the label h for *hypothetical formulas* to facilitate tracking reasoning paths during reasoning by cases. Thus, we work within the extended language $\mathscr{L}_h^{\mathsf{AC}} = \mathscr{L}^{\mathsf{AC}} \cup \mathscr{L}^h$. Recall from Definition 1 that an AC-argument is defined as a sequent of the form $\Gamma \Rightarrow \Delta$. We generalize sequents to be triple structures with a left-hand side Γ (LHS), a right-hand side Δ (RHS), and an additional component—a (possibly empty) set of sequents called the *hypothetical defeat information* (HDI) of the sequent, written as a subscript Θ on $\Gamma \Rightarrow_\Theta \Delta$. Intuitively, the HDI of an argument s contains sequents that were used in the hypothetically reasoning paths of reasoning by cases. Hence, if an argument t attacks any sequent in the HDI of s, then t also attacks s. This generalization is particularly important for representing argumentative defeat in Dung-style argumentation. This enables us to track defeat in arguments involving reasoning by cases. In the following, let $\mathsf{H} \in \{\mathsf{DAC}_4^{\mathbb{H}}, \mathsf{DAC}_4^{\mathbb{H}+}\}$.

$$\mathbf{Cut}^{\mathbb{H}②}\ \frac{\Gamma_1 \Rightarrow_\Theta \varphi^x \qquad \varphi^x, \Gamma_2 \Rightarrow_{\Theta'} \Delta}{\Gamma_1, \Gamma_2 \Rightarrow_{\Theta \cup \Theta'} \Delta} \qquad \mathbf{Def}^{\mathbb{H}}\ \frac{\Delta, (\varphi, \psi) \Rightarrow_\Theta}{\Delta \Rightarrow_\Theta \neg(\varphi, \psi)} \qquad \mathbf{HDet}_1\ \frac{\varphi^i, \Gamma \Rightarrow_\Theta \Xi}{\varphi^h, \Gamma \Rightarrow_\Theta \Xi}$$

$$\mathbf{HDet}_2{}^①\ \frac{\varphi^h, \Gamma \Rightarrow_\Theta \Xi}{\varphi^o, \Gamma \Rightarrow_\Theta \Xi} \qquad \mathbf{OR}^{\mathbb{H}③}\ \frac{s_1 : \varphi_1^h, \Gamma_1 \Rightarrow_{\Theta_1} \Delta \qquad s_2 : \varphi_2^h, \Gamma_2 \Rightarrow_{\Theta_2} \Delta}{(\varphi_1 \vee \varphi_2)^i, \Gamma_1, \Gamma_2 \Rightarrow_{\{s_1, s_2\} \cup \Theta_3} \Delta}$$

Fig. 4. The special rules of $\mathsf{DAC}_4^{\mathbb{H}} \backslash \mathsf{DAC}_4^{\mathbb{H}+}$ (Definition 4). Side-conditions ① and ② are as in Fig. 3, and ③ poses the following: $\Theta_3 = f(\Theta_1, \varphi_1^h) \cup f(\Theta_2, \varphi_2^h)$, where $f(\Omega, \varphi^h) = \{[\varphi^h, \Delta \Rightarrow \Gamma] \mid [\Delta \Rightarrow \Gamma] \in \Omega\}$, with Ω being an argument set and $\varphi^h \in \mathscr{L}^h$.

Definition 4. *An H-argument is of the form $\Gamma \Rightarrow_\Theta \Delta$, where $\Gamma \subseteq \mathcal{L}_h^{\mathsf{AC}}$ is a regular finite set, $\Delta \subseteq \mathcal{L}_h^{\mathsf{AC}}$ contains at most one element, and Θ is a set of H-arguments, which is typically omitted when empty. $\mathsf{DAC}_4^{\mathbb{H}}$ consists of the rules $\{\boldsymbol{Ax}, \boldsymbol{Det}, \boldsymbol{Cut}^{\mathbb{H}}, \boldsymbol{Def}^{\mathbb{H}}, \mathbf{OR}^{\mathbb{H}}, \mathbf{HDet}_1, \mathbf{HDet}_2\}$ and $\mathsf{DAC}_4^{\mathbb{H}+} = \mathsf{DAC}_4^{\mathbb{H}} \cup \{\boldsymbol{TP}\}$, as shown at Fig. 3 and Fig. 4. The initial sequent rules $\boldsymbol{Ax}$, $\boldsymbol{TP}$, and $\boldsymbol{Det}$ have empty HDIs (and so H-arguments are not circularly defined). An H-derivation of $\Gamma \Rightarrow_\Theta \Delta$ and $\vdash_\mathsf{H} \Gamma \Rightarrow_\Theta \Delta$ are defined in the usual way.*

---

[6] Recall that Disjunction Elimination in natural deduction expresses that, given a disjunction $\varphi \vee \psi$, we may assume both φ and ψ and reason with each separately. If both assumptions lead to the same conclusion γ, γ can be derived from $\varphi \vee \psi$.

The special rules of $\mathsf{DAC}_4^{\mathbb{H}} \backslash \mathsf{DAC}_4^{\mathbb{H}+}$ need commentary. Notably, they all share a common feature: they preserve the HDI of the premises in the conclusion. $\mathbf{Cut}^{\mathbb{H}}$ and $\mathbf{Def}^{\mathbb{H}}$ are simply HDI-augmented versions of $\mathbf{Cut}$ and $\mathbf{Def}$. $\mathbf{HDet}_1$ and $\mathbf{HDet}_2$ together enable successive detachment, thus replacing $\mathbf{SDet}$ (Fig. 3), allowing defaults triggered by an input φ^i or hypothesis φ^h to be triggered by a hypothesis φ^h or output φ^o, respectively.

The $\mathbf{OR}^{\mathbb{H}}$ rule is the most significant for nuanced disjunctive reasoning:

$$\mathbf{OR}^{\mathbb{H}③} \ \frac{s_1 : \varphi_1^h, \Gamma_1 \Rightarrow_{\Theta_1} \Delta \qquad s_2 : \varphi_2^h, \Gamma_2 \Rightarrow_{\Theta_2} \Delta}{s_3 : (\varphi_1 \vee \varphi_2)^i, \Gamma_1, \Gamma_2 \Rightarrow_{\{s_1,s_2\} \cup \Theta_3} \Delta}$$

It incorporates the two arguments s_1 and s_2, constituting the premises of the rule, directly into the conclusion s_3's HDI. Additionally, each of the HDIs of rule's premises is retained in the conclusion's HDI, with a modification: each argument in the premises's HDIs is appended with its corresponding hypothesis. That is, for the HDIs of s_1 and s_2 (i.e., Θ_1 and Θ_2) the HDI of the conclusion s_3 is given by $\{s_1, s_2\} \cup f(\Theta_1, \varphi_1^h) \cup f(\Theta_2, \varphi_2^h)$, where the function $f(\Omega, \varphi^h) = \{[\varphi^h, \Delta \Rightarrow \Gamma] \mid [\Delta \Rightarrow \Gamma] \in \Omega\}$. The main purpose of this is to record the hypotheses used in the disjunctive branching by each application of $\mathbf{OR}^{\mathbb{H}}$.

Example 3 (Example 1 cont.). Reconsider $\mathbb{K}_1$ (Example 1), the H-argument c (Fig. 5), concluding to "use cash," is derived via $\mathbf{OR}^{\mathbb{H}}$, using the defaults $(l, p \vee b)$, (b, mc), (p, lu), (mc, c), and (lu, c). The following $\mathsf{DAC}_4^{\mathbb{H}}$-derivation demonstrates this:

$$\mathbf{HDet}_1 \frac{\mathbf{Det} \dfrac{b^i, (b, mc) \Rightarrow mc^o}{b^h, (b, mc) \Rightarrow mc^o} \quad \mathbf{HDet}_2 \dfrac{\mathbf{HDet}_1 \dfrac{\mathbf{Det} \dfrac{}{mc^i, (mc, c) \Rightarrow c^o}}{mc^h, (mc, c) \Rightarrow c^o}}{mc^o, (mc, c) \Rightarrow c^o}}{h_1 : b^h, (b, mc), (mc, c) \Rightarrow c^o} \ \mathbf{Cut}^{\mathbb{H}}$$

$$\mathbf{HDet}_1 \frac{\mathbf{Det} \dfrac{p^i, (p, lu) \Rightarrow lu^o}{p^h, (p, lu) \Rightarrow lu^o} \quad \mathbf{HDet}_2 \dfrac{\mathbf{HDet}_1 \dfrac{\mathbf{Det} \dfrac{}{lu^i, (lu, c) \Rightarrow c^o}}{lu^h, (lu, c) \Rightarrow c^o}}{lu^o, (lu, c) \Rightarrow c^o}}{h_2 : p^h, (p, lu), (lu, c) \Rightarrow c^o} \ \mathbf{Cut}^{\mathbb{H}}$$

$$\mathbf{HDet}_1, \mathbf{HDet}_2 \frac{\mathbf{OR}^{\mathbb{H}} \dfrac{h_1 : b^h, (b, mc), (mc, c) \Rightarrow c^o \qquad h_2 : p^h, (p, lu), (lu, c) \Rightarrow c^o}{(p \vee b)^i, (b, mc), (mc, c), (p, lu), (lu, c) \Rightarrow_{\{h_1, h_2\}} c^o}}{(p \vee b)^o, (b, mc), (mc, c), (p, lu), (lu, c) \Rightarrow_{\{h_1, h_2\}} c^o}$$

Based on the arguments derived thus far, using l^i and $(l, p \vee b)$, argument c can be derived using the $\mathbf{Det}$ and $\mathbf{Cut}^{\mathbb{H}}$ rule:

$$c : \ [l^i, (l, p \vee b), (b, mc), (mc, c), (p, lu), (lu, c) \Rightarrow_{\{h_1, h_2\}} c^o],$$

where the arguments $h_1 : [b^h, (b, mc), (mc, c) \Rightarrow c^o]$ and $h_2 : [p^h, (p, lu), (lu, c) \Rightarrow c^o]$ are the only two disjunctive paths used in the application of $\mathbf{OR}^{\mathbb{H}}$.

Similar to $\mathsf{DAC}_4/\mathsf{DAC}_4^+$, $\mathsf{DAC}_4^{\mathbb{H}}/\mathsf{DAC}_4^{\mathbb{H}+}$ generate two central types of arguments: attacking arguments and arguments concluding output formulas. However, with the introduction of hypotheses, we now need to compare the implication relations between the hypotheses of the attacking and of the attacked arguments. In simple terms, the hypothesis of the attacking argument must be subsumed by that of the attacked argument in order for the attack to succeed, meaning that it uses the same or less hypotheses in order to attack. Furthermore, since each argument is now equipped with an HDI, if the HDI of an argument is attacked, the argument itself will also be attacked. Let us make this precise.

In Definition 2, it sufficed to consider only $\mathbb{K}$-based arguments. With the use of hypotheses, an H-argument may additionally be based on the hypothesis set $\mathcal{H}_{\mathbb{K}}$:

Definition 5. *The* hypothesis set *of* $\mathbb{K} = \langle \mathcal{F}, \mathcal{D} \rangle$ *is* $\mathcal{H}_{\mathbb{K}} = \{\varphi^h \mid \varphi \in \mathbf{Body}(\mathcal{D})\} \cup \{\gamma^h \mid \exists(\alpha \in \mathcal{F})\exists(m > 0)\exists(\beta_1, \ldots, \beta_m \in \mathbf{Body}(\mathcal{D}))$ s.t. $\nvdash \alpha$ and $\vdash \alpha \leftrightarrow (\bigvee_{i=1}^{m} \beta_i \vee \gamma)\}$, *where* $\mathbf{Body}(\mathcal{D}) = \{\varphi \mid (\varphi, \psi) \in \mathcal{D}\}$.*

The set of hypotheses $\mathcal{H}_{\mathbb{K}}$ of a knowledge base $\mathbb{K} = \langle \mathcal{F}, \mathcal{D} \rangle$ restricts the available options to build arguments based on hypothetical formulas (see Definition 6). The reason why we do not only consider the bodies of defaults is to account for knowlwedge bases such as $\mathcal{F} = \{(p \vee q)^i\}$ and $\mathcal{D} = \{(p, u)\}$ in the presence of **TP**, where we want to conclude $(u \vee q)^o$ and need to apply **HOR** to $[p^h, (p, u) \Rightarrow (u \vee q)^o]$ and $[q^h \Rightarrow (u \vee q)^o]$ to obtain $[(p \vee q)^i, (p, q) \Rightarrow (u \vee q)^o]$. In deontic applications without **TP** we can define $\mathcal{H}_{\mathbb{K}}$ by $\{\varphi^h \mid \varphi \in \mathbf{Body}(\mathcal{D})\}$.

For any argument $t = [\Gamma \Rightarrow_{\Theta} \Delta]$, we define the following functions: $\mathbf{HDI}(t) = \Theta$, $\mathbf{Con}(t) = \Delta$, $\mathbf{Pre}(t) = \Gamma$ and $\mathbf{Hyp}(t) = \Gamma \cap \mathscr{L}^h$.

Definition 6. *Let* $\mathbb{K} = \langle \mathcal{F}, \mathcal{D} \rangle$. *An* H-*instantiated argumentation framework, denoted by* $\mathscr{AF}_{\mathsf{H}}(\mathbb{K}) = \langle \mathsf{Arg}_{\mathsf{H}}, \mathsf{Att}_{\mathsf{H}} \rangle$, *is defined as follows:*

- $\Gamma \Rightarrow \Delta \in \mathsf{Arg}_{\mathsf{H}}$ *iff* a *is* H-*derivable and* $\Gamma \subseteq \mathcal{F} \cup \mathcal{D} \cup \mathcal{H}_{\mathbb{K}}$.
- $(a, b) \in \mathsf{Att}_{\mathsf{H}}$ *iff i)* $a, b \in \mathsf{Arg}_{\mathsf{H}}$, $\mathbf{Con}(a) = \neg(\varphi, \psi)$, $(\varphi, \psi) \in \mathbf{Pre}(b)$, *and* $\mathbf{Hyp}(a) \subseteq \mathbf{Hyp}(b)$, *or ii) there exists* $t \in \mathbf{HDI}(b)$ *such that* $(a, t) \in \mathsf{Att}_{\mathsf{H}}$.

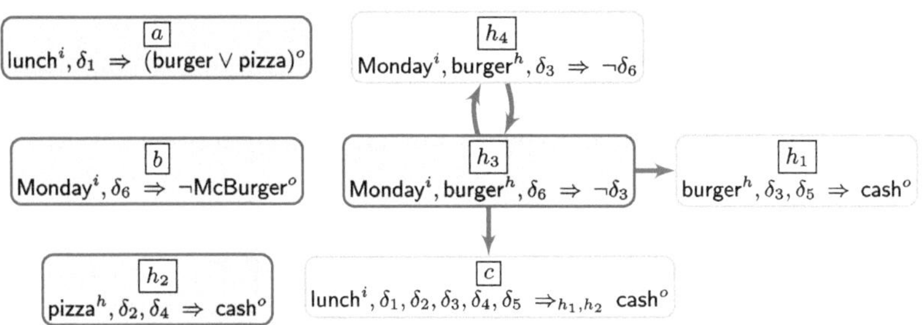

Fig. 5. A partial representation of $\mathscr{AF}_{\mathsf{H}}(\mathbb{K}_1)$ from Example 4, where $\mathsf{H} \in \{\mathsf{DAC}_4^{\mathbb{H}}, \mathsf{DAC}_4^{\mathbb{H}+}\}$. Arrows represent attacks. The arguments $\{a, b, h_2, h_3\}$ (in green) form the only stable extension of $\mathscr{AF}_{\mathsf{H}}(\mathbb{K}_1)$. (Color figure online)

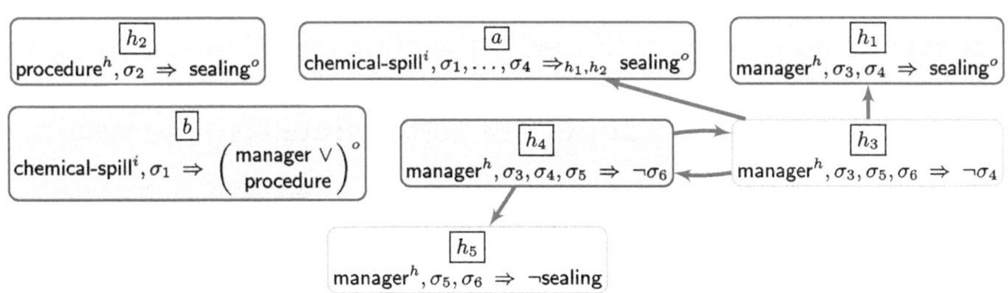

Fig. 6. A partial representation of $\mathscr{AF}_{\mathsf{H}}(\mathbb{K}_2)$ from Example 5, where $\mathsf{H} \in \{\mathsf{DAC}_4^{\mathbb{H}}, \mathsf{DAC}_4^{\mathbb{H}+}\}$. Arrows represent attacks. The arguments $\{a, b, h_1, h_2, h_4\}$ (in green) form a stable extension of $\mathscr{AF}_{\mathsf{H}}(\mathbb{K}_2)$. (Color figure online)

Example 4 (Example 1 cont.) Fig. 5 illustrates $\mathscr{A}\mathscr{F}_{\mathsf{H}}(\mathbb{K}_1)$, with H-arguments derived from $\mathbb{K}_1$ in Example 1. The set of arguments $\{h_1, h_2, h_3, h_4, a, b, c\} \subseteq \mathsf{Arg}_{\mathsf{H}}$ is shown, where $h_1, h_2 \in \mathbf{HDI}(c)$. Arrows represent attacks, e.g., the arrow from h_3 to h_1 denotes $(h_3, h_1) \in \mathsf{Att}_{\mathsf{H}}$. Since $h_1 \in \mathbf{HDI}(c)$, it follows that $(h_3, c) \in \mathsf{Att}_{\mathsf{H}}$.

Example 5 (Example 2 cont.) Fig. 6 illustrates $\mathscr{A}\mathscr{F}_{\mathsf{H}}(\mathbb{K}_2)$, with H-arguments derived from $\mathbb{K}_2$ in Example 2. The set of arguments $\{h_1, h_2, h_3, h_4, h_5, a, b\} \subseteq \mathsf{Arg}_{\mathsf{H}}$ is shown, where $h_1, h_2 \in \mathbf{HDI}(a)$. Arrows represent attacks, e.g., the arrow from h_3 to h_1 denotes $(h_3, h_1) \in \mathsf{Att}_{\mathsf{H}}$. Since $h_1 \in \mathbf{HDI}(a)$, it follows that $(h_3, a) \in \mathsf{Att}_{\mathsf{H}}$.

Definition 7. *Let $\mathscr{A}\mathscr{F}_{\mathsf{H}}(\mathbb{K}) = \langle \mathsf{Arg}_{\mathsf{H}}, \mathsf{Att}_{\mathsf{H}} \rangle$, $\mathcal{A} \subseteq \mathsf{Arg}_{\mathsf{H}}$ is conflict-free if for all $a, b \in \mathcal{A}$, if $\mathbf{Con}(a) = \neg(\varphi, \psi)$, then $(\varphi, \psi) \notin \mathbf{Pre}(b)$. Then:*

- *$\mathcal{A}$ is stable iff $\mathcal{A}$ is conflict-free, and for every $b \in \mathsf{Arg}_{\mathsf{H}} \setminus \mathcal{A}$, there exists an $a \in \mathcal{A}$ such that $(a, b) \in \mathsf{Att}_{\mathsf{H}}$.*

We define credulous (c) and skeptic (s) nonmonotonic inference accordingly:

- *$\mathbb{K} \hspace{1pt}\vdash^{c}_{\mathsf{H,sta}} \varphi$ iff there exists a stable set $\mathcal{A} \subseteq \mathsf{Arg}_{\mathsf{H}}$ such that there is an $a \in \mathcal{A}$ with $\mathbf{Con}(a) = \varphi$ and $\mathbf{Hyp}(a) = \emptyset$.*
- *$\mathbb{K} \hspace{1pt}\vdash^{s}_{\mathsf{H,sta}} \varphi$ iff for every stable set $\mathcal{A} \subseteq \mathsf{Arg}_{\mathsf{H}}$, there is an $a \in \mathcal{A}$ with $\mathbf{Con}(a) = \varphi$ and $\mathbf{Hyp}(a) = \emptyset$.*

As defined by Definition 7, hypotheses fulfill a central role in disjunctive reasoning but when we draw final conclusions from a given knowledge base, we are interested in fully committed arguments, i.e., not containing any hypotheses (cf. [13]).

Example 6 (Example 1 cont.). For $\mathbb{K}_1$, there is only one stable extension: $\mathcal{A} = \{a, b, h_2, h_3\} \subseteq \mathsf{Arg}_{\mathsf{H}}$ (colored green in Fig. 5). The set $\mathcal{A}' = \{h_1, h_2, h_4, c, a, b\} \subseteq \mathsf{Arg}_{\mathsf{H}}$ is not stable because $\mathbf{Con}(h_4) = \neg\delta_6$ and $\delta_6 \in \mathbf{Pre}(b)$. Although $h_2 \in \mathcal{A}$ and $\mathbf{Con}(h_2) = c^o$, we have $\mathbf{Hyp}(h_2) \neq \emptyset$. By Definition 7, $\mathbb{K}_1 \hspace{1pt}\vdash^{s}_{\mathsf{H,sta}} (p \vee b)^o$ but $\mathbb{K}_1 \hspace{1pt}\not\vdash^{s}_{\mathsf{H,sta}} c^o$.

Example 7 (Example 2 cont.). For $\mathbb{K}_2$, there are two stable extensions: $\mathcal{A}_1 = \{a, b, h_1, h_2, h_4\} \subseteq \mathsf{Arg}_{\mathsf{H}}$ (colored green in Fig. 6) and $\mathcal{A}_2 = \{h_2, h_3, h_5, b\} \subseteq \mathsf{Arg}_{\mathsf{H}}$. By Definition 7, $\mathbb{K}_2 \hspace{1pt}\vdash^{s}_{\mathsf{H,sta}} (m \vee p)^o$, but $\mathbb{K}_2 \hspace{1pt}\not\vdash^{s}_{\mathsf{H,sta}} s^o$.

In both examples, $\mathsf{H} \in \{\mathsf{DAC}_4^{\mathbb{H}}, \mathsf{DAC}_4^{\mathbb{H}+}\}$ yield the desired cautious outcome.

5 Meta-theoretical Properties for Disjunctive Reasoning

In this section, we evaluate Argumentation Calculi and $\mathsf{DAC}_4^{\mathbb{H}}/\mathsf{DAC}_4^{\mathbb{H}+}$ against meta-theoretical properties for disjunction. In the following, let $\mathsf{S} \in \mathsf{AC} \cup \{\mathsf{DAC}_4^{\mathbb{H}}, \mathsf{DAC}_4^{\mathbb{H}+}\}$. We refer to [3] for an overview of analyzing logical argumentation frameworks through postulates. We study properties of extensions (Definition 11) and of inference relations (Definition 12). We first define embedded arguments (Definition 8) and disjunctive defaults (Definition 9), where the latter are those defaults in arguments that cannot be triggered by any of their embedded arguments.

Definition 8 (Embedded Argument). *Let a and b be S-arguments. We say that b is an embedded argument of a, denoted as $b \in \mathbf{Embed}(a)$, if and only if the following conditions hold:* $\mathbf{Pre}(b) \subseteq \mathbf{Pre}(a)$ *and* $\mathbf{HDI}(b) \subseteq \mathbf{HDI}(a)$.

Definition 9 (Disjunctively Triggered Default). *Let a be an S-argument with* $\mathbf{Def}(a) = \mathbf{Pre}(a) \cap \mathcal{L}^d$ *the set of defaults occurring in the LHS of a. Then,* $\mathsf{D}_a = \{(\varphi, \psi) \in \mathbf{Def}(a) \mid \forall b \in \mathbf{Embed}(a), \mathbf{Con}(b) \neq \varphi\}$ *is the set of disjunctively triggered defaults in a.*

Intuitively, if $\mathsf{D}_a \neq \emptyset$, then a contains hypothetical subarguments.

Example 8 (Example 1 cont.). Take argument c in Fig. 5 and Example 3. We have $\mathsf{D}_c = \{\delta_3 : (b, mc), \delta_5 : (mc, c), \delta_2 : (p, lu), \delta_4 : (lu, c)\}$ (see Fig. 7). As another example, take argument a in Fig. 6, where $\mathsf{D}_a = \{\sigma_3 : (m, r), \sigma_4 : (r, s), \sigma_2 : (p, s)\}$.

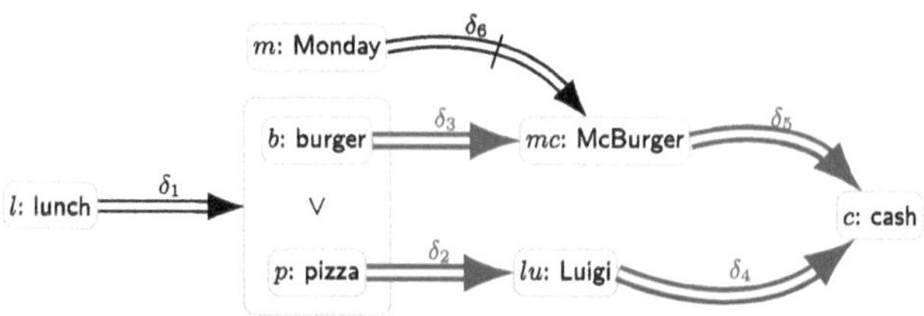

Fig. 7. The set of disjunctive defaults $\mathsf{D}_c = \{\delta_3, \delta_5, \delta_2, \delta_4\}$ of the argument $c = [\mathsf{lunch}^i, \delta_1, \delta_2, \delta_3, \delta_4, \delta_5 \Rightarrow_{h_1, h_2} \mathsf{cash}^\circ]$ (thick arrows).

Definition 10. *For an argument set $\mathcal{E}$, define:* $\mathcal{E}^* = \{a \in \mathcal{E} \mid \mathbf{Con}(a) \notin \overline{\mathcal{L}^n}$ *and* $\mathbf{Con}(a) \in \mathcal{L}^\circ\}$ *and* $\mathcal{E}^\Delta = \{a \in \mathcal{E}^* \mid \mathbf{Hyp}(a) = \emptyset\}$.

Definition 11 (Meta-Theoretical Properties I). *Let $\mathbb{K} = \langle \mathcal{F}, \mathcal{D} \rangle$ be a knowledge base and let $\mathscr{AF}_\mathsf{S}(\mathbb{K}) = \langle \mathsf{Args}_\mathsf{S}, \mathsf{Atts}_\mathsf{S} \rangle$. Suppose $\mathcal{E} \subseteq \mathsf{Args}_\mathsf{S}$ is stable, and let $a \in \mathcal{E}$ with $\mathsf{D}_a \neq \emptyset$. Let $\mathbb{K} \oplus \varphi = \langle \mathcal{F} \cup \{\varphi\}, \mathcal{D} \rangle$, for each $(\alpha, \beta) \in \mathsf{D}_a$, consider the updated argumentation framework $\mathscr{AF}_\mathsf{S}(\mathbb{K} \oplus \alpha) = \langle \mathsf{Args}_\mathsf{S}^{\oplus \alpha}, \mathsf{Atts}_\mathsf{S}^{\oplus \alpha} \rangle$. We define:*

- Conclusion Consistency: $\mathbf{Con}(\mathcal{E}^\Delta)$ *is* $\vdash_{\mathsf{CL}}$-*consistent.*
- Disjunctive Consistency: $\{\beta\} \cup \mathbf{Con}(\mathcal{E}^\Delta)$ *is* $\vdash_{\mathsf{CL}}$-*consistent.*
- Hypothetical Consistency: *If $a \in \mathcal{E}^*$ and there is no $b \in \mathsf{Args}_\mathsf{S}$ with $\mathbf{Con}(b) = \neg\mathbf{Con}(a)$ then there is no $b \in \mathsf{Args}_\mathsf{S}^{\oplus \alpha}$ with $\mathbf{Con}(b) = \neg\mathbf{Con}(a)$.*

Definition 11 specifies three interesting properties to be considered for stable extensions: *Conclusion consistency* states that the conclusions of all non-hypothetical arguments in a stable extension are consistent. *Disjunctive consistency* stipulates that the conclusions of all non-hypothetical arguments in a stable extension $\mathcal{E}$ are consistent with the head of any disjunctive norm in $\mathcal{E}$. *Hypothetical consistency* states that if there is no rebuttal of a for $\mathbb{K}$ (i.e., an argument with the opposite conclusion) then there is also no rebuttal of a for $\mathbb{K} \oplus \alpha$. Through the $\oplus$ operation, we incorporate the body α of a disjunctively triggered default of a into the knowledge base $\mathbb{K}$).

Proposition 1. *All* $S \in \mathsf{AC} \cup \{\mathsf{DAC}_4^{\mathbb{H}}, \mathsf{DAC}_4^{\mathbb{H}+}\}$ *satisfy conclusion consistency.*

Proposition 2. $\mathsf{DDC}, \mathsf{DAC}_4^{\mathbb{H}}$ *and* $\mathsf{DAC}_4^{\mathbb{H}+}$ *satisfy disjunctive consistency, while* $\mathsf{DAC}_4/\mathsf{DAC}_4^+$ *do not.*

Proposition 3. $\mathsf{DAC}_4^{\mathbb{H}}/\mathsf{DAC}_4^{\mathbb{H}+}$ *satisfy hypothetical consistency, while* DAC_4, DAC_4^+ *and* DDC *do not.*[7]

Definition 12 (Meta-Theoretical Properties II). *Let* $\mathbb{K} = \langle \mathcal{F}, \mathcal{D} \rangle$ *be a knowledge base and* $\mathbb{K} \oplus \varphi = \langle \mathcal{F} \cup \{\varphi\}, \mathcal{D} \rangle$ *its update. We define the following properties for nonmonotonic inference relations:*

- Introduction of Disjunction*: If* $\mathbb{K} \oplus \varphi \mathrel{|\!\sim} \alpha$ *and* $\mathbb{K} \oplus \psi \mathrel{|\!\sim} \alpha$, *then* $\mathbb{K} \oplus (\varphi \vee \psi) \mathrel{|\!\sim} \alpha$.
- Disjunctive Rationality [19]*: If* $\mathbb{K} \oplus (\varphi \vee \psi) \mathrel{|\!\sim} \alpha$, *then* $\mathbb{K} \oplus \varphi \mathrel{|\!\sim} \alpha$ *or* $\mathbb{K} \oplus \psi \mathrel{|\!\sim} \alpha$.

Proposition 4. $\mathrel{|\!\sim}^s_{\mathsf{DAC}_4^{\mathbb{H}},\mathsf{sta}}$, $\mathrel{|\!\sim}^s_{\mathsf{DAC}_4^{\mathbb{H}+},\mathsf{sta}}$, $\mathrel{|\!\sim}^s_{\mathsf{DAC}_4,\mathsf{sta}}$, *and* $\mathrel{|\!\sim}^s_{\mathsf{DAC}_4^+,\mathsf{sta}}$ *satisfy the introduction of disjunction, but* $\mathrel{|\!\sim}^c_{\mathsf{DAC}_4^{\mathbb{H}},\mathsf{sta}}$ *and* $\mathrel{|\!\sim}^c_{\mathsf{DAC}_4^{\mathbb{H}+},\mathsf{sta}}$ *do not.*

Proposition 5. $\mathrel{|\!\sim}^s_{\mathsf{DAC}_4^{\mathbb{H}},\mathsf{sta}}$, $\mathrel{|\!\sim}^s_{\mathsf{DAC}_4^{\mathbb{H}+},\mathsf{sta}}$, $\mathrel{|\!\sim}^s_{\mathsf{DAC}_4,\mathsf{sta}}$, $\mathrel{|\!\sim}^s_{\mathsf{DAC}_4^+,\mathsf{sta}}$, *and* $\mathrel{|\!\sim}^s_{\mathsf{DDC},\mathsf{sta}}$ *do not satisfy the disjunctive rationality.*

Let us show that disjunctive rationality fails for $\mathrel{|\!\sim}^s_{\mathsf{DAC}_4^{\mathbb{H}},\mathsf{sta}}$ and $\mathrel{|\!\sim}^s_{\mathsf{DAC}_4^{\mathbb{H}+},\mathsf{sta}}$ (abbreviated as $\mathrel{|\!\sim}^s$). (The case for DDC is analogous.) Let $\mathscr{AF}(\mathbb{K} \oplus \varphi) = \langle \mathsf{Arg}_\varphi, \mathsf{Att}_\varphi \rangle$. Consider $\mathbb{K} = \langle \mathcal{F}, \mathcal{D} \rangle$ with $\mathcal{F} = \{g_1 \vee g_2\}$ and $\mathcal{D} = \{(g_1, c_1), (g_2, c_2), (c_1, s), (c_2, s), (p, \neg c_1), (q, \neg c_2)\}$. Intuitively, in $\mathscr{AF}(\mathbb{K} \oplus p \vee q)$ the argument $a : [(g_1 \vee g_2)^i, (g_1, c_1), (c_1, s), (g_2, c_2), (c_2, s) \Rightarrow s^o]$ seems problematic since we have $p \vee q \in \mathcal{F}$ and both p and q give rise to potential defeaters of the two respective arms of the disjunctive argument underlying a. But the only stable extension for $\mathscr{AF}(\mathbb{K} \oplus p \vee q)$ contains a and $\mathbb{K} \oplus (p \vee q) \mathrel{|\!\sim}^s s$. In contrast, the only stable extension in Arg_φ contains the argument $[p^i, g_1^h, (p, \neg c_1) \Rightarrow \neg(g_1, c_1)]$ and the only stable extension in Arg_ψ contains the argument $[q^i, g_2^h, (q, \neg c_2) \Rightarrow \neg(g_2, c_2)]$. So, $\mathbb{K} \oplus p \mathrel{|\!\not\sim}^s s$ and $\mathbb{K} \oplus q \mathrel{|\!\not\sim}^s s$. Thus, disjunctive rationality fails.

Table 1 summarizes Propositions 1–5. The question whether the "introduction of disjunctions" property holds for DDC is still open, as well as the question whether there is a system that satisfies all discussed properties, including disjunctive rationality, a property that recently attracted renewed attention [9].[8]

---

[7] We illustrate why hypothetical consistency does not hold for $S \in \{\mathsf{DAC}_4, \mathsf{DAC}_4^+, \mathsf{DDC}\}$. Consider $\mathbb{K}_2$ in Example 2. Let $\mathscr{AF}_S(\mathbb{K}_2) = \langle \mathsf{Arg}_S, \mathsf{Att}_S \rangle$. Let $y = [c^i, \sigma_1, \ldots, \sigma_4 \Rightarrow s^o]$. There are no arguments $e \in \mathsf{Arg}_S(\mathbb{K}_2)$ with $\mathbf{Con}(e) = \neg s^o$. Then, $\sigma_3 = (m, r) \in \mathbf{D}_y$. Since $c = [m, \sigma_5, \sigma_6 \Rightarrow \neg s^o] \in \mathsf{Arg}_S^{\oplus m}$, hypothetical consistency does not hold. In contrast, $[m^h, \sigma_5, \sigma_6 \Rightarrow \neg s^o] \in \mathsf{Arg}_S$ for $S \in \{\mathsf{DAC}_4^{\mathbb{H}}, \mathsf{DAC}_4^{\mathbb{H}+}\}$.

[8] We conjecture that disjunctive rationality can be achieved by generalizing $\mathbf{OR}^{\mathbb{H}}$ to

$$\frac{s_1 : \varphi_1^h, \Gamma_1 \Rightarrow_{\Theta_1} \neg(\varphi_1, \psi_1) \qquad s_2 : \varphi_2^h, \Gamma_2 \Rightarrow_{\Theta_2} \neg(\varphi_2, \psi_2)}{(\varphi_1 \vee \varphi_2)^h, \Gamma_1, \Gamma_2 \Rightarrow_{\{s_1, s_2\} \cup f(\Theta_1, \varphi_1^h) \cup f(\Theta_2, \varphi_2^h)} \neg(\varphi_1, \psi_1), \neg(\varphi_2, \psi_2)}$$

and allow an argument $\Gamma \Rightarrow_\Theta \neg(\varphi_1, \psi_1), \ldots, \neg(\varphi_n, \psi_n)$ to attack an argument $\Delta, (\varphi_1, \psi_1), \ldots, (\varphi_n, \psi_n) \Rightarrow_{\Theta'} \Xi$ (subject to additional restrictions as in Definition 6). In this case $[(p \vee q)^i, (g_1 \vee g_2)^i, (p, \neg c_1), (q, \neg c_2) \Rightarrow \neg(g_1, c_1), \neg(g_2, c_2)]$ defeats a.

Table 1. Consistency properties and properties of nm inferences of $H \in AC \cup \{DAC_4^H, DAC_4^{H+}\}$ (Proposition 1–5).

Consistency properties	DAC_4/DAC_4^+	DDC	DAC_4^H/DAC_4^{H+}
conclusion consistency	✓	✓	✓
disjunctive consistency	–	✓	✓
hypothetical consistency	–	–	✓
Properties for inferences $\mathrel{\vert\!\sim}_H^s$			
introduction of disjunction	✓	?	✓
disjunctive rationality	–	–	–

6   Related and Future Work

Disjunctive reasoning with default rules is fundamentally challenging. While Sect. 2 highlights the limitations of I/O logic in handling disjunctive examples, such shortcomings are prevalent in many nonmonotonic systems for defeasible rules, e.g., Gelfond and Lifschitz's Disjunctive Default Logic [15]. Furthremore, the Disjunctive Default Calculus (DDC) [8] also fails to address conflicts triggered by individual disjuncts in Example 2 due to its lack of hypothetical consistency. Our work proposes a cautious perspective and demonstrates that sequent-based logical argumentation is quite suitable for exploring variations of disjunctive reasoning. In particular, our proof-theoretic approach provides a refined treatment of disjunctive reasoning by making hypothetical reasoning paths explicit.

Future directions include integrating Inquisitive Logic [11] for systematic hypothesis generation and exploring Bayesian probability for uncertainty quantification in disjunctive reasoning chains (cf. [16]). Furthermore, our rule-based labeled approach, containing explicit hypothetical information for argumentative defeat, promotes a transparency that serves explanation purposes [6, 7]. This connection remains to be investigated. Last, a comparison to related work employing explicit hypotheses and suppositions in the reasoning process is also warranted, e.g., think of dialectical argumentation frameworks [13].

In summary, our key contributions are: (i) a hypothesis-driven approach for disjunctive reasoning in logical argumentation that explicitly tracks reasoning paths to preserve defeasibility, (ii) a specialized disjunction rule for Argument Calculi and hypothesis-based attack relations in argumentation frameworks, and (iii) novel rationality postulates (disjunctive consistency and hypothetical consistency) for disjunctive reasoning and their study with respect to nonmonotonic inference based on Argument Calculi.

Acknowledgments. This work is supported by the Major Project of the National Social Science Foundation (19ZDA041) and partially funded by the Deutsche Forschungsgemeinschaft (DFG) 511915728 and the Austrian Science Fund (FWF) 10.55776/COE12.

192   Z. Zhou et al.

Disclosure of Interests. The authors have no competing interests to declare that are relevant to the content of this article.

Appendix: Selected Proofs

In the context of the following proofs, let $\mathbb{K} = \langle \mathcal{F}, \mathcal{D} \rangle$ denote an arbitrary knowledge base, $\vdash_\mathsf{H}$ be the derivability relation for $\mathsf{H} \in \{\mathsf{DAC}_4^\mathbb{H}, \mathsf{DAC}_4^{\mathbb{H}+}\}$, and $\vdash_\mathsf{CL}$ be the derivability relation of classical logic (the subscript is omitted when the context disambiguates). Due to space constraints, some proofs are abbreviated or omitted. For brevity, we typically exclude argument HDIs unless explicitly required.

*Proof (**Proposition** 1).* We prove the case for DAC. Assume there are formulas $\varphi_1, \ldots, \varphi_n \in \mathbf{Con}(\mathcal{E}^\triangle)$ such that $\vdash_\mathsf{CL} \neg(\varphi_1 \wedge \cdots \wedge \varphi_n)$. Then, there exist $[\Delta_1 \Rightarrow \varphi_1^{x_1}], \ldots, [\Delta_n \Rightarrow \varphi_n^{x_n}] \in \mathcal{E}$. Without loss of generality, assume that $x_1 = \cdots = x_n = o$ and that $\exists (\alpha, \beta) \in \Delta_1$. By **Ax** and multiple applications of **Cut**, we obtain $\vdash_\mathsf{DAC} [\Delta_1, \ldots, \Delta_n \Rightarrow (\varphi_1 \wedge \cdots \wedge \varphi_n)^o]$. On the other hand, by **Ax**, we have $\vdash_\mathsf{DAC} [\Rightarrow \neg(\varphi_1 \wedge \cdots \wedge \varphi_n)^o]$. Applying **Cut**, we conclude $\vdash_\mathsf{DAC} [\Delta_1, \ldots, \Delta_n \Rightarrow]$. Moreover, by **Def**, $a = [\Delta_1 \setminus (\alpha, \beta), \ldots, \Delta_n \Rightarrow \neg(\alpha, \beta)]$ is DAC-derivable. Clearly, $a \in \mathcal{E}$, but a attacks $[\Delta_1 \Rightarrow \varphi_1^o]$, contradicting the conflict-freeness of $\mathcal{E}$.

*Proof (**Proposition** 2).* We prove the case for $\mathsf{H} \in \{\mathsf{DAC}_4^\mathbb{H}, \mathsf{DAC}_4^{\mathbb{H}+}\}$. Assume that $\varphi_1, \ldots, \varphi_n \in \{\beta\} \cup \mathbf{Con}(\mathcal{E}^\triangle)$ such that $\vdash_\mathsf{CL} \neg(\varphi_1 \wedge \cdots \wedge \varphi_n)$. Since H satisfies conclusion consistency, $\beta \in \{\varphi_1, \ldots, \varphi_n\}$. It is easy to see that $d_1 = [\alpha^h, (\alpha, \beta) \Rightarrow \beta^o] \in \mathcal{E}$.

Without loss of generality, assume that $\{\varphi_1, ..., \varphi_k\} \cup \beta = \{\varphi_1, ..., \varphi_n\}$. Therefore, $\varphi_1, ..., \varphi_k \in \mathbf{Con}(\mathcal{E}^\triangle)$. For $i \in \{1, ..., k\}$, we have $\vdash_\mathsf{H} [\Delta_i \Rightarrow \varphi_i^o] \in \mathcal{E}$. By applying $\mathbf{Cut}^\mathbb{H}$ multiple times, we get $\vdash_\mathsf{H} d_2 = [\Delta_1, \ldots, \Delta_k \Rightarrow (\varphi_1 \wedge \cdots \wedge \varphi_k)^o]$. Using d_1 and d_2, we obtain $\vdash_\mathsf{H} d_3 = [\alpha^h, (\alpha, \beta), \Delta_1, \ldots, \Delta_k \Rightarrow (\varphi_1 \wedge \cdots \wedge \varphi_n)^o]$. Since $\vdash_\mathsf{CL} \neg(\varphi_1 \wedge \cdots \wedge \varphi_n)$, we have $\vdash_\mathsf{H} d_4 = [\Rightarrow \neg(\varphi_1 \wedge \cdots \wedge \varphi_n)^o]$. By $\mathbf{Cut}^\mathbb{H}$ and $\mathbf{Def}^\mathbb{H}$, from d_4 and d_3, we get $\vdash_\mathsf{H} d_5 = [\alpha^h, \Delta_1, \ldots, \Delta_k \Rightarrow \neg(\alpha, \beta)]$. We claim that $d_5 \in \mathcal{E}$. For the sake of contradiction, assume there exists an argument $d_6 \in \mathcal{E}$ such that $\mathbf{Con}(d_6) = \neg(\xi, \theta)$ and $(\xi, \theta) \in \mathbf{Pre}(d_5)$. Therefore, $(\xi, \theta) \in \bigcup_{i=1}^k \Delta_i$, which implies $\exists i \in \{1, ..., k\}$ such that $(\xi, \theta) \in \mathbf{Pre}(\Delta_i \Rightarrow \varphi_i^o)$. This is a contradiction.

Finally, since $d_5, a \in \mathcal{E}$ and $(\alpha, \beta) \in \mathbf{Pre}(a)$, this contradicts the conflict-freeness of $\mathcal{E}$.

*Proof (**Proposition** 3).* We show that $\mathsf{DAC}_4^\mathbb{H}/\mathsf{DAC}_4^{\mathbb{H}+}$ satisfies hypothetical consistency. Let $\mathsf{H} \in \{\mathsf{DAC}_4^\mathbb{H}, \mathsf{DAC}_4^{\mathbb{H}+}\}$. Assume there exists $b \in \mathsf{Arg}_\mathsf{H}^{\oplus \alpha}$ such that $\mathbf{Con}(b) = \neg\mathbf{Con}(a)$. Without loss of generality, assume $b = [\alpha^i, \Gamma \Rightarrow \Delta]$. Then, we can derive $b' = [\alpha^h, \Gamma \Rightarrow \Delta]$ in H (by the rule $\mathbf{HDet}_1$). It is easy to see that $b' \in \mathsf{Arg}_\mathsf{H}$. However, $\mathbf{Con}(b') = \neg\mathbf{Con}(a)$.

*Proof (**Proposition** 4).* We prove that $\mathrel|\joinrel\sim_{\mathsf{DAC}_4^\mathbb{H}, \mathtt{sta}}^s$ and $\mathrel|\joinrel\sim_{\mathsf{DAC}_4^{\mathbb{H}+}, \mathtt{sta}}^s$ (abbreviated as $\mathrel|\joinrel\sim^s$) satisfy the introduction of disjunction. Let $\mathscr{AF}(\mathbb{K} \oplus \varphi) = \langle \mathsf{Arg}_\varphi, \mathsf{Att}_\varphi \rangle$,

$\mathscr{A}\mathscr{F}(\mathbb{K} \oplus \psi) = \langle \mathsf{Arg}_\psi, \mathsf{Att}_\psi \rangle$, and $\mathscr{A}\mathscr{F}(\mathbb{K} \oplus (\varphi \vee \psi)) = \langle \mathsf{Arg}_{\varphi\vee\psi}, \mathsf{Att}_{\varphi\vee\psi} \rangle$. Let $\mathtt{sta}_\varphi$, $\mathtt{sta}_\psi$, and $\mathtt{sta}_{\varphi\vee\psi}$ denote the stable extensions of Arg_φ, Arg_ψ, and $\mathsf{Arg}_{\varphi\vee\psi}$, respectively.

For an argument set $\mathscr{E}$, define $\Omega = \{a \in \mathscr{E} \mid \varphi^i \in \mathbf{Pre}(a)\}$ and $\Theta = \{[(\varphi \vee \psi)^i, \Delta \Rightarrow \Gamma] \mid [(\varphi \vee \psi)^h, \Delta \Rightarrow \Gamma] \in \mathscr{E}\}$. Define $\mathscr{E}^{\downarrow} = (\mathscr{E} \setminus \Omega) \cup \Theta$. The set $\mathscr{E}^{\downarrow}$ removes all arguments in $\mathscr{E}$ that contain φ^i in their premises and adds arguments that contain $(\varphi \vee \psi)^i$ in their premises. Let

$$\mathtt{sta}_\varphi^* = \{\mathscr{E}^{\downarrow} \mid \mathscr{E} \in \mathtt{sta}_\varphi \text{ and } \forall t \in \mathsf{Arg}_{\varphi\vee\psi} \setminus \mathscr{E}^{\downarrow}, \exists t' \in \mathscr{E}^{\downarrow} \text{ s.t } (t', t) \in \mathsf{Att}_{\varphi\vee\psi}\}$$

Define $\mathtt{sta}_\psi^*$ similarly. We now assert that $\mathtt{sta}_\varphi^* = \mathtt{sta}_{\varphi\vee\psi}$ and $\mathtt{sta}_\psi^* = \mathtt{sta}_{\varphi\vee\psi}$.

We prove that $\mathtt{sta}_\varphi^* = \mathtt{sta}_{\varphi\vee\psi}$. The case for $\mathtt{sta}_\psi^* = \mathtt{sta}_{\varphi\vee\psi}$ is analogous. The left-to-right direction is obvious. We now prove the right-to-left direction.

Suppose $\mathscr{E} \in \mathtt{sta}_{\varphi\vee\psi}$. Let $\Xi = \{[\varphi^i, \Delta \Rightarrow \Gamma] \mid [\varphi^h, \Delta \Rightarrow \Gamma] \in \mathscr{E}\}$. We now assert that $\mathscr{E} \cup \Xi \in \mathtt{sta}_\varphi$. Consistency is obvious. For the sake of contradiction, suppose there exists $b \in \mathsf{Arg}_\varphi \setminus (\mathscr{E} \cup \Xi)$ such that for all $a \in \mathscr{E} \cup \Xi$, $(a, b) \notin \mathsf{Att}_\varphi$. It is easy to see that $\varphi^i \in \mathbf{Pre}(b)$ (if $\varphi^i \notin \mathbf{Pre}(b)$, then $b \in \mathsf{Arg}_{\varphi\vee\psi}$, hence $b \in \mathsf{Arg}_{\varphi\vee\psi} \setminus \mathscr{E}$ or $b \in \mathscr{E}$, both of which clearly contradict the assumption).

Therefore, argument b has the form $b = [\varphi^i, \Delta \Rightarrow \Gamma]$. Let $b' = [\varphi^h, \Delta \Rightarrow \Gamma]$. Clearly, $b' \in \mathsf{Arg}_{\varphi\vee\psi}$ and $b' \notin \mathscr{E}$ (if $b' \in \mathscr{E}$, then $b \in \Xi$, contradicting the assumption). Since $\mathscr{E}$ is stable, there exists $z \in \mathscr{E}$ such that $(z, b') \in \mathsf{Att}_{\varphi\vee\psi}$. Therefore, $\mathbf{Hyp}(z) \subseteq \mathbf{Hyp}(b')$. Without loss of generality, let z have the form $z = [\varphi^h, \Delta_z \Rightarrow \Gamma_z]$. Let $z' = [\varphi^i, \Delta_z \Rightarrow \Gamma_z]$. Clearly, $z' \in \mathscr{E} \cup \Xi$ and $\mathbf{Hyp}(z') \subseteq \mathbf{Hyp}(b)$. Therefore, $(z', b) \in \mathsf{Att}_\varphi$, which leads to a contradiction. Hence, $\mathscr{E} \cup \Xi \in \mathtt{sta}_\varphi$. It is easy to see that $(\mathscr{E} \cup \Xi)^{\downarrow} = \mathscr{E}$. Therefore, $\mathscr{E} \in \mathtt{sta}_\varphi^*$. Since we have shown that $\mathscr{E} \in \mathtt{sta}_{\varphi\vee\psi}$ implies $\mathscr{E} \in \mathtt{sta}_\varphi^*$, we obtain $\mathtt{sta}_{\varphi\vee\psi} \subseteq \mathtt{sta}_\varphi^*$. Thus $\mathtt{sta}_\varphi^* = \mathtt{sta}_{\varphi\vee\psi}$.

Since $\mathbb{K} \oplus \varphi \mid\!\sim^s \alpha$, for any $\mathscr{E} \in \mathtt{sta}_\varphi$, there exists $t \in \mathscr{E}$ such that $\mathbf{Con}(t) = \alpha$ and $\mathbf{Hyp}(t) = \emptyset$. Without loss of generality, let t have the form $t = [\varphi^i, \Delta_t \Rightarrow \alpha]$. It is easy to see that $t' = [\varphi^h, \Delta_t \Rightarrow \alpha] \in \mathscr{E}$. Therefore, for any $\mathscr{E}^* \in \mathtt{sta}_\varphi^*$, there exists $b \in \mathscr{E}^*$ such that $\mathbf{Con}(b) = \alpha$ and $\mathbf{Hyp}(b) = \varphi^h$. A similar result can be proved for $\mathtt{sta}_\psi^*$. Therefore, for each $\mathscr{E} \in \mathtt{sta}_{\varphi\vee\psi}$, there exist $b = [\varphi^h, \Delta_b \Rightarrow \alpha] \in \mathscr{E}$ and $b' = [\psi^h, \Delta_{b'} \Rightarrow \alpha] \in \mathscr{E}$. Using the rule $\mathbf{OR}^{\mathbb{H}}$, we can derive $k = [(\varphi \vee \psi)^i, \Delta_b, \Delta_{b'}] \Rightarrow_{b,b'} \alpha$. It is easy to see that $k \in \mathscr{E}$ and $\mathbf{Hyp}(k) = \emptyset$.

References

1. Aloni, M.: Disjunction. In: Zalta, E.N., Nodelman, U. (eds.) The Stanford Encyclopedia of Philosophy. Metaphysics Research Lab, Stanford University, Winter 2024 edn. (2024)
2. Arieli, O., Borg, A., Heyninck, J., Straßer, C.: Logic-based approaches to formal argumentation. In: Handbook of Formal Argumentation, vol. 2, pp. 1793–1898. College Publications (2021)

3. Arieli, O., Borg, A., Straßer, C.: A postulate-driven study of logical argumentation. Artif. Intell. **322**, 103966 (2023)
4. Baroni, P., Gabbay, D., Giacomin, M., van der Torre, L. (eds.): Handbook of Formal Argumentation. College Publications, London, England (2018)
5. Beirlaen, M., Heyninck, J., Straßer, C.: Reasoning by cases in structured argumentation. In: Proceedings of the Symposium on Applied Computing - SAC '17 (2017)
6. van Berkel, K., Straßer, C.: Towards deontic explanation through dialogue. In: Proceedings of ArgXAI 2024, the 2nd International Workshop on Argumentation for eXplainable AI (2024)
7. van Berkel, K., Straßer, C.: Reasoning with and about norms in logical argumentation. In: Toni, F., Polberg, S., Booth, R., Caminada, M., Kido, H. (eds.) Computational Models of Argument, Proceedings, pp. 332–343. IOS Press (2022)
8. van Berkel, K., Straßer, C., Zhou, Z.: Towards an argumentative unification of default reasoning. In: Chris Reed, Matthias Thimm, T.R. (ed.) Computational Models of Argument, proceedings. pp. 313–324. IOS Press (2024)
9. Booth, R., Varzinczak, I.: Conditional inference under disjunctive rationality. In: Proceedings of the AAAI Conference on Artificial Intelligence, vol. 35, no. 7, pp. 6227–6234 (2021)
10. Caminada, M.: On the issue of contraposition of defeasible rules. In: Proceedings of COMMA2008, pp. 109–115. IOS Press, NLD (2008)
11. Ciardelli, I., Roelofsen, F.: Inquisitive logic. J. Philos. Log. **40**, 55–94 (2011)
12. Čyras, K., Rago, A., Albini, E., Baroni, P., Toni, F., et al.: Argumentative xai: a survey. In: Proceedings of the Thirtieth International Joint Conference on Artificial Intelligence, pp. 4392–4399 (2021)
13. D'Agostino, M., Modgil, S.: Classical logic, argument and dialectic. Artif. Intell. **262**, 15–51 (2018)
14. Dung, P.M.: On the acceptability of arguments and its fundamental role in nonmonotonic reasoning, logic programming and n-person games. Artif. Intell. **77**(2), 321–357 (1995)
15. Gelfond, M., Lifschitz, V., Przymusinska, H., Truszczynski, M.: Disjunctive defaults. In: Proceedings of the Second International Conference on Principles of Knowledge Representation and Reasoning, pp. 230–237 (1991)
16. Goldszmidt, M., Pearl, J.: Qualitative probabilities for default reasoning, belief revision, and causal modeling. Artif. Intell. **84**(1–2), 57–112 (1996)
17. Grice, H.P.: Logic and conversation. In: Speech Acts, pp. 41–58. Brill (1975)
18. Kraus, S., Lehmann, D., Magidor, M.: Nonmonotonic reasoning, preferential models and cumulative logics. Artif. Intell. **44**(1–2), 167–207 (1990)
19. Lehmann, D.J., Magidor, M.: What does a conditional knowledge base entail? Artif. Intell. **55**(1), 1–60 (1992)
20. Makinson, D., van der Torre, L.: Constraints for input/output logics. J. Philos. Log. **30**(2), 155–185 (2001)
21. Negri, S., Von Plato, J.: Structural Proof Theory. Cambridge university press, Cambridge (2008)
22. Reiter, R.: A logic for default reasoning. Artif. Intell. **13**(1–2), 81–132 (1980)

A Finitary Axiomatization of Arbitrary Social Announcement Logic

Rui Zhu[(✉)] [iD]

Arts, University of Auckland, Auckland, New Zealand
`zrui956@aucklanduni.ac.nz`

Abstract. This paper presents a finitary axiomatization of Arbitrary Social Announcement Logic (ASAL), a dynamic epistemic logic modeling belief diffusion in social networks. ASAL extends Social Announcement Logic (SAL) with arbitrary announcement operators, resembling those in Arbitrary Public Announcement Logic (APAL). Unlike APAL, ASAL is based on belief rather than knowledge and allows inconsistent beliefs and local information flow. While ASAL shares structural similarities with Boolean APAL (BAPAL), our approach differs by avoiding the necessity form technique and instead using a novel model transformation method. We prove the soundness of the finitary axiomatization, including a derivation rule for arbitrary announcements, and establish weak completeness via Henkin-style canonical models. Illustrative examples clarify the semantic distinctions from PAL-based systems and display the model transformation process. We conclude with potential extensions of ASAL and directions for future research.

Keywords: Dynamic Epistemic Logic · Arbitrary Announcement · Social Networks

1 Introduction

Arbitrary announcement operators, which quantify over all possible propositional or epistemic statements, offer a powerful means of analyzing information updates. These operators have been extensively studied in the context of Public Announcement Logic (PAL), particularly in its extension known as Arbitrary Public Announcement Logic (APAL), which allows reasoning about knowledge updates after any possible public announcement. The standard technique used in axiomatizing APAL and its variants is *necessity form*, introduced by Goldblatt [11] and applied in [3]. However, it was shown in [12] that a key inference rule for arbitrary epistemic quantification is unsound, casting doubt on the soundness of certain axiomatizations.

Due to its high expressivity, APAL is known to be undecidable [9]. While the subsequent works [2,4] provide APAL with an infinitary axiomatization, a finitary axiomatization remains unknown. The only known fragment of APAL that admits a finitary axiomatization is Boolean Arbitrary Public Announcement Logic (BAPAL), which restricts arbitrary announcements to Boolean formulas [15]. Another approach to achieving finitary axiomatizability is to extend the

V. Goranko et al. (Eds.): LORI 2025, LNCS 16010, pp. 195–212, 2026.
https://doi.org/10.1007/978-981-95-2481-5_14

model with a memory mechanism, resulting in Arbitrary Public Announcement Logic with Memory (APALM) [6].

The finitary axiomatizability of original APAL remains an open problem. Recent studies have further explored the boundaries of this issue: Ågotnes and Galimullin [1] investigated APAL extended with a *common knowledge* operator (APALC), and Galimullin and Kuijer [10] demonstrated that APALC is not finitarily axiomatizable.

Distinct from public announcements, which update the knowledge of all agents simultaneously, several logical frameworks have been proposed for epistemic reasoning in social networks [5, 7, 8, 13, 14]. These frameworks emphasize that information updates are *local* rather than global, and are shaped by the structure of the underlying social network. In such settings, an agent's announcement reaches only their direct followers, allowing for reasoning about belief diffusion and the structure of the social network. To formalize such belief updates directly, Xiong et al. [18] introduced *Propositional Network Announcement Logic* (PNAL), in which agents sincerely announce propositional beliefs to their followers, and beliefs evolve through the network accordingly.

Although PNAL may appear similar to PAL when the network is a complete digraph, it is neither a syntactic nor a semantic fragment of PAL. The belief-based semantics of PNAL, which allows agents to hold inconsistent beliefs, marks a significant departure from the knowledge-based semantics of PAL. Furthermore, sincerity in PNAL is optional: PNAL uses Boolean valuations rather than point Kripke models, and hence does not have preconditions for epistemic updates. These differences motivate a broader framework. We propose that PNAL and its extensions form a general family called *Social Announcement Logic* (SAL), which highlights the role of social network in information flow and reflects epistemic assumptions distinct from those in PAL.

Subsequent work has extended this framework to reason about structural properties [16]. Within this family, Xiong and Ågotnes [17] introduced arbitrary announcement operators into SAL, resulting in an infinitary system similar in spirit to APAL. However, their work left the open problem of constructing a finitary axiomatization for this logic.

Given that BAPAL admits a finitary axiomatization, and that the network structure in SAL does not interfere with the quantification over announcements, one might expect that similar technique could suffice. However, in this work, we provide a finitary axiomatization for SAL with arbitrary announcement operators, referred to as *Arbitrary Social Announcement Logic* (ASAL), based on a novel model transformation technique. Unlike BAPAL [15], which relies on the necessity form technique, our approach avoids this method altogether and instead offers a more standard and transparent proof. This alternative approach contributes to a clearer conceptual understanding of the completeness argument.

There are strong motivations for pursuing this approach. First, offering an independent proof ensures conceptual clarity and serves as a cross-verification of the finitary system. We observe that many steps in the completeness proof of BAPAL [15] reduce to arguments from APAL, which itself lacks a finitary axiomatization [4]. Furthermore, necessity-form-based technique is often presented in a sketchy manner, and it remains unclear at which step soundness fails in earlier attempts [3]. Second, our model transformation technique is conceptually more standard and direct. As we show, the Henkin-style construction is sufficient for dealing with arbitrary announcements, provided the model is appropriately pre-processed. Third, although SAL with a complete network may resemble PAL, the logic allows for announcement types that are not definable within PAL. This conceptual divergence justifies the need for distinct proof techniques and semantic interpretations.

This paper begins by formally introducing the syntax and semantics of ASAL. We then focus on the soundness of the rule for arbitrary announcements, which constitutes the main technical contribution. Based on this result, we present a finitary axiomatization and establish its weak completeness using a Henkin-style canonical construction. The paper concludes with a discussion of potential extensions and directions for future work.

2   Preliminaries

The language of arbitrary social announcement logic is an expansion of the language of propositional network announcement logic given in [18] by adding arbitrary operators.

Definition 1 (Language of ASAL). *Let* A *be a nonempty set of agents,* Prop *a countable set of propositional variables. A message* θ *and a formula* ϕ *in the language of ASAL, denoted* L_{ASAL}, *are defined as follows:*

$$\theta ::= p \mid \neg\theta \mid (\theta \wedge \theta) \qquad \phi ::= B_a\theta \mid \neg\phi \mid (\phi \wedge \phi) \mid \langle a!\theta \rangle\phi \mid \langle a! \rangle\phi$$

where $p \in$ Prop, $a \in$ A.

The standard logical connectives $\vee$, $\rightarrow$, $\leftrightarrow$ and the dual operators $[a!\theta]$ and $[a!]$ are defined as usual. The dynamic modalities $\langle a!\theta \rangle\phi$ and $\langle a! \rangle\phi$ represent epistemic changes induced by social announcements. The former means "after agent a sincerely announces θ to their followers, ϕ holds," while the latter expresses that "after some sincere announcement by a to their followers, ϕ holds." Following the notation in [18], we write $\langle \vec{c}! \rangle$ to denote a sequence of diamond sincere social announcement operators (e.g., $\langle c!\theta \rangle$), and $[\vec{c}!]$ for the corresponding box modalities.

Note that ASAL does not allow nested epistemic modalities such as $B_a B_b q$, and thus does not express higher-order beliefs. This reflects our main aim to focus on belief propagation rather than higher-order reasoning.

Definition 2 (Denotation). *Let* $V = \wp(\text{Prop})$ *be the* belief space. *The denotation function* $[\![\,]\!]$ *assigns to each message* θ *a subset* $[\![\theta]\!]$ *of* V, *called the* denotation *of* θ, *as follows:*

$$[\![p]\!] = \{x \in V \mid p \in x\}, \quad [\![\neg\theta]\!] = V \backslash [\![\theta]\!], \quad [\![(\theta_1 \wedge \theta_2)]\!] = [\![\theta_1]\!] \cap [\![\theta_2]\!]$$

Definition 3 (Satisfaction). *A model $m = (f, k)$ consists of a network function f, which assigns to each agent a a set $f(a) \subseteq \mathsf{A}$ of followers, and an epistemic distribution k, assigning to each agent a a belief state $k(a) \subseteq V$. Given a message θ, the update $[a : \theta]k$ is defined:*

$$[a : \theta]k(b) = \begin{cases} k(b) \cap [\![\theta]\!] & \text{if } b \in f(a), \\ k(b) & \text{otherwise.} \end{cases}$$

We define the satisfaction relation $\vDash$ between models and formulas as follows:

$$
\begin{aligned}
m &\vDash B_a\theta \iff k(a) \subseteq [\![\theta]\!] && (\text{agent } a \text{ believes } \theta) \\
m &\vDash \neg\phi \iff m \nvDash \phi \\
m &\vDash (\phi \wedge \psi) \iff m \vDash \phi \text{ and } m \vDash \psi \\
m &\vDash \langle a!\theta\rangle\phi \iff k(a) \subseteq [\![\theta]\!] \text{ and } (f, [a : \theta]k) \vDash \phi \\
m &\vDash \langle a!\rangle\phi \iff \exists\theta \text{ s.t. } k(a) \subseteq [\![\theta]\!] \text{ and } (f, [a : \theta]k) \vDash \phi
\end{aligned}
$$

We write $(f, [a : \theta]k)$ as $[a : \theta]m$ to denote the updated model. A formula ϕ is said to be satisfiable *if there exists a model m such that $m \vDash \phi$, and* valid *if it holds in all models. A set Σ of formulas semantically entails ϕ, written $\Sigma \vDash \phi$, if every model satisfying all formulas in Σ also satisfies ϕ.*

As shown in [18], self-following (i.e., whether $a \in f(a)$) cannot be expressed in the language of PNAL. Thus, we impose no restrictions on $f(a)$, allowing agents to broadcast messages to themselves, although this may seem counterintuitive in a social network context.

Example 1. We illustrate the semantics of ASAL using a diagrammatic model (f, k) and its update $[a : r]k$. In Fig. 1a, each agent is assigned a unique color, and arrows indicate following relationships. Nodes represent elements of the belief space V.

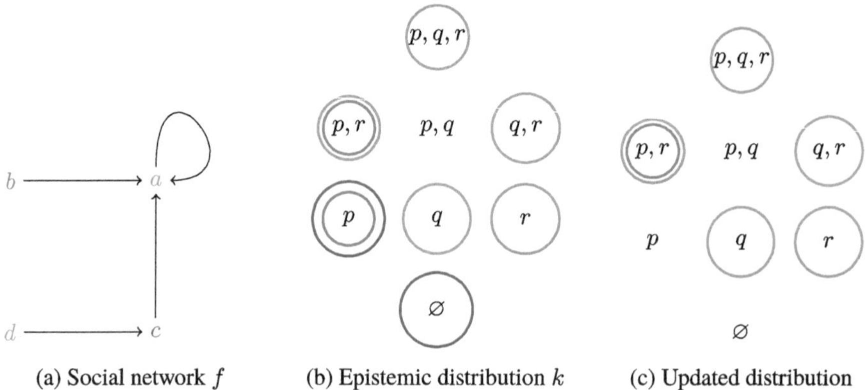

(a) Social network f (b) Epistemic distribution k (c) Updated distribution

Fig. 1. A model (f, k) and its update $[a : r]k$

Figure 1b illustrates the epistemic distribution k. A node circled in an agent's color belongs to that agent's belief state. It is easy to check that $(f, k) \vDash B_a(p \wedge r) \wedge \neg B_a q$ because all nodes circled in orange (agent a) satisfy $p \wedge r$, while the point $\{p, r\}$ does not belong to $[\![q]\!]$. Figure 1c displays the updated epistemic distribution after agent a announces r. In this updated model, we have $[a : r]m \vDash B_b r \wedge B_c p \wedge \neg B_d r$.

This example also illustrates a key distinction between SAL and PAL. In particular, $[a : r]m \vDash B_c \bot$, indicating that agent c holds an inconsistent belief after receiving the announcement. Such inconsistency is not allowed in PAL. As a logic of knowledge, PAL is based on Kripke structures enforcing consistency via equivalence relations.

Furthermore, our model structure naturally accommodates a more flexible dynamic operator known as a *free social announcement*, defined as:

$$m \vDash [a : \theta]\phi \iff [a : \theta]m \vDash \phi$$

if the language is expanded with $[a : \theta]$. Free social announcements allow epistemic updates without requiring any preconditions such as sincerity. This concept does not have a counterpart in PAL, where public announcements are interpreted over pointed Kripke models and require a condition that the designated world cannot be eliminated during the updates. Free announcements not only extend expressivity but also simplify completeness proofs. However, since the focus of this paper is to solve the open problem concerning arbitrary *sincere* announcements in [17], we focus on arbitrary sincere announcements and do not explore free announcements further.

Building on [17,18], we recall existing validities for explicit social announcement operators and introduce those involving arbitrary operators.

Proposition 1 (Arbitrary Validities). *The following schemes are valid and the rule preserves validity.*

$K_!$ $[a!](\phi \rightarrow \psi) \rightarrow ([a!]\phi \rightarrow [a!]\psi)$ $Arb_!$ $\langle a!\theta\rangle\phi \rightarrow \langle a!\rangle\phi$
$Dual_!$ $\neg\langle a!\rangle\phi \leftrightarrow [a!]\neg\phi$ $Nec_!$ *From* $\vdash \phi$, *infer* $\vdash [a!]\phi$

Proof. The proofs for $K_!$, $Dual_!$, *and* $Nec_!$ *are straightforward. We focus on proving* $Arb_!$. *Assume there exists a model* $m = (f, k)$ *such that* $m \vDash \langle a!\theta\rangle\phi$. *By definition 3,* $k(a) \subseteq [\![\theta]\!]$ *and* $[a : \theta]m \vDash \phi$. *Then, there exists a message* γ *such that* $k(a) \subseteq [\![\gamma]\!]$ *and* $[a : \gamma]m \vDash \phi$ *when we consider* $\gamma = \theta$. *By definition 3 we have* $m \vDash \langle a!\rangle\phi$. $\square$

The standard Henkin construction method requires a rule that ensures every formula with an arbitrary announcement operator has a witness in the maximal consistent set built through the construction. As is customary, formulas with new variables occurring in explicit announcements serve as witnesses. Therefore, we must establish that the following rule preserves validity.

Arb_D *From* $\langle a!p\rangle\phi \rightarrow \psi$, infer $\langle a!\rangle\phi \rightarrow \psi$ where p does not occur in ϕ or ψ.

The following section is devoted to the soundness proof of Arb_D. This forms the most technical part in this paper, as it requires constructing an appropriate model transformation that preserves validity under arbitrary social announcements.

3   Soundness

3.1 p-Free Models

First, we analyze the difficulty in proving $\mathsf{Arb_D}$. Suppose $\langle a!\rangle\phi \to \psi$ is not valid; we must then demonstrate that $\langle a!p\rangle\phi \to \psi$ is invalid. A common approach is to transform the model that falsifies $\langle a!\rangle\phi \to \psi$ to obtain a new model that meets our requirements. The key condition is that in the finally transformed model, agent a must believe the fresh proposition p, and the announcement of p must ensure that ϕ holds while ψ does not.

However, simply introducing a new message is insufficient, as its semantics might already be implicitly constrained by the agents' belief states. For instance, in a given model, some agents may already believe p while others do not. If p is implicitly used in an arbitrary announcement, using it as the announcement could lead to unintended consequences.

To address this, we first transform the given model into a form where no agent's belief state containing any point in the denotation of p. In other words, every agent should believe $\neg p$, ensuring that any announcement involving p is equivalent to making the same announcement but replacing p with $\bot$. To achieve this transformation systematically, we introduce an enumeration process that helps regulate the introduction of fresh propositions.

Definition 4 (Agreement function). *Let $\Pi = (p, p_1, p_2, ...)$, where $p_0 = p$ be an injective enumeration of an infinite subset of Prop.*

(a) The successor *function of Π is a function $g : \mathsf{Prop} \to \mathsf{Prop}$ defined by*

$$g(q) = \begin{cases} p_{n+1} & \text{if } q = p_n \text{ for } n \geq 0; \\ q & \text{otherwise} \end{cases}$$

(b) Given a point $w \subseteq \mathsf{Prop}$, we say $g[w] = \{g(q) \mid q \in w\}$ is the image *of w under g.*
(c) Then an operation of message $g^*\theta$ over function g is recursively defined:

$$g^*q = g(q) \qquad g^*\neg\theta = \neg g^*\theta \qquad g^*(\theta \wedge \gamma) = g^*\theta \wedge g^*\gamma$$

Note that by definition $g^*\bot = \bot$ because $\bot$ stands for formulas in form of $\theta \wedge \neg\theta$, and then we can see $g^*\bot = g^*\theta \wedge \neg g^*\theta = \bot$.
(d) Given a set $\Sigma \subseteq V$, the image of Σ is defined by $g[\Sigma] = \{g[w] \mid w \in \Sigma\}$.
(e) Given a function k, The image of k is a function $g[k] : \mathsf{A} \to \wp(V)$ defined by $g[k](a) = g[k(a)]$ for all $a \in \mathsf{A}$.

The successor function maps each propositional variable to its successor according to the enumeration. The image function applies this function to every propositional variable in a set, ensuring that no element contains p. Apart from a function for successors, we also introduce a function for preceders, termed as *predecessor functions*. These enable us to retrieve the original messages when they do not contain p.

Definition 5 (Preceder). *Given an enumeration $\Pi = (p_0, p_1, p_2, ...)$ with $p_0 = p$, we define a* predecessor function $\bar{g} : \mathsf{Prop} \setminus \{p\} \to \mathsf{Prop}$ *by saying*

$$\bar{g}(q) = \begin{cases} p_{n-1} & \text{if } q = p_n \text{ for } n > 0; \\ q & \text{otherwise} \end{cases}$$

We also have an operation of message $\bar{g}^*\theta$ over function $\bar{g}$ which is defined:

$$\bar{g}^*q = \bar{g}(q) \qquad \bar{g}^*\neg\theta = \neg\bar{g}^*\theta \qquad \bar{g}^*(\theta \wedge \gamma) = \bar{g}^*\theta \wedge \bar{g}^*\gamma$$

Lemma 1 (Identities). *For any message θ not containing p and any message γ, we can prove that*

$$\bar{g}^*g^*\gamma = \gamma \tag{s.p.}$$
$$g^*\bar{g}^*\theta = \theta \tag{p.s.}$$

Now we define our first target: p-free models where announcing p in a p-free model is equivalent to announcing $\bot$, as every agent believes $\neg p$. In this model, the role of each variable is substituted by its successor. The infinite set Prop allows us to shift propositions via g, similar to introducing fresh constants in First-Order Logic.

Definition 6 (p-free model). *A model $m = (f, k)$ is a p-free model if*

$$\bigcup_{i \in \mathsf{A}} k(i) \subseteq \llbracket \neg p \rrbracket$$

Lemma 2 (Substitution of p-free model). *Given a p-free model $m = (f, k)$, it is provable that for every message θ,*

$$[a : \theta]m \vDash \phi \iff [a : \theta[p/\bot]]m \vDash \phi \tag{p.sub}$$

Proof. We show this result by induction on θ.
Base case *when $\theta = q \neq p$ we know $\theta = \theta[p/\bot]$, the result is obtained trivially. When $\theta = p$ consider for every agent $i \in A$, if $i \notin f(a)$ we have $k(i) = [a : p]k(i) = [a : p[p/\bot]]k(i)$ by def.3. If $i \in f(a)$, we know $[a : p]k(i) = k(i) \cap \llbracket p \rrbracket = k(i) \cap \llbracket \bot \rrbracket = [a : p/\bot]k(i)$ because in a p-free model it holds that $k(i) \cap \llbracket p \rrbracket = \varnothing$ for every agent i.*
Inductive case *when $\theta = \neg\gamma$ or $\theta = \gamma \wedge \chi$, the result is covered by the I.H..* $\square$

Proposition 2 (Value Reservation). *The following equivalence is provable:*

$$w \in \llbracket \theta \rrbracket \iff g[w] \in \llbracket g^*\theta \rrbracket \tag{v.res}$$

Lemma 3 (Value Agreement). *Given a set $\Sigma \subseteq V$, $g[\Sigma]$ the image of Σ under a successor function g, we can show that for every message θ,*

$$\Sigma \subseteq \llbracket \theta \rrbracket \iff g[\Sigma] \subseteq \llbracket g^*\theta \rrbracket \tag{v.agr}$$

Now we can show that for every formula that does not contain any message of Π, the model transformation will preserve its truth value. For some p that may be implicitly used in an arbitrary announcement, we can replace it according to quantification in the new model.

Furthermore, we need to note that although Π has countably infinite variables, each formula ϕ has only finitely many messages, and it cannot be co-finite. Therefore, we can always define a Π.

Lemma 4 (Changing p-free model). *Given a model $m = (f, k)$, it is provable that for every message θ and Π-free formula ϕ*

$$g[[a : \theta]m] \vDash \phi \Longrightarrow [a : g^*\theta]g[m] \vDash \phi \qquad \text{(outwards)}$$
$$[a : \theta]g[m] \vDash \phi \Longrightarrow g[[a : \bar{g}^*\theta[p/\bot]]m] \vDash \phi \qquad \text{(inwards)}$$

Proof. The proof of outwards follows from v.agr and g's distribution over set-intersection that for every agent $i \in f(a)$ that $g[k(i) \cap [\![\theta]\!]] = g[k(i)] \cap [\![g^*\theta]\!]$. To show inwards we only need to prove

$$[a : \theta[p/\bot]]g[m] \vDash \phi \Longrightarrow g[[a : \bar{g}^*\theta[p/\bot]]m] \vDash \phi \qquad \text{(back)}$$

because $g[m]$ is a p-free model and we can apply p.sub to the antecedent. The proof of back also follows from g's distribution over set-intersection. Just note $\bar{g}^*\theta[p/\bot]$ exists because there is no p in $\theta[p/\bot]$ and $\theta[p/\bot] = g^*\bar{g}^*\theta[p/\bot]$ due to p.s.. □

Theorem 1 (Model Reservation). *Given a model $m = (f, k)$ and an infinite enumeration of atomic messages $\Pi = (p, p_1, p_2, ...)$ and its successor function g. It is provable that for every Π-free formula ϕ,*

$$m \vDash \phi \Longleftrightarrow g[m] \vDash \phi \qquad \text{(m.res)}$$

After establishing m.res, we conclude that a p-free formula is satisfiable if and only if it is satisfied in a p-free model. Since every formula contains only finitely many propositional variables, it is always possible to construct an infinite enumeration Π and define a corresponding successor function g accordingly.

Example 2. We illustrate here the model transformation to a p-free model using diagrammatic representations. Suppose a social network where b follows a but does not follow c. Let Π be a fixed enumeration such that $p = p_0$, $r = p_1$, $s = p_2$, and so on, and suppose that q does not occur in Π.

Figure 2a shows the epistemic distribution in the original model, while Fig. 2b displays the corresponding p-free model obtained via our model transformation technique. Nodes circled with dashed lines represent belief states generated through transformation, whereas solid-circled nodes remain unchanged during the process.

For any Π-free formula, the original and transformed models agree on satisfaction. For example, both models satisfy the formula $\neg B_a q \wedge \neg B_c \neg q$. Moreover, the original model satisfies $\langle a! \rangle B_b \neg q$, since a can sincerely announce $p \vee \neg q$; and in the p-free model, this remains true because a can instead announce $s \vee \neg q$ sincerely.

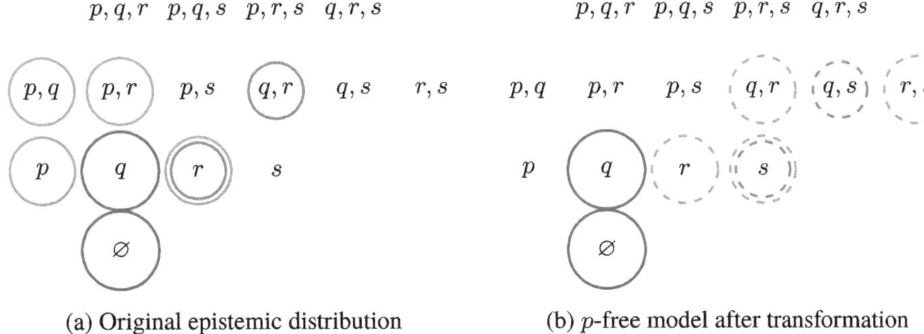

(a) Original epistemic distribution   (b) p-free model after transformation

Fig. 2. Model transformation to a p-free model

The first step to construct a p-free model ends here. In our plan, we treat p as a controlled message, aiming to ensure that whenever a formula ϕ is satisfied after announcing some θ in a given model m, there exists a model, obtained from the p-free model, in which ϕ is satisfied by announcing p in place of θ. This requires a second model transformation which will be addressed in the following subsection.

3.2   Soundness of Derivation Rule

To ensure that the social announcement of p has the same effect as announcing θ, we need to ensure that the model satisfies the formula $B_b(p \leftrightarrow \theta)$ for all $b \in A$. However, we cannot change each belief state arbitrarily. The model transformation must preserve the satisfaction of p-free formulas. This requires establishing invariance properties for p-free formulas.

Proposition 3 (Value Invariant).   *Given $p \in$ Prop and $w, v \in V$: we can show that the equivalence $v \in [\![\theta]\!] \iff w \in [\![\theta]\!]$ implies that for any message γ not containing p and $w \in V$,*

$$w \in [\![\gamma]\!] \iff w \cup \{p\} \in [\![\gamma]\!] \iff w \setminus \{p\} \in [\![\gamma]\!] \qquad \text{(o.agr)}$$

Next, we introduce a modulo-based technique along with a modified satisfaction relation. This refinement allows us to streamline the subsequent soundness proof, making it more concise and transparent.

Definition 7 (Invariant Equivalence).   *For $\Sigma \subseteq V$ and $p \in$ Prop, let*

$$\Sigma + p = \{w \cup \{p\} \mid w \in \Sigma\}$$

We say models $m_1 = (f_1, k_1)$ and $m_2 = (f_2, k_2)$ agree modulo p, write $m_1 \sim_p m_2$ if $k_1(a) + p = k_2(a) + p$ for all $a \in A$. We also say m_1 and m_2 are p-invariant equivalent if they agree modulo p.

Example 3. We now illustrate the notion of invariant equivalence using diagrammatic models. The two epistemic distributions below are derived from the model of Fig. 2b by joining the proposition p into some or all points of each agent's belief state. Nodes obtained via this modification are marked with dotted circles.

For instance, in Fig. 3a, the belief state $\{p, q, r\}$ of agent a results from joining p to $\{q, r\}$; likewise, b's belief state $\{p, q, s\}$ comes from extending $\{q, s\}$, and c's $\{p, q\}$ from $\{q\}$. In Fig. 3b, p is added to every point of each agent's belief state.

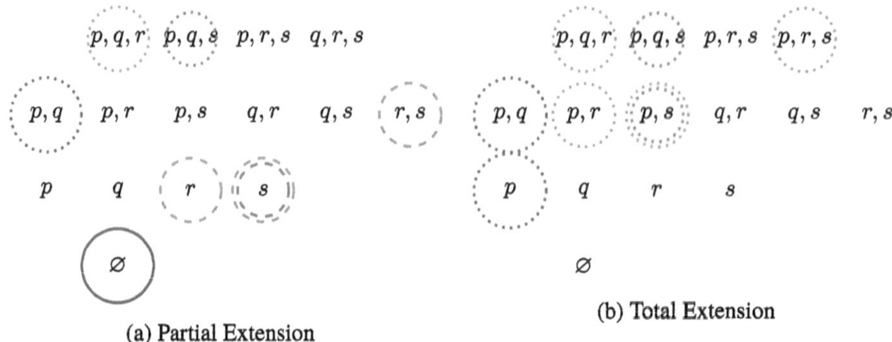

(a) Partial Extension

(b) Total Extension

Fig. 3. Invariance Equivalence Between Different Extensions

Given the same social network structure, the three models of Figs. 2b, 3a and 3b are pairwise p-invariant. That is, for every agent i, the set $k(i) + p$ is identical across the three models as the case represented by Fig. 3b.

Definition 8 (Unreference Satisfaction). *For a formula ϕ not containing p, we define satisfaction without reference to p, denoted $m \vDash_p \phi$, as follows:*

$m \vDash_p \phi \iff m \vDash \phi$ *if ϕ is not an arbitrary announcement formula*

$m \vDash_p \langle a! \rangle \phi \iff$ *there exists a θ that does not contain p such that $k(a) \subseteq \llbracket \theta \rrbracket$ and $[a : \theta] m \vDash_p \phi$*

This satisfaction will play a transitional role in our proof as $\vDash_p$ ensures $\langle a! \rangle \phi$ is equiped with adequate p-free announcements, aiding $\mathsf{Arb_D}$ proof. Next, we will examine some special equivalent models under this satisfaction. Before doing so, we need to note the following result.

Proposition 4 (Model Dynamics). *If $k_1(i) + p = k_2(i) + p$ for all $i \in \mathsf{A}$ then for any message θ not containing p and any $a \in \mathsf{A}$*

$$[a : \theta] k_1(i) + p = [a : \theta] k_2(i) + p \qquad \text{(mod.dy)}$$

Lemma 5 (Invariant Preservation). *If $m_1 \sim_p m_2$ then for every ϕ not containing p, it is provable that*

$$m_1 \vDash_p \phi \iff m_2 \vDash_p \phi$$

The next step is to show that under certain conditions, the two satisfaction relationships are equivalent. To show it, we need the following proposition first.

Proposition 5 (Substitution Equations). *If for some message θ not containing p and agent $a \in A$, $k(a) \subseteq [\![p \leftrightarrow \theta]\!]$ then it is provable that*

$$k(a) \cap [\![\gamma]\!] = k(a) \cap [\![\gamma[p/\theta]]\!].$$

Proof. To show this result, since we know $k(a) \subseteq [\![p \leftrightarrow \theta]\!]$, we only need to show

$$[\![p \leftrightarrow \theta]\!] \cap [\![\gamma]\!] = [\![p \leftrightarrow \theta]\!] \cap [\![\gamma[p/\theta]]\!]$$

This can be shown by induction on γ. The basic case is covered by propositional logic and set theory. The Boolean cases can be obtained by I.H. easily because the substitution $[p/\gamma]$ is defined recursively so it can distribute into each atomic subformula. $\square$

Lemma 6 (Equivalent Satisfaction). *If for some message θ not containing p and all $i \in A$, $m \vDash B_i(p \leftrightarrow \theta)$ then for all ϕ not containing p, it is provable that*

$$m \vDash \phi \Longleftrightarrow m \vDash_p \phi$$

Proof. We show this result by induction on ϕ. If ϕ is not an arbitrary announcement formula, the result is obvious so we only consider the arbitrary case. Let $m = (f, k)$, we have

$$
\begin{aligned}
(f, k) \vDash \langle a! \rangle \psi &\Leftrightarrow (f, [a : \gamma]k) \vDash \psi \ \text{ and } \ (f, k) \vDash B_a \gamma && \textit{for some } \gamma \textit{ by def.3.} \\
&\Leftrightarrow [a : \gamma[p/\theta]](f, k) \vDash \psi \ \text{ and } \ (f, k) \vDash B_a \gamma[p/\theta] && \textit{by prop.5} \\
&\Leftrightarrow [a : \gamma[p/\theta]](f, k) \vDash_p \psi \ \text{ and } \ (f, k) \vDash_p B_a \gamma[p/\theta] && \textit{by the I.H.} \\
&\Leftrightarrow (f, k) \vDash_p \langle a! \rangle \psi && \textit{by def.8}
\end{aligned}
$$

This completes the induction and the proof. $\square$

To determine the condition under which two messages are interchangeable in updating belief states, we introduce the following lemma.

Lemma 7 (Interchangeable Messages). *If $m \vDash B_i(\theta \leftrightarrow \delta)$ for all $i \in A$, then it is provable that for all a,*

$$m \vDash \langle a!\theta \rangle \phi \leftrightarrow \langle a!\delta \rangle \phi$$

Proof. To show this result, we only prove that $[a : \theta]m = [a : \delta]m$. Let $[a : \theta]m = (f_1, k_1)$ and $[a : \delta]m = (f_2, k_2)$. It is clear $f_1 = f_2$, we show that for each agent $i \in A$, $k_1(i) = k_2(i)$. For agent i that not follow a, the result is trivial. We only consider the case that i follows a. Since $k(i) \subseteq [\![\theta \leftrightarrow \delta]\!]$, the result is covered by $[\![\theta \leftrightarrow \delta]\!] \cap [\![\theta]\!] = [\![\theta \leftrightarrow \delta]\!] \cap [\![\delta]\!]$. $\square$

We now proceed to show that the derivation rule preserves validity, which is the final step in establishing soundness. The lemmas we have proved so far will play significant roles in this proof.

Proposition 6 (Arbitrary Derivation). Arb_D *preserves validity.*

Proof. Suppose $\langle a!p\rangle\phi \to \psi$ is valid and then there is a model $m = (f,k)$ such that $(f,k) \vDash \langle a!\rangle\phi$. Then there exists some θ such that $k(a) \subseteq [\![\theta]\!]$ and $[a:\theta]m \vDash \phi$. By outwards, we know $(f,[a:g^*\theta]g[k]) \vDash \phi$ and by v.agr $g[k](a) \subseteq [\![g^*\theta]\!]$.

Moreover, since $g[\Sigma] \cap [\![p]\!] = \varnothing$ for every $\Sigma \subseteq V$, we have $(f,g[k]) \vDash B_b(p \leftrightarrow \bot)$ for all $b \in \mathsf{A}$. So by def.6, we have $(f,g[k]) \vDash_p \langle a!g^*\theta\rangle\phi$. Just note that $\langle a!g^*\theta\rangle\phi$ does not contain p. Now define a new model $m' = (f,k')$ by specifying

$$k'(b) = (g[k(b)] \setminus [\![g^*\theta]\!]) \cup ((g[k(b)] \cap [\![g^*\theta]\!]) + p)$$

This construction guarantees that $m' \sim_p g[m]$ as $k'(a)+p = g[k](a)+p$ for all $a \in \mathsf{a}$. Then by lemma 5, we have $m' \vDash_p \langle a!g^*\theta\rangle\phi$. Also, since $m' \vDash B_b(p \leftrightarrow g^*\theta)$ for all b and $g^*\theta$ does not contain p, by def.6 we have $m' \vDash \langle a!g^*\theta\rangle\phi$, and by lemma 7 we have $m' \vDash \langle a!p\rangle\phi$. Thus, by def.3 we have $m' \vDash \langle a!p\rangle\phi$.

Since $\langle a!p\rangle\phi \to \psi$ is valid, we have $m' \vDash \psi$. Additionally, in m', announcing p is equivalent to announcing $g^*\theta$, which does not contain p, so $m' \vDash_p \psi$. Given $m' \sim_p g[m]$ and ψ lacking p, by Lemma 5, we deduce $g[m] \vDash_p \psi$ and $g[m] \vDash_p g^*\psi$. Note that $g^*\psi = \psi$.

Finally, by 6, we know $g[m] \vDash g^*\psi$ and then by m.res $m \vDash \psi$. Therefore, we show that for arbitrary model m, we have $m \vDash \langle a!\rangle\phi \to \psi$, and hence that $\langle a!\rangle\phi \to \psi$ is valid. □

The construction in this part is the core result of this paper, as it establishes not only the soundness of the derivation rule $\mathsf{Arb_D}$ in social announcement logic but also provides a new model transformation method for dynamic epistemic logic with arbitrary operators.

4   Completeness

Definition 9 (System of ASAL). *Let* sal *denote the formal system for* L_{ASAL}, *consisting of all formulas that can be derived from the axioms and inference rules specified below. A formula is said to be a theorem of* sal *if it can be obtained by a finite application of these axioms and rules.*

Axioms and Inference Rules of sal

Taut	*All subs. ins. of prop. taut.*	$\mathsf{K_!}$	$[a!](\phi \to \psi) \to ([a!]\phi \to [a!]\psi)$
$\mathsf{K_B}$	$B_a(\theta \to \gamma) \to (B_a\theta \to B_a\gamma)$	$\mathsf{Dual_!}$	$\neg\langle a!\rangle\phi \leftrightarrow [a!]\neg\phi$
$\mathsf{K_\top}$	$B_a\theta$ *(when θ is a taut.)*	$\mathsf{Arb_I}$	$\langle a!\theta\rangle\phi \to \langle a!\rangle\phi$
$\mathsf{K_\theta}$	$[a!\theta](\phi \to \psi) \to ([a!\theta]\phi \to [a!\theta]\psi)$	$\mathsf{Nec_\theta}$	*From $\vdash \phi$, infer $\vdash [a!\theta]\phi$*
Cnsv	$B_b\chi \to [a!\theta]B_b\chi$	$\mathsf{Nec_!}$	*From $\vdash \phi$, infer $\vdash [a!]\phi$*
Null	$\langle a!\theta\rangle\phi \to \phi$ *(when θ is a taut.)*	MP	*From $\vdash (\phi \to \psi)$ and $\vdash \phi$, infer $\vdash \psi$*
Sinc	$[a!\theta]\phi \leftrightarrow (B_a\theta \to \langle a!\theta\rangle\phi)$	$\mathsf{Arb_D}$	*From $\vdash \langle a!p\rangle\phi \to \psi$, infer $\vdash \langle a!\rangle\phi \to \psi$*
Rat	$\langle a!\theta\rangle B_b\gamma \to B_b(\theta \to \gamma)$		(where p is a fresh variable not in ϕ or ψ)
Foll	$\langle \vec{c}!\rangle(\neg B_b\chi \wedge \langle a!\theta\rangle B_b\chi) \to [\vec{c}!][a!\theta]B_b\theta$		

These axioms and rules without an arbitrary operator have been introduced in [18]. Briefly, Cnsv represents *conservatism*, which states that agents never abandon their previously held beliefs. Null asserts that if an announcement is a tautology, it does not induce any change in beliefs. Sinc ensures that when the precondition is met, the operator $\langle a!\theta \rangle$ and its dual are equivalent. Rat captures the principle that agents only adopt beliefs that logically follow from their prior beliefs and the social announcements they receive. Foll expresses a conditional relationship regarding the follower relation. The remaining axioms are standard in modal logic. Next, it is straightforward to establish the following theorems for sal.

Theorem 2 (Equivalence Replacement). *Let ϕ be a formula of L_{ASAL}, ϕ' a formula obtained from ϕ by replacing some occurrences of $\psi_1, ..., \psi_n$ by $\psi'_1, ..., \psi'_n$ respectively. It is provable*

$$\vdash_{\mathsf{sal}} \psi_1 \leftrightarrow \psi'_1, ..., \vdash_{\mathsf{sal}} \psi_n \leftrightarrow \psi'_n \implies \vdash_{\mathsf{sal}} \phi \leftrightarrow \phi'$$

Theorem 3 (Soundness of SAL). *The system* sal *is sound, i.e. for every ϕ in L_{ASAL}.*

$$\vdash_{\mathsf{sal}} \phi \implies \models \phi$$

Proof. The soundness of validities without an arbitrary operator has been established by [18]. The rest are covered by Propositions 1 and 6. □

4.1 Non-compactness

Given the inclusion of arbitrary operators in our language, our logic cannot be compact. This result can be illustrated through the subsequent counterexample.

Theorem 4 (Non-Compactness of SAL). *ASAL is not compact.*

Proof. A counter-example is given in [17] to show the result. Here we construct a simpler counter-example:

$$\{\langle a! \rangle B_b p \wedge \neg B_b p\} \cup \{\neg B_a \theta \mid \theta \ \text{is not a tautology}\}$$

Observe that in any model satisfying the set $\{\langle a! \rangle B_b p \wedge \neg B_b p\}$, we have $b \in f(a)$. Each finite subset of this set is satisfiable, since for any such subset, we can find a suitable message γ and construct a model in which $k(a) \subseteq [\![\gamma]\!]$ and $k(b) \subseteq [\![\gamma \rightarrow p]\!]$.

However, the entire infinite set is not satisfiable. Intuitively, there exists no non-tautological message that a can sincerely announce: in such a model, a is assumed to believe nothing except tautologies. As a result, the infinite set has no model. □

This result indicates that we can only establish the weak completeness of any finitary axiomatization of ASAL. Consequently, in constructing the maximal consistent set (MCS), we require countably many fresh propositions in Prop to serve as witnesses for arbitrary announcements.

208   R. Zhu

4.2 Non-greedy Set

We will establish our proof as usual by constructing a model based on the given consistent formula using the Henkin construction. This construction requires witnesses, which we define below.

Definition 10 (Witness of ASAL). *Let Σ be a set of formulas of L_{ASAL}. We say Σ is* witnessed *if for any formula $\langle a!\rangle\phi \in \Sigma$, there exists some message γ such that $\langle a!\gamma\rangle\phi \in \Sigma$. We call γ a* witness.

While expanding the given set by adding a formula with fresh propositional variables, we merely need to ensure that the complement of the set of atomic messages in Prop remains infinite. This implies we can invariably identify an atomic message to function as a witness, allowing an agent to have a message to announce. Therefore, we introduce the subsequent definition and formulation:

Definition 11 (Greedy Set). *We say a set of formulas is* greedy *if its formulas contain a cofinite set of atomic messages. Otherwise, it is* non-greedy.

Given our objective to show weak completeness, the original set must be non-greedy. Subsequently, we will apply the conventional Lindenbaum construction. We aim to show that every non-greedy sal-consistent set is satisfiable via a Henkin-style construction.

4.3 Weak Completeness
Lemma 8 (Lindenbaum for ASAL). *Let ϕ be a consistent formula, there exists a witnessed maximal sal-consistent set containing ϕ.*

Definition 12 (Canonical Model for SAL). *[18]   For any maximal sal-consistent set of formulas Σ, we construct a model $m_\Sigma = (k_\Sigma, f_\Sigma)$ by letting*

$$k_\Sigma(a) = \bigcap\{\llbracket\theta\rrbracket \mid B_a\theta \in \Sigma\}$$
$$f_\Sigma(a) = \{b \mid [\vec{c!}][a!\theta]B_b\theta \in \Sigma \ \text{ for all } \ \vec{c} \ \text{ and } \ \theta \in \Sigma\}$$

Lemma 9 (Belief Characterization). *[18] $k_\Sigma(a) \subseteq \llbracket\theta\rrbracket \Leftrightarrow B_a\theta \in \Sigma$ for any a and θ.*

Definition 13 (Update MCS). *[18] Let Σ be a maximal consistent set. The epistemic update resulting from agent a announcing θ, denoted as $\langle a!\theta\rangle\Sigma$, is defined as follows:*
$$\langle a!\theta\rangle\Sigma = \{\phi \mid \langle a!\theta\rangle\phi \in \Sigma\}$$

Furthermore, we define the relation $\preceq$ between sets of formulas as follows: $\Sigma \preceq \Sigma'$ iff $B_a\theta \in \Sigma$ and $\Sigma' = \langle a!\theta\rangle\Sigma$ for some a and θ. Also, we define $\leq$ be the transitive closure of $\preceq$.

Lemma 10 (Maximal Closure). *[18] If Σ is a maximal consistent set and $\Sigma \leq \Sigma'$ then*

1. Σ' is also a maximal consistent set.
2. there is a $\overrightarrow{c}$ such that: (a) $\Sigma' = \langle \overrightarrow{c!} \rangle \Sigma$, and (b) $[\overleftarrow{c!}]\phi \in \Sigma \Leftrightarrow \phi \in \Sigma'$ for all ϕ, where $\overleftarrow{c}$ is the reversal of $\overrightarrow{c}$.

The closure-based method can now handle fragments with formulas only involving explicit social announcements. To extend this to formulas with arbitrary announcement operators, we can reduce them to the explicit cases using the following lemma.

Lemma 11 (Explicit Lemma). *Let $\phi[\epsilon/\langle b! \rangle \psi]$ denote ϕ with all occurrences of subformula ϵ replaced by $\langle b! \rangle \psi$. Then for all MCSs Σ,*

$$\phi[\epsilon/\langle b! \rangle \psi] \in \Sigma \Leftrightarrow \text{ there exists some } \gamma \text{ such that } \phi[\epsilon/\langle b! \gamma \rangle \psi] \in \Sigma \quad \text{(sinc. ins)}$$

Proof. *we show this result by induction on ϕ.*
Base case *when $\phi = B_a\theta \neq \epsilon$, the result holds for $B_a\theta[\epsilon/\langle \gamma \rangle \psi] = B_a\theta$ trivially. When $\phi = \epsilon$, by witnessed property of MCSs, given in Lemma 8, we can derive sinc. ins.*
Inductive step *Boolean cases can be covered by the I.H..*
When $\phi = \langle a!\theta \rangle \chi[\epsilon/\langle b \rangle \psi]$, we have the following:

$$
\begin{aligned}
\langle a!\theta \rangle \chi[\epsilon/\langle b! \rangle \psi] \in \Sigma &\Leftrightarrow \chi[\epsilon/\langle b! \rangle \psi] \in \langle a!\theta \rangle \Sigma && \text{by def.13} \\
&\Leftrightarrow \chi[\epsilon/\langle b! \gamma \rangle \psi] \in \langle a!\theta \rangle \Sigma && \text{by I.H.} \\
&\Leftrightarrow \langle a!\theta \rangle \chi[\epsilon/\langle b! \gamma \rangle \psi] \in \Sigma && \text{by def.13}
\end{aligned}
$$

Note that $\langle a!\theta \rangle \Sigma$ is a MCS by Lemma 10.
When $\phi = \langle a! \rangle \chi[\epsilon/\langle b! \rangle \psi]$, we have the following:

$$
\begin{aligned}
\langle a! \rangle \chi[\epsilon/\langle b! \rangle \psi] \in \Sigma &\Leftrightarrow \langle a!\theta \rangle \chi[\epsilon/\langle b! \rangle \psi] \in \Sigma && \text{by Lemma 8} \\
&\Leftrightarrow \chi[\epsilon/\langle b! \rangle \psi] \in \langle a!\theta \rangle \Sigma && \text{by def.13} \\
&\Leftrightarrow \chi[\epsilon/\langle b! \rangle \psi] \in \langle a!\theta \rangle \Sigma && \text{by I.H.} \\
&\Leftrightarrow \langle a!\theta \rangle \chi[\epsilon/\langle b! \rangle \psi] \in \Sigma && \text{by def.13}
\end{aligned}
$$

Note that $\langle a! \rangle \chi[\epsilon/\langle b! \rangle \psi]$ has a witness in Σ by Lemma 8. $\square$

By Lemma 11, we can ensure that every formula in a MCS corresponds to a formula without an arbitrary announcement operator. Consequently, it suffices to address only the case within propositional network announcement logic.

Lemma 12 (Dynamic Characterization). *[18] For any MCS Σ, if $\Sigma \leq \Sigma'$ and $B_a\theta \in \Sigma'$ then $[a : \theta]k_{\Sigma'} = k_{\langle a!\theta \rangle \Sigma'}$.*

Theorem 5 (Completeness of SAL). *The system* sal *is weakly complete, i.e. for any finite* sal*-consistent set Σ, there exists some model m such that $m \vDash \Sigma$.*

Proof. We firstly construct a maximal consistent set Σ^+ based on Σ, which can be done and guaranteed by Lemma 8. Now we can show the truth lemma by an induction on ϕ.

$$\phi \in \Sigma^+ \Leftrightarrow m_{\Sigma^+} \vDash \phi \qquad \text{(truth lemma)}$$

Base case *when ϕ is an atomic formula, the result can be guaranteed by def.12.*
Inductive step *The boolean cases can be covered by the I.H..*

$$
\begin{aligned}
\textbf{when}\quad \phi = \langle a!\theta\rangle\psi \quad & \langle a!\theta\rangle\psi \in \Sigma^+ \Leftrightarrow \psi \in \langle a!\theta\rangle\Sigma^+ \ \ and \ \ B_a\theta \in \Sigma^+ && by\ Lemma10 \\
& \Leftrightarrow m_{\langle a!\theta\rangle\Sigma^+} \vDash \psi \ \ and \ \ m_{\Sigma^+} \vDash B_a\theta && by\ I.H. \\
& \Leftrightarrow [a:\theta]m_{\Sigma^+} \vDash \psi \ \ and \ \ m_{\Sigma^+} \vDash B_a\theta && by\ Lemma\ 12 \\
& \Leftrightarrow m_{\Sigma^+} \vDash \langle a!\theta\rangle\psi && by\ def.3 \\
\textbf{when}\quad \phi = \langle a!\rangle\psi \quad & \langle a!\rangle\psi \in \Sigma^+ \Rightarrow \langle a!\gamma\rangle\psi \in \Sigma^+ && for\ some\ \gamma\ by\ sinc.ins \\
& \Rightarrow m_{\Sigma^+} \vDash \langle a!\gamma\rangle\psi && as\ case\ above \\
& \Rightarrow m_{\Sigma^+} \vDash \langle a!\rangle\psi && by\ def.3
\end{aligned}
$$

Note that Σ^+ is witnessed. Conversely, we have

$$
\begin{aligned}
m_{\Sigma^+} \vDash \langle a!\rangle\psi & \Rightarrow [a:\theta]m_{\Sigma^+} \vDash \psi \ and\ m_{\Sigma^+} \vDash B_a\theta && for\ some\ \theta\ by\ def.3 \\
& \Rightarrow m_{\langle a!\theta\rangle\Sigma^+} \vDash \psi \ and\ m_{\Sigma^+} \vDash B_a\theta && by\ Lemma12 \\
& \Rightarrow \psi \in \langle a!\theta\rangle\Sigma^+ \ and\ B_a\theta \in \Sigma^+ && by\ I.H. \\
& \Rightarrow \langle a!\theta\rangle\psi \in \Sigma^+ \ and\ B_a\theta \in \Sigma^+ && by\ def.13 \\
& \Rightarrow \langle a!\rangle\psi \in \Sigma^+ && by\ \mathsf{Arb_I}
\end{aligned}
$$

This completes the induction and the proof. $\qquad\qquad\square$

This establishes weak completeness of sal, ensuring that every sal-consistent formula is satisfiable in a model with epistemic distributions and social networks. This axiomatization connects ASAL's syntax and semantics, enabling the validation of theorems via logical inference and enhancing our understanding of agents' beliefs and social relationships.

5 Discussion and Future Work

This paper introduces the first finitary axiomatization of ASAL, extending PNAL by incorporating arbitrary announcement operators. Our key innovation lies in a modulo-based model transformation technique that enables a sound and complete formal system. This paper solved the open problem left by [17], and leads to a Henkin-style weak completeness result for ASAL.

Compared to PAL and its extensions, ASAL emphasizes local and asymmetric information updates in social networks and accommodates inconsistent belief states. While BAPAL [15] also achieves a finitary axiomatization by restricting announcement quantification to Boolean formulas, it remains grounded in knowledge-based semantics. In contrast, ASAL operates in a belief-based setting without higher-order modalities, and its announcement dynamics are intrinsically social and directional. Furthermore, our approach avoids the necessity-form technique commonly used in APAL and its variants. Instead we employ a direct model-theoretic transformation which results in a more standard completeness proof. This not only improves conceptual clarity but also provides an independent validation of finitary completeness results in this area.

Future work may explore extending ASAL with nested epistemic modalities to capture higher-order beliefs, or incorporating belief revision mechanisms and constraints on network structure. We also suggest investigating computational complexity, and potential applications in modeling deceptive or strategic communication via free social announcements.

Acknowledgments. I would like to thank Jeremy Seligman for his long-term support and guidance on this line of research. I also thank the anonymous reviewers for their valuable suggestions, which improved the paper's clarity and quality.

Disclosure of Interests. The author has no competing interests to declare that are relevant to the content of this article.

References

1. Ågotnes, T., Galimullin, R.: Quantifying over information change with common knowledge. Auton. Agent. Multi-Agent Syst. **37**, 19 (2023)
2. Balbiani, P.: Putting right the wording and the proof of the truth lemma for apal. J. Appl. Non-Classical Logics **25**, 2–19 (2015)
3. Balbiani, P., Baltag, A., Ditmarsch, H., Herzig, A., Hoshi, T., Lima, T.: Knowable 'as' known after an announcement. Rev. Symbolic Logic **1**, 305–334 (2008)
4. Balbiani, P., Ditmarsch, H.: A simple proof of the completeness of apal. Stud. Logic **8**, 65–78 (2015)
5. Baltag, A., Christoff, Z., Rendsvig, R.K., Smets, S.: Dynamic epistemic logics of diffusion and prediction in social networks. Stud. Logica. **107**, 489–531 (2019)
6. Baltag, A., Özgün, A., Sandoval, A.L.V.: Arbitrary public announcement logic with memory. J. Philosophical Logic, 1–58 (2022)
7. Christoff, Z.: A logic for social influence through communication. In: EUMAS, pp. 31–39. Citeseer (2013)
8. Christoff, Z., Hansen, J.U.: A logic for diffusion in social networks. J. Appl. Log. **13**, 48–77 (2015)
9. French, T., Ditmarsch, H.: Undecidability for arbitrary public announcement logic. Adv. Modal Logic **7**, 23–42 (2008)
10. Galimullin, R., Kuijer, L.B.: Satisfiability of arbitrary public announcement logic with common knowledge is σ_1^1-hard. Electron. Proc. Theoretical Comput. Sci. **379**, 260–271 (2023)
11. Goldblatt, R.: Axiomatising the Logic of Computer Programming, Springer, Heidelberg (1982)
12. Kuijer, B.: Unsoundness of r$\square$ (2015). https://personal.us.es/hvd/APAL_counterexample.pdf
13. Seligman, J., Liu, F., Girard, P.: Logic in the community. In: Logic and Its Applications, pp. 178–188. Springer (2011)
14. Seligman, J., Liu, F., Girard, P.: Facebook and the epistemic logic of friendship. In: Schipper, B.C. (eds.) TARK 2013, pp. 229–238 (2013)

15. van Ditmarsch, H., French, T.: Quantifying over boolean announcements. Logical Methods Comput. Sci. **18**, 20:1–20:22 (2022)
16. Xiong, Z., Ågotnes, T.: On the logic of balance in social networks. J. Logic Lang. Inform. **29**, 53–75 (2019)
17. Xiong, Z., Ågotnes, T.: Arbitrary propositional network announcement logic. In: International Workshop on Dynamic Logic, pp. 277–293. Springer (2020)
18. Xiong, Z., Ågotnes, T., Seligman, J., Zhu, R.: Towards a logic of tweeting. In: International Workshop on Logic, Rationality and Interaction, pp. 49–64. Springer (2017)

Author Index